Springer Series in
Surface Sciences

8

Editor: Gerhard Ertl

Springer Series in **Surface Sciences**

Editors: Gerhard Ertl and Robert Gomer

Kinetics of Interface Reactions

Proceedings of a Workshop on
Interface Phenomena, Campobello Island, Canada,
September 24–27, 1986

Editors: M. Grunze, H. J. Kreuzer

With 152 Figures

Springer-Verlag Berlin Heidelberg New York
London Paris Tokyo

Professor Dr. Michael Grunze

Laboratory for Surface Science & Technology, Department of
Physics, University of Maine at Orono, Orono, ME 04469, USA

Professor Dr. Hans Jürgen Kreuzer

Department of Physics, Dalhousie University, Halifax,
Nova Scotia, Canada, B3H3J5

Series Editors

Professor Dr. Gerhard Ertl

Fritz-Haber-Institut der Max-Planck-Gesellschaft, Faradayweg 4–6
D-1000 Berlin 33

Professor Robert Gomer

The James Franck Institute, The University of Chicago, 5640 Ellis Avenue,
Chicago, IL 60637, USA

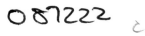

ISBN 3-540-17821-X Springer-Verlag Berlin Heidelberg New York
ISBN 0-387-17821-X Springer-Verlag New York Berlin Heidelberg

Offsetprinting: Druckhaus Beltz, 6944 Hemsbach/Bergstr.
Bookbinding: J. Schäffer GmbH & Co. KG., 6718 Grünstadt
2153/3150-543210

Preface

This book contains the proceedings of the first Workshop on Interface Phenomena, organized jointly by the surface science groups at Dalhousie University and the University of Maine. It was our intention to concentrate on just three topics related to the kinetics of interface reactions which, in our opinion, were frequently obscured unnecessarily in the literature and whose fundamental nature warranted an extensive discussion to help clarify the issues, very much in the spirit of the Discussions of the Faraday Society. Each session (day) saw two principal speakers expounding the different views; the session chairmen were asked to summarize the ensuing discussions.

To understand the complexity of interface reactions, paradigms must be formulated to provide a framework for the interpretation of experimental data and for the construction of theoretical models. Phenomenological approaches have been based on a small number of rate equations for the concentrations or mole numbers of the various species involved in a particular system with the relevant rate constants either fitted (in the form of the Arrhenius parametrization) to experimental data or calculated on the basis of microscopic models. The former procedure can at best serve as a guide to the latter, and is, in most cases, confined to ruling out certain reaction pathways rather than to ascertaining a unique answer.

Gases react at the surface of solids; the adsorption and desorption processes themselves are to be understood as "surface bond" making and breaking mechanisms, respectively. If the energy transfer is fast enough, then these processes can be understood in terms of thermodynamic arguments such as formulated in transition state theory. However, in most situations such a simple treatment is not sufficient and the details of the microscopic dynamics must be invoked. Unfortunately, this point is very frequently overlooked in the analysis of kinetic data of gas-surface reactions, leading to rather murky discussions in the literature. We therefore devoted the first day of the workshop to a thorough exposé of the dynamics versus thermodynamics controversy in adsorption-desorption kinetics. We are very fortunate in having secured the clear and stimulating contributions by Menzel and Brenig on this subject.

The phenomenological analysis of surface reactions in terms of kinetic rate equations quite often has to invoke precursor states as reaction intermediates to fit experimental data. However, because such an analysis rarely leads to a unique answer, independent evidence must be brought forward if precursors in a given reaction are to be accepted as more than just mythical "mis-fits". In particular, because precursors are reaction transients, their lifetime should surely be of interest in deciding whether they are states (i.e., local minima on some potential energy surface) or mere resonances in a scattering channel. The session on "Precursors: Myth or Reality?" was a stormy one, as the contributions by Weinberg, Auerbach, and others in Part II attest, foreshadowing more controversies in the analyses of specific systems.

The equilibrium properties of surface phase transitions have been studied for many decades, and they exhibit a wealth of fascinating detail. The exploration of their kinetics, on the other hand, had to await the advent of time resolved surface analysis techniques. In particular, video-LEED has made it possible to study the kinetics of surface reconstruction, as is reviewed comprehensively by Heinz in the third part of these proceedings. It was our intention at this workshop to take early stock, also, of the theoretical approaches, as outlined by Gunton, and to identify some of the interesting features of surface phase transitions in order to stimulate the interest of experimentalists and theorists alike.

The workshop was held at the Roosevelt Campobello International Park on Campobello Island, New Brunswick. The superb accommodation in the various "cottages" within the Park and the serene beauty of the Bay of Fundy provided an ideal setting for intensive discussions. Speaking on behalf of all participants, we would like to express our gratitude to the staff of the Park for making our stay there so pleasant.

Orono and Halifax,   *Michael Grunze*
December 1986   *H. Jürgen Kreuzer*

Acknowledgements

This workshop was supported by the Air Force Office of Scientific Research, The Army Research Office, the Office of Naval Research, the Surface Science Division of the Canadian Association of Physicists, Dalhousie University, and the University of Maine. The views, opinions, and/or findings contained in this report are those of the authors and should not be construed as an official position, policy, or decision by any of the above agencies, unless so designated by other documentation.

We would also like to thank the following companies for their donations to help sponsor graduate student attendance:

American Cyanamid Company
IBM Almaden Research Center, San Jose
ICI America, Inc.
Leybold-Hereaus
Perkin Elmer/Physical Electronics Division
Sanders Associates

Solar Energy Research Institute
SPECS, GmbH
Sperry Corporation
Standard Oil of Ohio
Varian Associates

Contents

Part I

Adsorption-Desorption Kinetics:
Dynamics versus Thermodynamics

Equilibrium and Non-equilibrium Effects in Adsorption-Desorption Kinetics: Influences of Interaction Dynamics, Reaction Kinetics, and Statistical Mechanics of the Adlayer

D. Menzel

Physik-Department, Technische Unversität München,
D-8046 Garching, Fed. Rep. of Germany

The various meanings of non-equilibrium in our context (irreversibility of experimental situation; deviation from equilibrium within the layer; non-equilibration in energy and/or momentum transfer during the surface collision, leading to non-equipartition among the various degrees of freedom) are clarified first. If quasi-equilibrium persists (no influence of irreversibility; internal equilibrium of the layer), then the expression for desorption rates can be split into an equilibrium factor and a dynamic factor; the latter can be identified with the sticking coefficient and contains the non-equipartition effects. It can then be expected that, depending on circumstances, one or the other will govern the rate behaviour. Examples of experimental results from physisorption as well as chemisorption systems are discussed which show that effects of the statistical mechanics of the adsorbate layer, the internal equilibration of surface layers, and of the microscopic dynamics of the gas-surface interaction can be clearly seen in many cases, even though they can mix and interfere as well. Thermodynamic influences become most obvious if detailed coverage dependences of desorption rate parameters are investigated which can then be used to derive the statistics of the layer. Dynamic effects can be most easily seen in state-resolved beam experiments, but are also obvious in more classical experiments, in particular of sticking coefficients. Some remarks about experimental difficulties are made in an appendix.

1. INTRODUCTION

Adsorption and desorption are, together with surface diffusion, the simplest surface kinetic processes which form the basis of more complicated reaction processes [1-3]. Strong theoretical efforts have been made, therefore, to come to a detailed, hopefully microscopic, understanding of them [4-8]. Also, partly because of the relative ease of experimentation, a large body of results exists about thermal desorption and sticking, although a critical evaluation shows that quite a large part of it is not usable for comparison with these theoretical expectations. In recent years, increasing use has been made of energy, angle, and state-resolved studies to investigate details of the microscopic interactions [9,10]. This survey intends to show that while the latter type of experiments can give more direct access to dynamic effects, the more conventional rate studies (which are also justified by their intrinsic interest and connections to other properties of the adlayer) can be used in many cases to distinguish thermodynamic and dynamic effects in ad- or desorption kinetics, if they are carried out with sufficient care. For linking kinetics to the interactions within the adsorbate layer, which are most obvious in coverage dependences, the classical experiments are more informative. The main emphasis of this survey will, therefore, be on the conventional type of adsorption, desorption and equilibrium measurements, although selective reference to beam or other state-selected results will be made.

2

The discussion will be based on ref. 3, some conclusions of which will be repeated here for clarity, but which will in general be assumed as known. The main topic to be treated is the distinction of dynamic from thermodynamic (or "equilibrium" in a broad sense), and kinetic influences in surface kinetics. Since discussions with colleagues and referees as well as reading of the literature have suggested to this author that the basic concepts of this distinction are not always too clear, we will start with a (partly admittedly trivial) clarification of terms which leads to a well-defined nomenclature. Then a few examples will be shown from physisorption and chemisorption systems, in which all aspects and their coupling show up.

2. SOME BASIC CONCEPTS

The terms "equilibrium" "and non-equilibrium" effects in surface kinetics are being used in several meanings. While there are definitely connections between these, it appears helpful to distinguish them clearly first. This will be attempted in the following:

A) In true thermodynamic equilibrium, the integral rates of adsorption and desorption, R_a and R_d, as well as the differential fluxes $R(j)$ along any individual path, however defined, have to be opposed and equal as a consequence of the second law of thermodynamics (detailed balance, or reciprocity [8]):

$$R_a = R_d \quad \text{and} \quad R_a(j) = R_d(j). \tag{1}$$

Considering for simplicity only one adsorbing gas phase species, the equilibrium between gas and surface will depend only on one (gas or surface) concentration of this species and on temperature. The normalized integral rates

$$s = R_a/N_g v_T \quad \text{(sticking coefficient)} \quad \text{and}$$

$$k_d = R_d/N_a \quad \text{(specific desorption rate)}$$

(where N_g and N_a are the gas and surface concentrations and v_T the thermal velocity [3]) depend only on (average) coverage and on temperature. (In equilibrium, the average coverage will be a sufficient parameter even for a complicated adlayer; see below). The equilibrium parameters (e.g. $N_a = f(N_g,T)$ can then be measured in a suitable way and expressed with any of the various cuts through parameter space, isotherms, isobars, or isosters; isosteric heats and entropies of adsorption as function of coverage can be obtained from these [3] (see also Appendix).

The actual measurement of rates requires a deviation from equilibrium. While relaxation methods could be used to make this deviation very small, this is rarely done in classical surface kinetics. (Relaxation methods are more often used in beam experiments, but these represent non-equilibrium situations in themselves). Here one usually measures under strongly or totally irreversible conditions: to make the analysis unambiguous, adsorption is usually carried out at low enough temperatures to make desorption negligible, and desorption at low enough pressure and high enough temperatures that re-adsorption is negligible. Because of the small windows of measurement (see Appendix), this means that adsorption, equilibrium (or stationary state), and desorption measurements are carried out at low, medium, and high temperature, with overlap between each two adjacent ranges but not of the extremes (see ref. 11 for a notable exception). These experimental aspects are interesting for our topic in the following sense. On

3

going from equilibrium to irreversible conditions even for the same T, N_a values, one cannot be absolutely sure whether the remaining flux is the same as under equilibrium. Even for the simplest case of desorption, adatoms of near zero coverage climbing the vibrational ladder in their surface potential, the upper levels may be depleted if the incoming adsorbing flux is removed. This will certainly happen if the heat flow from the surface which leads to refilling of these levels is so severely impeded that it proceeds on the same or a longer time scale as desorption itself (KRAMERS-SUHL low friction case [2,3]). More complicated mechanisms can lead to additional changes (see below). The so-called quasi-equilibrium assumption excludes such cases, so that detailed balance holds even for this irreversible (but isothermal) case. When people talk about "testing detailed balance for surface kinetics", they usually mean the test of this assumption. To do this without ambiguity, k_d and (isothermal) s values and equilibrium results must be available for the same temperature. Such a test [11] will be discusssed below. If only s and $\overline{k_d}$ values at strongly differing temperatures (as described above) are available, extrapolation to make them overlap in the range of equilibrium results is necessary which requires very good and extensive results. If good agreement between kinetic and equilibrium parameters is reached, then this not only proves the applicability of the quasi-equilibrium assumption, but also that adsorption and desorption proceed via the same dominant channels at high and low temperatures [12,13]. This is a more far-reaching conclusion than the "validity of detailed balancing", and can be expected with much less confidence. If adsorption (as well as desorption) can proceed via several alternative paths (e.g. via precursors or not; via a 2d-gas into a 2d solid or directly), then the weighting of these paths will depend on their relative rates. If one then compares the mechanisms at, say, different temperatures, then this weighting and the effective mechanisms could be very different. Nevertheless, at each temperature the reversible and irreversible rates could still be equal, so "detailed balance holds".

B) Another type of non-equilibrium effect can occur in such cases of alternative mechanisms. When discussing the variables necessary for characterization of rate parameters, we assumed that the adlayer is in internal equilibrium during adsorption and desorption. If this is not the case, one coverage parameter is not enough to characterize the layer; if only one is used (or known), then the rates may appear to depend on the prehistory, i.e. the way the (average) coverage measured has been reached. In terms of mechanism, this means that instead of an equilibrium preceding the rate-determining step (for instance between a 2D condensed and a 2D gas phase preceding desorption), now two (or more) rates of a sequence are of comparable order. Even if we always prepare the same starting layer, the effect will show up in the temperature dependence in the case of desorption which will not be Arrhenius-like and, if forced to such a form, will yield an unphysical "desorption energy". The situation can become simple again if the first rate becomes the slowest; then an Arrhenius form will usually work again, but the derived energy will correspond to some other energy than the binding energy of the adsorbate. Similar situations can arise in adsorption, if precursor kinetics persist [2,3,14]. All these cases can be characterized by the fact that a new slow variable is to be introduced [8].

C) The most interesting "non-equilibrium" effects consist in anisotropies in the dependences of s and k_d on the dynamical variables, as compared to Boltzmann distributions. In desorption, angular dependences can be far from cosine (either strongly directional or wider), and translational and/or inner energies can deviate from the Boltzmann distribution; the corresponding anisotropies can be found in the sticking coefficient. Such deviations

4

must be caused by the details of the particle-surface interaction potential, i.e. are of microscopic dynamic nature. They signify a "non-equilibrium" of energy distribution over the degrees of freedom of the product
in desorption and conversely variations of the sticking probability for
different initial states of the gas molecules. To distinguish them clearly
from the "irreversibility" effects discussed in A) and B), I shall call
them "non-equipartition" effects. Obviously, such dynamical effects will
occur both under reversible and irreversible conditions, as they only depend on the dynamics of the gas-surface interaction which must be time-
reversible. In fact, the most dramatic test for the applicability of the
quasi-equilibrium assumption (the "validy of detailed balance", see above)
is the finding that the same anisotropies or non-equipartition effects
exist for s and k_d [12]. We see that "non-equilibrium" (i.e. dynamic non-
equipartition) effects can exist in equilibrium (i.e. under reversible
conditions), while equipartition can easily be found under irreversible
conditions. Irreversible conditions have to be applied to find dynamic
effects, but these can be used to prove that the quasi-equilibrium assumption is valid.

D) Many authors (see e.g. refs. 3, 15-16) have shown that the dynamic
effects in surface kinetics can be factorized from the thermodynamic influences, if quasi-equilibrium exists as discussed above (detailed balance
as well as internal equilibrium). Then the reversible and irreversible
rates are equal (even though they may contain strong dynamic effects);
adsorption can be described by an isothermal sticking coefficient which
only depends on (average) coverage and temperature, and desorption follows
an Arrhenius law. If we take a non-activated, nondissociative desorption as
example, so that it makes sense to normalize (see above) the rate to the
coverage, we get

$$R_d = N_a \; k_o^{(1)} \; \exp \; (-E_d/RT) \tag{2}$$

(where E_d and $k_o^{(1)}$ are still coverage- and possibly T-dependent [3]). On the
other hand the equilibrium condition demands that

$$N_g/N_a = (F_g(3D)/F_a) \; \exp \; (\; E_{eq}/RT) \, , \tag{3}$$

where the F's are the partition functions of the three-dimensional gas and
the adsorbate, respectively, and E_{eq} the equilibrium energy of adsorption.
Combining eqs. 2 and 3 yields (if $E_d = E_{eq}$)

$$k_o = s \; v_T \; (F_g(3D)/F_a) = s(kT/h) \; (F_g(2D)/F_a) = A(s/F_a) \, , \tag{4}$$

where $A = (kT/h) \; F(2D)$ is a (weakly T-dependent) constant. As has been
shown in ref. 3, exactly the same result follows from transition state
theory (TST) [17], if s is identified with the TST transmission coefficient. This is not surprising, since TST is a quasi-equilibrium treatment.
We see that the pre-exponential of desorption depends on the sticking coefficient and on the partition function of the layer. The sticking coefficient will contain two kinds of contributions. If no dynamic effects exist
there will be a coverage dependence, $s \sim s_o f(\Theta)$, due to the available sites
on the surface. In the simplest, Langmuir-type case, this will be $s \sim (1-\Theta)$,
but could be more complicated; but it will be connected to the statistics
of the layer. If $s_o \neq 1$, then there is no complete equilibration upon impact
with an empty site, and dynamic effects exist. It is seen that the specific
desorption rate can be split into an equilibrium part determined by the
equilibrium statistics of the layer, and a dynamic part. The latter shows
up directly in the sticking coefficient; it is equally contained in the

desorption rate, but because of the overwhelming influence of the exponential, it will generally be difficult to clearly observe it there. A better chance of seeing it even there exists if not the absolute values, but relative changes (e.g. anisotropies) are measured. Extension to other cases (e.g. dissociative adsorption, activated adsorption) is straightforward. The treatment obviously breaks down when the quasi-equilibrium assumption and/or internal equilibrium do not apply.

To summarize this chapter:

Dynamic effects are due to the microscopic molecule-surface interaction which can lead to anisotropies in the rate parameters s and k_d; these show up in non-equipartition effects.

Measurements of rates are always carried out under irreversible conditions. As long as equilibration within the layer and with the substrate is fast compared to desorption times, the rates will depend only on dynamic and equilibrium properties of the adlayer. If the equilibration rates become comparable to desorption rates, complicated kinetics will be the consequence which cannot be described by simple equations (deviation from the Arrhenius form of temperature dependence, or from coverage dependences based on equilibrium statistical mechanics of the layer).

Finally, if the equilibration rates become slowest, i.e. are rate-determining, kinetics can follow simple equations again, but the derived rate parameters cannot be connected to thermodynamic or dynamic parameters.

As long as quasi-equilibrium (detailed balance even under irreversible conditions) and internal equilibrium can be assumed, a simplified treatment can be applied which allows the factorization of influences of collision dynamics from the equilibrium statistics of the layer. The sticking coefficient then carries the dynamic information most clearly, while in the desorption rate the strong exponential tends to obscure such influences. On the other hand, desorption rates show the clearer connection to the thermodynamics of the layer.

There are obvious connections of these three realms to the three time-scales introduced by BRENIG [8].

3. EXAMPLES

In this chapter some examples from physisorption and chemisorption experiments will be examined in the light of the preceding section to extract influences of layer statistics, kinetic mechanisms and interaction dynamics. It will become obvious that clear cases of each of these are available at least qualitatively, and that they can also mix. Emphasis will be on nondissociative adsorption (atomic adsorption of rare gases, nonactivated molecular adsorption), with examples mainly from the work of the author's group. One example for dissociative adsorption will be discussed, and inferences from beam results will also be made. Because of the limited space, the discussion will be qualitative.

3.1 Adsorption of Rare Gases

Adsorption of rare gas atoms which are bound by dispersion forces only (or at least predominantly) and have no internal degrees of freedom, might be

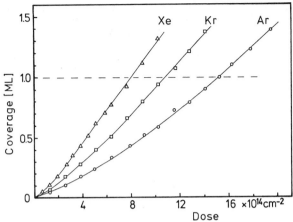

Fig. 1: Evolution of the coverage of Ar, Kr, Xe on Ni(111) as function of exposure (in 10^{14} coll/cm²). Dosing from a capillary source (gas incidence 30° from the surface normal, T_g = 295 K); T_s = 25 K. Note the exponential increase of coverage with exposure, approaching linear behaviour above one monolayer. Coverages are from thermal desorption; XPS gave similar behaviour. After [20].

expected to be simple, but is actually surprisingly complicated. This is mainly due to the fact that there is no big difference between the atom-surface (the "vertical" interaction) and the atom-atom (the "lateral") interactions.

To start with dynamic effects: Sticking is strongly influenced by incomplete equilibration of energy upon collision with a (clean) surface. This would be expected from the known values of accommodation coefficients [5,7,18] (the a.c. is an upper limit of the sticking coefficient; under certain conditions [19] the two are equal). In fact, recent measurements for Ar, Kr, Xe on Ni(111) [20] and for Ne and Ar on Ru(001) [21] have borne this out in a very dramatic way. What is found (figs. 1 and 2) is that coverage increases <u>exponentially</u> with exposure, so that the sticking coefficient increases <u>linearly</u> with coverage according to $s = s_0 + \Theta . s_1$ (where Θ is the fractional coverage with $\Theta = 1$ corresponding to 1 monolayer). The values found for s_0 and s_1 are not drastically different for the heavy rare gases (s_0 about 0.23, 0.30 and 0.46 for Ar, Kr, Xe/Ni(111); s_1/s_0 about 2.8, 2.2 and 1.4 for the same) and the s_0 values are in the range expected from a.c. values [18]. Similar results were found for physisorbed N_2.

For Ne/Ru(001), however, the effect is really dramatic: s_1/s_0 reaches values of almost 10^3, while s_0 can become as small as 1.10^{-3}. A range is indicated here, because it is found that the s_0 values depend strongly on surface cleanness and perfection as well as on gas temperature. The obvious explanation for these observations must be that the energy transfer to the <u>clean</u> surface is dynamically impeded (severely for Ne). While there is no other way to stick at zero or low coverage (leading to the low s_0 values), sticking can proceed via collision with rare gas surface atoms (with efficiency s_1) once a fraction of the surface is covered. The very high s_1/s_0 value for Ne is then mainly due to the action of the effective mass ratio for the collision partners (which is a near [22] perfect match for collision with adatoms, while the unfavourable mass ratio for the clean surface is further enhanced by the coupling among metal surface atoms which leads

7

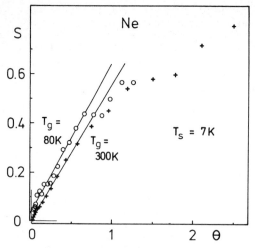

Fig. 2:
Sticking coefficient of Ne on Ru(001) for T_s =7 K and T_g=295 K or 80 K (thermal beam, normal incidence), as a function of coverage. Sticking is seen to increase linearly for most of the first monolayer. Asymptotic values at $\Theta \to 0$ were 0.0022 (295 K) and 0.024 (80 K) for the runs shown; the lowest value obtained for the supposedly best surface preparation was 0.0012 (295 K). After [21].

to an effective collision mass of surface atoms corresponding to at least four atoms [23]). The influence of impurities should then mainly be a mass effect again, possibly with some enhancement by changed coupling to the substrate. Disorder can be effective via the latter effect (metal adatoms) as well as via improvement of transfer of parallel momentum (at steps and the like). It is stressed that all the measurements were carried out at non-isothermal conditions. Qualitatively the observed influence of the gas temperature is expected [18], but a quantitative analysis has not yet been made. It should be noted that the measurements are quite difficult [20,21].

Desorption measurements for the same systems were carried out both with temperature-programmed and isothermal desorption (TPD and ITD, respectively). In all cases, several subsequent layers (2 to 4) can be distinguished, before a bulk phase develops (Fig.3). It is found in all cases and with

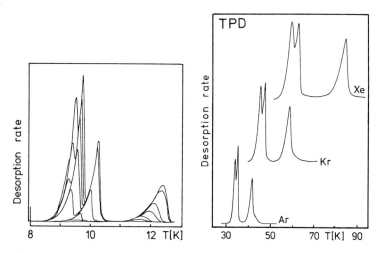

Fig. 3: Programmed desorption of Ne/Ru(001) [20] (left) and Ar, Kr, Xe/-Ni(111) [21] (right). First, second, and higher layers can be distinguished.

both methods that there exists a rather broad coverage range where the rate changes very little with coverage, even for the first layer, i.e. the formal order is close to zero (Figs.4 to 6 which show Xe as example). At very low coverage, first order behaviour is observed, but in between,com-

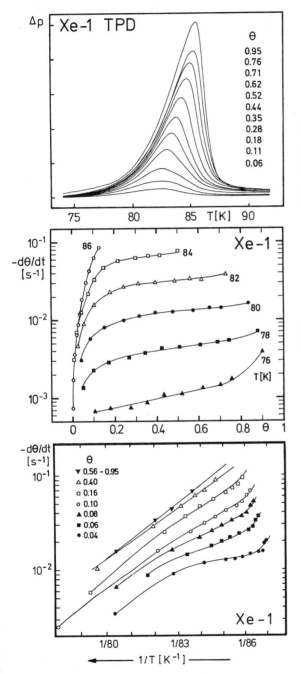

Fig. 4:
Temperature - programmed desorption traces for Xe/Ni(111) for various initial coverages up to one monolayer (0.5 K/s). From ref. 20

Fig. 5:
Replotting of the data of Fig. 4 as desorption isotherms. Close to zero order behaviour is most obvious in the medium coverage-temperature range. From ref. 20

Fig. 6:
Isosteric evaluation of the data of Figs. 4 and 5. Note the curvature of traces at low coverages and high temperatures. From ref. 20

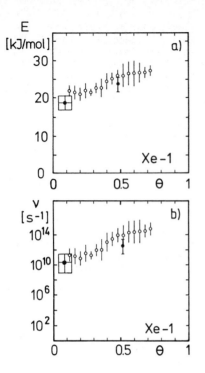

E
[kJ/mol]

a)

Xe-1

v
[s⁻¹]

b)

Xe-1

Fig. 7:
Isosteric desorption energies
and preexponentials for Xe/
Ni(111), extracted from Figs.
4-6 (open symbols). Two results from isothermal desorption (full symbols) are given
for the zero order range (at
$\Theta \sim 0.5$) and for first order
desorption from the 2D gas
alone (0.15> Θ >0.05). After
ref. 20

plicated coverage dependences result. Evaluation by desorption isosters
works well down to about $\Theta \sim 0.2$ (for Xe), but curved isosters result below,
with varying slope over most of the range. Isosteric desorption energies
and prefactors extracted from the well-behaved region are shown for Xe in
Fig.7. The qualitative interpretation for the zero order range is quite
obvious and has been given before by other authors (e.g. ref. 24). At high
coverage a two-dimensional condensed phase, 2D-cond., exists because of
lateral interactions among adatoms; it is in equilibrium with a 2D-gas
phase, both on the free surface part and on top of the islands. Because of
the higher binding energy on the free surface (i.e. in the first layer),
the coverage or density of 2D-gas is much lower on top of the islands (in
the second layer); under our conditions for Xe by factors between 10^2 and
10^3. Nevertheless, desorption proceeds through both of these regions, which
is seen to correspond to reversal of the adsorption paths. Besides, desorption can also proceed directly from the condensed phase. As long as there
is internal equilibrium among the surface phases, the area-specific rates
along these paths are the same, or more accurately have the same ratios as
the corresponding isothermal sticking coefficients, according to detailed
balance, and their contributions in the total desorption rate vary in proportion to the surface fractions covered by them. It should be noted that
all paths have to exist to give close to zero order behaviour. The temperature dependence of the rate in the corresponding range will give the
binding energy of the 2D-condensed phase, i.e. the sum of the binding energy of the 2D-gas phase concerned and its energy difference from the 2D-
condensate; the preexponential of desorption is that expected for desorption from
a condensate. This is the situation prevailing down to $\Theta \sim 0.4$. The increase of order and change of apparent desorption energy occurring below
seems to indicate that phase equilibrium is not maintained any more. Equi-

librium of the condensed islands with the first layer gas will break down because of increasing distances between islands of condensate. Diffusion will enter as a rate - controlling step, and concentration gradients will develop in the 2D-gas. Complex rate behaviour is the consequence for which kinetic models can be constructed with all their intrinsic ambiguity [20]. Nevertheless, Θ is still an essentially sufficient parameter to characterize the layer, as shown by the well-behaved isosters down to $\Theta \sim 0.2$. Only the effective desorption energy decreases due to admixture of other kinetic steps. At low coverage and low temperature, first order kinetics results corresponding to desorption of the 2D-gas remaining after removal of the condensate; this is most clearly seen in the isothermal experiments. At the higher temperatures reached for the same coverages in TPD, however, the isosters become very ill-behaved (Fig. 6). There may well be a contribution from impurities and/or surface inhomogeneities with a range of binding energies. However, the well-behaved isothermal results as well as the results for Ar and Kr [20] and for the second layers make it more likely that in this range a strong kinetic non-equilibrium persists which makes the composition of the layer dependent on prehistory, so that the average coverage is not a sufficient variable. We are in the process of carrying out numerical simulations to test this hypothesis [25]. There may also be a contribution from a range of sites with variable properties (inhomogeneities and/or impurities).

This shows that thermodynamic as well as non-equilibrium effects exist in desorption from these systems. Dynamic effects which showed up in sticking are not strong enough for the heavy rare gases to be visible in desorption, besides the exponentials and besides the complex kinetic features. Their anisotropy effect - a decrease of the mean translational energy or a widening of angular distributions in desorption, as predicted by TULLY [16] and found by AUERBACH et al. [27] - need more selective measurements for their detection. In the case of Ne, where sticking showed dynamic effects to be very strong, they are expected to be detectable much more easily. Here measurements are in preparation.

While the gas temperatures were quite different in adsorption and desorption in these measurements, out of experimental necessities, so that a direct test of detailed balance is not possible, the discussion above has suggested that adsorption and desorption at least at high coverages proceed along the same paths (although the anisotropy between first and second layer sticking will certainly be smaller under isothermal than under non-isothermal conditions). When the internal phase equilibrium breaks down, the rate determining step can shift strongly. Then detailed balance certainly breaks down, at least one new slow variable is necessary to describe the behaviour, and there may be a dependence of the kinetics on prehistory. Measurements with strongly varying heating rates would be necessary to test this, but these are very difficult in this temperature range. An indication of such effects, however, comes from the different behaviour of TPD and ITD data as mentioned (see refs. 20 and 21 for details).

3.2 Molecular Chemisorption

The example we will consider here is CO on Ru(001), for which detailed information exists. Beam experiments, in particular with state selection, are mostly available for NO, because of the ease of applying laser methods to this molecule.

Extensive adsorption and desorption measurements have been done for CO/Ru(001) [13], with various methods (programmed and isothermal desorption, with gas phase and in-situ monitoring). The desorption measurements have shown very clear connections with the lateral interactions and the 2D-phase transitions which are their consequence [13,28]. CO is adsorbed standing up, with the C-end attached; it prefers to sit on top of Ru atoms [29]. The main lateral interaction is a strong repulsion between CO molecules on next neighbour Ru atoms which effectively excludes such a tight packing. As there is a weak attraction for near-next neighbours, islands of a $\sqrt{3}$-phase are formed from low coverages on; at coverage 1/3 (referred to Ru substrate atoms) which is about half the saturation coverage of CO, a rigid ordered layer exists with all molecules identical. At lower coverage, an order-disorder transition occurs in the range 150 to 340 K, depending on coverages [28,30]. At coverages above $\sqrt{3}$, complicated low temperature ordered phases exist which, however, all disorder below 150 K [30]. The lateral interactions also show up in high resolution infrared spectra [31]. The influence of the completion of the $\sqrt{3}$ phase is seen most easily in the drop of the binding or desorption energies (which are identical within errors of measurements) and of the pre-exponentials of desorption (fig. 8). The quasi-equilibrium assumption was tested by comparing the directly measured pre-exponentials of desorption to those obtained from equilibrium energies and sticking coefficients extrapolated to the temperature of equilibrium measurements. The best test of the overriding effect of the statistical mechanics of the layer on desorption kinetics can be seen in the high values and strong coverage dependences of k_0 (from $\sim 10^{16}$ s^{-1} at low coverage to $\sim 10^{19}$ at the $\sqrt{3}$ coverage, where it drops to $\sim 10^{14}$s^{-1} to slowly rise again towards saturation). These values, as well as the precipitous rise towards $\Theta = 1/3$ are easily explained by the low partition function (or entropy) of the rigid $\sqrt{3}$-layer [3,13,32]. The attractive interactions in the ordering layer are also seen in the increase of energy towards $\Theta = 1/3$. When more molecules are squeezed into this layer, not only the energy drops by the repulsive interaction, but the entropy increases because of the mobility of these disturbances, yielding a low and almost normal (for mobile layers) value of k_0. The k_0 values can be quantitatively accounted for by statistics. This is documented in the literature so well [3,13,33] that we can omit a discussion here. Similar behaviour has been seen for CO on Ni(111) [34]. Desorption kinetics is thus seen to be fully understandable in terms of thermodynamics in this case.

Comparison with the behaviour of sticking coefficients shows, however, that dynamic effects do exist even in this system which are not seen in desorption because the variations caused by them are comparatively small. Sticking at zero coverage is high, but not perfect ($s_0 = 0.7$ independent of surface temperature) and decreases as $\sim(1-\Theta)$ for the range where no islands of ordered $\sqrt{3}$ are formed ($T_S \sim 200$ K) to decrease sharply where this happens. At lower surface temperatures, sticking decreases more slowly with coverage and remains high to higher coverages. These results show that even in sticking, thermodynamics pushes through, so that dynamics and thermodynamics mix. The reason for this is that the properties of the precursors involved are strongly influenced by the structure of the layer, which in turn depends on the lateral interactions. We have termed this a mixing of lateral and vertical interactions [13,33]. The coverage dependence of s and its sensitivity to temperature has been explained by different types of precursors (2 different extrinsic ones on islands of the $\sqrt{3}$ layer; intrinsic on the clean surface). The influence of the surface phases on sticking stems from the existence (or not) of $\sqrt{3}$ islands. Of course, the various competing rates for precursor formation, integration and destruction also introduce temperature dependences. The s -value which clearly deviates from unity has been explained by imperfect accommodation of rotation. Strongly rotating molecules can be trapped in a weakly bound "dynamic precursor"

12

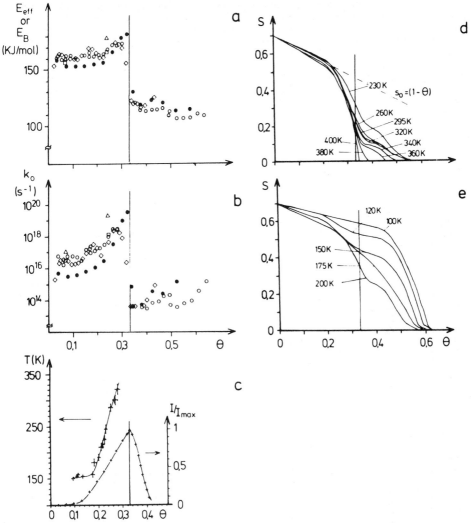

Fig. 8: Correlation of kinetic and thermodynamic data for CO/Ru(001), as function of coverage (after [33]).
a) Binding energy from quasiequilibrium measurements (full points), and desorption energies from various methods of thermal desorption (open symbols).

b) Preexponentials of desorption from the same kinetic data (open symbols), and from binding energies and sticking coefficient given in d,e) (open points).

c) Phase boundary of the √3 superstructure due to NNN occupation (left ordinate), and √3 intensity during filling of the layer (right ordinate).
d,e) Coverage dependence of the sticking coefficient at the given surface temperature, at Tg=300 K. Note the correlation to data of c).

[12,35] which does not feel the chemisorption potential because of the lacking orientation, so that molecules skid around on the surface in this state. As they have a chance to desorb before they lose their rotation, s_0 is less than unity. There may also be a conversion of rotational to translational energy upon impact, so that the chance of sticking is decreased. Both effects would lead to an underrepresentation of higher rotational excitations in desorbing molecules under quasi-equilibrium conditions ("rotational cooling") which has indeed been seen in state-resolved scattering [36] and desorption [37] of NO. We clearly see that there are again dynamic, kinetic, and equilibrium statistics effects in this system. Dynamics mostly shows up in the rotational behaviour; without it, the small deviation of s_0 from unity might not have been believed (the measurement is accurate to about 5%, though). One may ask why the kinetic effects which must exist through the complicated precursor systems do not show up. Formally, there are strong formal similarities between the phase coexistence in rare gases (see 3.1) and the precursor systems, as the same type of reaction sequence equations are valid in both cases. Why, then, are there no kinetic non-equilibrium effects seen here? The important difference is that precursors are energetically much closer to the gas phase than to the stable adsorbate state, while the opposite is true for co-existing surface phases. In the first case the relevant equilibrium (which may break down to lead to kinetic effects) involves the gas phase, in the second the stable adsorbate; equilibrium precedes (or not) adsorption for precursors, but desorption for phase co-existence. Therefore no effect of equilibrium break down results for the former in desorption.

3.3 Dissociative Adsorption

Here we only examine the simplest case, dissociative adsorption of H_2 into 2H atoms. Very interesting data exist for the three close-packed faces on nickel, most extensively for Ni(111), by Rendulic and coworkers [11,12]. Comparing equilibrium isotherms with isotherms constructed from (isothermal) sticking and desorption measurements starting at the same temperature, they have attained very good agreement which proves detailed balance for this system [11]. Furthermore, the angular dependences of sticking coefficients [12,38] for a wide range were shown to exhibit the same anisotropies which proves that essentially the same mechanism predominates under low and higher temperature conditions in this system. These anisotropies can be well understood with an activated adsorption model which assumes no or weak accommodation of parallel momentum. This is in good agreement with conclusions from translational energy distributions of desorbing H_2 [39]. Measurements of sticking and desorption from Ni(110) show no deviation from isotropy, which must be ascribed to precursor - mediated mechanisms, while a direct path over a barrier dominates for Ni(111) which makes dynamic effects strong, as already suggested by the low value of the sticking coefficient. Ni(100) appears to be intermediate [40]. We note that detailed balance applies for Ni(111) and ad- and desorption paths are the same for low and high temperatures; the system is seen to be governed by dynamic effects. The precursor kinetics on the other close-packed Ni faces, which lead to strong coupling of the molecule to the surface and thus high s_0-values, suppresses the dynamic effects seen in integral and even angular-dependent measurements. Whether there is a rotational anisotropy as in CO and NO, is an interesting question. On Cu(100), rotational influences of molecular H_2 adsorption at low temperature (corresponding to the adsorption of the precursor for dissociative adsorption) have been found [41]. On the other hand, a dramatic effect of surface thermodynamics has been observed in "explosive" associative desorption of H from Ni(110) which is coupled to the relief of H-induced surface reconstruction [42].

14

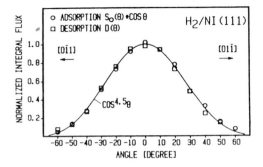

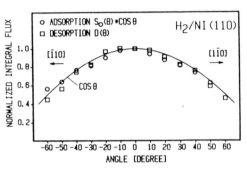

Fig. 9:
Angular dependence of sticking probability and desorption flux for H_2 on Ni(111) (top) and Ni(110) (bottom). Adsorption with Maxwellian beam of $T_g=300K$, for surface temperatures of 190 and 220 K, respectively. After Steinrück et al. [12], with permission.

3.4 Discussion

These examples have shown that thermodynamic effects (influences of the equilibrium statistics of the adlayer) most readily show up in coverage-dependent desorption measurements, while dynamic effects are most easily discerned in sticking measurements. As suggested by the decomposition of the rate equation into these two influences, dynamic effects are most likely to predominate in systems with sticking coefficients deviating strongly from unity. But even in cases where s is close to 1, dynamic effects can be seen in accurate sticking measurements; for the same systems desorption is often governed by thermodynamics, because of the overwhelming influence of the exponential term. Dynamic effects can also be seen well if a particular anisotropy is investigated selectively (angular or energy distributions, translational or internal).

Dynamic non-equipartition effects have been shown to exist under quasi-equilibrium conditions (c.f. $H_2/Ni(111)$), while equipartition can exist under irreversible conditions (c.f. angular distributions of desorbing CO, or $H_2/Ni(110)$). Strong non-equipartition effects, being due to the microscopic dynamics of molecule-surface collision, can be expected to persist in a wide temperature range, so that the mechanisms of interaction are not different at high and low temperature (c.f. $H_2/Ni(111)$). Isotropic systems, expected to involve precursors, may be much more sensitive to temperature, so that predominant mechanisms can easily change. If equilibrium within the layer or with precursors is an important feature, then other rates than adsorption or desorption can become rate-determining; additional variables can exist, the state of the system may depend on its prehistory, and simple analysis of rate data may not be possible any more. This shows that it makes sense to distinguish dynamic and kinetic nonequilibrium features, even though "dynamics" could be carried far enough to include the potential surfaces of all intermediate steps, or alternatively a kinetic treatment of parallel and consecutive reaction steps could be widened to include energy and momentum transfer. It appears more advantageous, however, to distinguish the influences as proposed.

Acknowledgments

This work has greatly profitted from valuable discussions with W. Brenig and H.J. Kreuzer as well as with M.J. Breitschafter, P. Feulner, H. Pfnür, H. Schlichting, and E. Umbach whom I also want to thank for their excellent cooperation. I thank H.J. Kreuzer and Dalhousie University for their hospitality during preparation of this manuscript, which was aided by a NATO travel grant. Our mentioned experiments were supported by the Deutsche Forschungsgemeinschaft through SFB 128.

Appendix

Here we make a few additional remarks which amplify the discussion of Sect. 2 in connection with experimental problems.

Isosteric heats and entropies are in almost all cases not obtained under equilibrium, in particular not under isothermal conditions. Because of the difficulty of changing the temperature of an isotropic gas, most measurements have been carried out with constant T_g, varying only T_s; i.e. under steady state conditions. A notable exception is ref. 11. The usual argument is that this only introduces an error of order kT in the isosteric heat. However, if the sticking coefficient is strongly dependent on gas temperature, then a much larger error could be introduced.

Furthermore, isosteric heats are determined from temperature dependences of the equilibrium pressure for constant coverage. As only about 4-5 powers of ten are accessible for the pressure, the temperature range of measurement necessarily changes with coverage, with high coverages being measured at low T and low coverages at high T. In a plot of E_{iso} (Θ) this influence does not show up but nevertheless exists. In particular, if 2-dimensional phase changes take place in the system concerned - which should express themselves in the E_{iso} behaviour - this must not be forgotten in their discussion. This is even more severe in Arrhenius plots of isosteric desorption rates where the window of measurement is often even much narrower (rarely more than 2-3 powers of 10). We re-emphasize here (see also ref. 3) the importance of adequate evaluation of desorption rates. Any analysis assuming a fixed order and/or the coverage-independence of a kinetic parameter (E_d, k_0, s) is totally inadequate for the derivation of coverage dependences, as are methods which only use peak maxima and widths. The preferred analysis is the construction of desorption isotherms with subsequent isosteric analysis for TPD (temperature-programmed desorption), and direct isosteric analysis for isothermal desorption measurements (see refs. 3, and earlier references therein). Such analyses are only sensible, of course, for sufficiently accurate and detailed data: TPD with variation of initial coverage and, even more preferable, of heating rate over a range as wide as possible; suitable construction of apparatus to ensure high signal/noise and signal/background (see the references given in ref. 3) and accurate temperature measurement and control (which is very difficult below 60 K). In cases of kinetic nonequilibrium effects, problems exist for ITD (isothermal desorption) measurements with temperature jump, as then the starting conditions may be difficult to define.

REFERENCES

1. J.H. deBoer: The Dynamical Character of Adsorption (Clarendon Press, Oxford 1968)
2. See for instance:
    D. Menzel: In Interactions on Metal Surfaces, ed. by R. Gomer (Springer, Berlin 1975) p.101;
    D.A. King: CRC Crit. Rev. Solid Mater. Sci 7, 167 (1978);
    J.T. Yates, Jr.: In Solid State Physics: Surfaces, Methods of Experimental Physics, vol. 22, ed. by R.L. Park and M.G. Lagally (Academic Press, Orlando 1985), p. 425
3. D. Menzel: In Chemistry and Physics of Solid Surfaces, ed. by R. Vanselow and R. Howe (Springer Berlin 1982), p.389
4. J.E. Lennard-Jones: Trans. Faraday Soc. 28, 333 (1932)
5. F.O. Goodman, Y. Wachman: Dynamics of Gas-Surface Scattering (Academic Press, New York 1976)
6. J.C. Tully: Ann. Rev. Phys. Chem. 31, 319 (1980)
7. H.J. Kreuzer, Z.W. Gortel: Physisorption Kinetics (Springer, Berlin 1986).
8. W. Brenig: this volume, and references therein.
9. J.A. Barker, D.J. Auerbach: Surface Sci. Reports 4, 1 (1985)
10. G. Comsa, R. David: Surface Sci. Reports 5, 145 (1985)
11. K.D. Rendulic, A. Winkler: J. Chem. Phys. 79, 5151 (1983)
12. H.P. Steinrück, M. Luger, A. Winkler, K.D. Rendulic: Phys. Rev. B 32, 5032 (1985);
    H.P. Steinrück, K.D. Rendulic, A. Winkler, Surface Sci. 152/3, 323 and 154, 99 (1985)
13. H. Pfnür, P. Feulner, H.A. Engelhardt, D. Menzel: Chem. Phys. Lett., 59, 481 (1978);
    H. Pfnür, D. Menzel: J. Chem. Phys. 79, 2400 (1983);
    H. Pfnür, P. Feulner, D. Menzel: J. Chem. Phys. 79, 4613 (1983)
14. W.H. Weinberg: this volume.
15. J.C. Keck: Disc. Faraday Soc. 33, 173 (1962);
    G. Iche, Ph. Nozieres: J. Physique 37, 1313 (1976);
    W. Brenig, K. Schönhammer: Z. Physik B24, 91
16. J.C. Tully: Surface Sci. 111, 461 (1981), and this volume
17. S. Glasstone, K.J. Laidler, H. Eyring: The Theory of Rate Processes (McGraw-Hill, New York 1941)
18. J. Kouptsidis, D. Menzel: Ber. Bunsenges. Phys. Chem. 74, 512 (1970)
19. W. Brenig: Z. Physik B48, 127 (1982)
20. M.J. Breitschafter: Ph.D. Thesis (T.U. München 1985);
    M.J. Breitschafter, E. Umbach, D. Menzel: in preparation
21. H. Schlichting: Ph.D. Thesis (T.U. München 1987);
    H. Schlichting, P. Feulner, D. Menzel: in preparation
22. Interestingly, s continues to rise above 1 monolayer and saturates only above 3 monolayers for Ne (fig. 2). This shows that the first 2 layers still feel the influence of weak coupling to the metal
23. D. Menzel, J. Kouptsidis: In Gas-Oberflächen-Wechselwirkung (Dornier, Friedrichshafen 1973) p. 282
24. R. Opila, R. Gomer: Surface Sci. 36, 1 (1981)
25. H.J. Kreuzer and D. Menzel, in preparation
26. U. Leuthäusser: Z. Phys. B50, 65 (1983)
27. J.E. Hurst, C.A. Becker, J.P. Cowin, K.C. Janda, L. Wharton, D.J. Auerbach: Phys. Rev. Letters 43, 1175 (1979)

28. H. Pfnür, D. Menzel: Surface Sci. 148, 411 (1984)
29. G. Michalk, W. Moritz, H. Pfnür, D. Menzel: Surface Sci. 129, 92 (1983)
30. E.D. Williams, W.H. Weinberg: Surface Sci. 82, 93 (1979);
    H. Pfnür, H.J. Heier: Ber. Bunsenges. phys. chem. 90, 272 (1986)
31. H. Pfnür, D. Menzel, F.M. Hoffmann, A. Ortega, A.M. Bradshaw: Surface Sci. 93, 431 (1980)
32. U. Leuthäusser: Z. Physik B37, 65 (1980)
33. D. Menzel, H. Pfnür, P. Feulner: Surface Sci. 126, 374 (1983)
34. H. Ibach, W. Erley, H. Wagner: Surface Sci. 92, 28 (1980)
35. D. Menzel: In Proceedings of the 3S'83 Conference Obertraun, ed. by P. Braun et al. (HTU Wien 1983) p. 218
36. J. Segner, H. Robota, W. Vielhaber, G. Ertl, F. Frenkel, J. Häger, W. Krieger, H. Walter: Surface Sci. 131, 273 (1983)
37. R.R. Cavanagh, D.S. King: Phys. Rev. Letters 46, 1829 (1981)
38. D.O. Hayward, A.O. Taylor: Chem. Phys. Letters 124, 264 (1986)
39. G. Comsa, R. David, B.J. Schumacher: Surface Sci. 85, 45 (1979)
40. H.P. Steinrück: private communication
41. S. Andersson, J. Harris: Phys. Rev. B27, 9 (1983);
    S. Andersson, L. Wilzen, J. Harris: Phys. Rev. Letters 55, 2591 (1985)
42. K. Christmann, F. Chehab, V. Penka, G. Ertl: Surface Sci. 152/53, 356 (1985)

18

Kinetics and Dynamics of Gas-Surface Interaction: The Principles of Detailed Balance and Unitarity

W. Brenig

Physik-Department, Technische Universität München,
D-8046 Garching, Fed. Rep. of Germany

Some results of a quantum theory for inelastic scattering, sticking and desorption of gas molecules at surfaces are reviewed. The theory is valid if the time scale $\tau_{life}=1/r$ introduced by the relaxation rates r of the vibrational states of adsorbates is large compared to their vibrational periods τ_{vib}. For sufficiently large activation energies of the adsorbates another time constant τ_{res}, the residence time of adsorbates can be determined from the theory. One then can distinguish essentially three different regimes corresponding to the three time scales τ_{vib}, τ_{life} and τ_{res}. There are two steps leading from one regime to the next one: The solution of the Schrödinger equation leading to the relaxation rates r of the kinetic equation, and the solution of the kinetic equation leading to sticking coefficients, accommodation coefficients, desorption rates and angular and energy distributions of desorbing particles. In all cases the principles of detailed balance and unitarity are very helpful in reducing the number of independent parameters of the theory and as a check of the results of calculations and experiments.

I. Introduction

Molecules which are scattered by surfaces quite often occur in two groups: A so-called direct scattering fraction and a trapping-desorption fraction. While the direct part can be described by quantum mechanical scattering theory alone the trapping-desorption part usually needs a kinetic treatment. The long time behavior of the corresponding kinetic equations then can be described approximately by the rate equations of chemistry.

Chemical processes usually occur at time scales which are large compared to typical microscopic times of molecular dynamics namely those of the order of the periods τ_{vib} of atomic vibrations in molecules and solids. This is due to the occurrence of Arrhenius factors exp(D/T) in chemical reactions with potential well depths D usually being large as compared to temperature T. The residence times $\tau_{res} = \tau_{vib}exp(D/T)$ therefore are usually large as compared to molecular vibrational periods.

From the point of view of molecular dynamics chemical processes such as desorption are tedious. Suppose the Arrhenius exponent exp(D/T) is of the order of 10^4 then the adsorbed particle moves back and forth in the attractive well for about 10^4 vibrational periods before it gets desorbed. During this time because of the coupling to the heat bath the motion is rather complicated in detail but boring on the average. Only during a small fraction of the time the motion differs from that in thermal equilibrium, namely if the particle energetically has come close to the top of the attractive well and is on its way to desorption. In order to calculate energy and angular distributions of desorbing particles one has to go through such boring calculations many (say hundred) times.

The obvious way to save computing time in such a situation is to "average first and then calculate" instead of "first calculate and then average".

In the following sections we are going to consider situations in which the coupling of the particle to the heat bath is weak enough that a successful scheme can be developed which works with distribution functions rather than individual particle trajectories and which leads to a simple description of kinetic and inelastic processes at surfaces.

II.   Kinetic Theory

Table 1 exhibits some typical time scales encountered in the interaction of gas particles with surfaces as well as the equations and parameters used to describe the dynamics in the corresponding regimes.

At zero time resolution (integrating the intensity of outgoing particles over all times) a generalized reflectivity P_{fi} is sufficient to describe the experimental situation completely [1,2]. P_{fi} is the probability that a particle with initial quantum numbers i occurs with final quantum numbers f after interaction with the surface or else in an experiment with constant (time independent) fluxes j_i, j_f one has $j_f = \sum P_{fi} j_i$. For atoms one may choose the components p and $\underline{P}$ of the momentum perpendicular and parallel to the surface as quantum numbers (or the energy and two polar angles), for molecules one needs the internal (vibrational and rotational) quantum numbers in addition.

Table 1:   Time scales in gas-surface interactions. The Schrödinger eq. is, of course, valid for all times, but used only for short times.

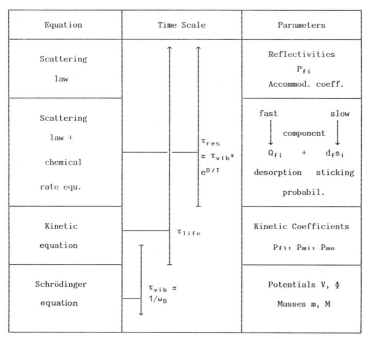

Equation	Time Scale	Parameters
Scattering law		Reflectivities P_{fi} Accommod. coeff.
Scattering law + chemical rate equ.	τ_{res} $\approx \tau_{vib}{}^x$ $e^{D/T}$	fast slow component Q_{fi} + $d_f s_i$ desorption sticking probabil.
Kinetic equation	τ_{life}	Kinetic Coefficients P_{fi}, P_{mi}, P_{mn}
Schrödinger equation	$\tau_{vib} \approx 1/\omega_D$	Potentials V, Φ Masses m, M

P_{fi} has to obey three general conditions:

1. Particle number conservation ("Unitarity")

$$\sum_f P_{fi} = 1 \tag{1}$$

2. The "equilibrium condition"

$$\sum_i P_{fi}(T_s) \, J_i(T_s) = J_f(T_s); \tag{2}$$

where

$$J_i(T_s) = |p_i| N_i(T_s) \sim |p_i| \exp(-\epsilon_i/T_s)$$

(and similarly for the final state) which guarantees that a Maxwellian gas of temperature T_s in contact with a surface at temperature T_s is in equilibrium.

3. Time reversal symmetry ("reciprocity")

$$P_{fi}(T_s) \, J_i(T_s) = P_{\underleftarrow{i}\,\underleftarrow{f}} \, J_f(T_s) \tag{3}$$

where $\underleftarrow{i}$ and $\underleftarrow{f}$ are the quantum numbers of the time reversed states. If the surface has spatial inversion symmetry (or is amorphous) the arrows of time inversion can be ommitted in (3). The resulting relation then usually is called "detailed balance". We shall use the same notation for (3) always. The equilibrium condition (2) is valid even without reciprocity (3). But if (1) and (3) hold (2) can be derived by summing (3) over i.

With increasing time resolution in most cases one can distinguish two different components in the outgoing beam: a fast and slow one. The fast component can be described by a contribution Q_{fi} to P_{fi} while the time behavior of the slow one is described in terms of a sticking coefficient s_i and a desorption rate r_d. The sticking coefficient is defined as the fraction of the incident flux $j_i(t)$ which is not scattered immediately but gets trapped for some time at the surface leading to an increase of the number $n^a(t)$ of adsorbed particles. The adsorbed particles then after some time are desorbed with a rate r_d. Both these phenomena are contained in the chemical rate equation

$$dn^a(t)/dt = \sum s_i j_i(t) - r_d n^a(t). \tag{4}$$

The energy and angular distribution of desorbing particles can be described by

$$P_{fi} = Q_{fi} + d_f s_i . \tag{5}$$

Such an equation is intuitively expected as expressing the statistical independence of the processes of adsorption and desorption. It can be derived from a time dependent master equation [2]. Since the total number of desorbing particles has to be equal to the number of adsorbed particles one has

$$\sum d_f = 1 \tag{6}$$

which, combined with total particle number conservation (1) yields

$$\sum Q_{fi} = 1 - s_i . \tag{7}$$

The equilibrium condition (2) combined with (4) leads to

$$\Sigma s_i J_i(T_s) = r_d N^a(T_s) \,. \tag{8}$$

Reciprocity then allows to express d_f in terms of the sticking coefficient. Combining (3) with (6) one finds

$$d_f = s_f J_f(T_s)/\Sigma s_i J_i(T_s) \,. \tag{9}$$

The situation is particularly simple if the sticking coefficient is close to one, then the angular dependence of desorbing particles is determined by the angular dependence of $J_f(T_s)$ which according to (2) is proportional to the <u>vertical</u> component p_f of the momentum, i.e. proportional to $\cos\Theta$ (Θ the angle between the surface normal and the direction of the outgoing beam). In many cases therefore the angular distribution of scattered particles consists of two groups: The fast component described by Q_{fi} in (5) (often called "directly scattered") producing a "lobe" in a polar diagram of scattered intensities near the specular direction and the slow component (the "trapping-desorption" part) which quite often is nearly circular in a polar diagram (see figures 1 to 3). The identification of the two components from the angular distribution alone is only hypothetical. It has to be supplemented by the energy distribution (and its angular dependence). Fig. 4 exhibits such measurements using time of flight techniques [5]. As one can see the energy dependence of the intensity of scattered particles is Maxwellian outside the lobular region, where one would expect the trapping-desorption component to dominate. Deviations from "cos Θ times Maxwellian" distributions occur if the sticking coefficient depends on angle of incidence and on energy. In this case the simultaneous measurement of sticking coefficients and distribution functions of desorbing particles provides an important cheque of "detailed balance" (3) [6].

Collecting everything what has been said from eq. (4) on one may say that particle number conservation, equilibrium condition and reciprocity allows to express everything in terms of two quantities s_i and $N^a(T_s)$ provided there is just one slowly varying variable $n^a(t)$ besides the (time

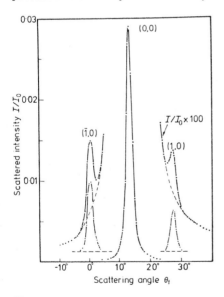

Fig. 1   Scattered intensity of He diffracted from Ag(111), in the ⟨112⟩ azimuth, $\Theta_i = 13.5°$. The intensity is normalized to the full beam intensity I_0. The first-order diffraction peaks are seen on a magnified scale (× 100). The diffraction peaks after substracting the background are also shown, [3].

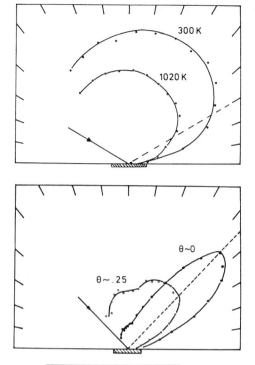

Fig. 2 Angular scattering distribution in the scattering plane for CO scattered from Pd (111) for substrate temperatures of 300 and 1020K. Angle of incidence = 60°, [4].

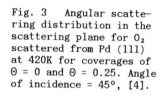

Fig. 3 Angular scattering distribution in the scattering plane for O_2 scattered from Pd (111) at 420K for coverages of $\Theta = 0$ and $\Theta = 0.25$. Angle of incidence = 45°, [4].

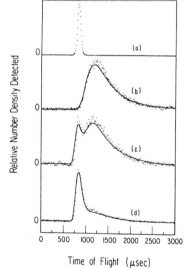

Fig. 4 Time-of-flight spectrum at $T_s = 185K$, angle of incidence 75° from normal. Curve (a): incident Xe beam with $\langle E_{kin}\rangle_i/k_B = 1615$ K; Curve (b): Xe scattered at 0° (normal); curve (c): 45°; curve (d): 75° (specular). Solid lines are from a model as described in [5].

integrated) fluxes. In principle, as we shall see, the quantity $N^a(T_s)$ is not an equilibrium property, but, in practice in most cases it can be identified with the total number of adsorbed particles in equilibrium in the so-called deep well limit. Then the only kinetic quantity for the time resolution we are considering so far is the sticking coefficient s_i. A microscopic theory therefore should concentrate on this quantity.

23

Unfortunately there seems to be no way which leads directly from the Schrödinger equation to the sticking coefficient, except for $T_s = 0$. Quite often, however, one has to deal with a situation with essentially one further well-separated time scale in the game. This occurs for light (or weakly bound) particles on heavy substrates. In this case the adsorbate has well defined vibrational levels with a life time τ_{life} large compared to the vibrational period and small compared to the residence time $\tau_{res} = 1/r_d$. The transitions between the vibrational and continuum states occur with certain transition rates which can be calculated directly from the Schrödinger equation. The calculation of the sticking coefficient then is reduced to the solution of a set of kinetic equations containing these transition rates as parameters. From Schrödinger's equation one can calculate certain so-called prompt sticking probabilities p_{mi} from an initial continuum state i to a bound state with quantum numbers m. This is the fraction of particles out of an incident particle in state i occurring in a bound state m after about a vibrational period (or one "round-trip" of the incident particle in the attractive potential well. From these prompt differential sticking coefficients one may determine a total prompt sticking probability

$$p_i = \Sigma p_{mi} . \tag{10}$$

As a first orientation the sticking coefficient s_i may be identified with p_i. This is correct only at $T_s = 0$. At nonzero T_s only a certain fraction, say, s_m of particles trapped after one round trip stays in the attractive well for a long time after cascading down the ladder of bound states. The fraction $1 - s_m$ reappears in the continuum after a few round trips by thermal excitations.

The true sticking coefficient then is given by

$$s_i = \Sigma s_m \, p_{mi} . \tag{11}$$

This equation may be supplemented by a similar one for the P_{fi}, namely [2,7]

$$P_{fi} = p_{fi} + \Sigma P_{fm} \, p_{mi} . \tag{11a}$$

Here the first term on the r.h.s. is the contribution of the fast (direct) continuum-continuum transitions which can be calculated directly from the Schrödinger equation. The second term contains all contributions from particles which intermediately have gone through bound states. The sticking-desorption part is one of these contributions. The quantity P_{fm} obeys an equation similar to (11a) but with the initial state i being replaced by a bound state quantum number, say, n.

$$P_{fn} = p_{fn} + \Sigma P_{fm} \, p_{mn} . \tag{12a}$$

The matrix elements p_{fn} and p_{mn} occurring in this equation can again be determined directly from the Schrödinger equation. P_{fm} is the probability of all transitions from a bound state to a continuum state. The p_{fi} and p_{fm} obey conditions in analogy to (1,2,3). In approximate solutions of the Schrödinger equation these conditions often are more or less violated. For instance if one uses the first order distorted wave Born approximation unitarity is violated. The violation is particularly bad for strong coupling and can lead to sticking coefficients larger than one [8]. If one uses classical trajectory approximations without recoil detailed balance is violated. Then the heating of cold beams by hot surfaces is not described properly [14].

The equation for s_m can, in general, be derived from the kinetic equation describing the transitions between the bound states. It is the eigenvalue equation for the lowest eigenvalue of the kinetic equation, which is the desorption rate r_d. This equation becomes particularly simple in the "deep well limit" mentioned above. The m-sum occuring in (12a) then can be extended to "minus infinity" and the probability P_{fm} for transitions from very low lying bound states into the continuum equals the probability for desorption d_f. In this case the equation for s_m approaches the homogeneous counterpart of (12a) which is related to (12a) just as (11) is related to (11a) [2]:

$$s_n = \sum s_m \, P_{mn} \qquad \text{with} \qquad s_m \to 1 \quad \text{for} \quad m \to -\infty \ . \tag{12}$$

The resulting equation is known from the theory of games as "gambler's ruin equation" [9]. The solution of this equation is rather simple since the range of r_{mn} usually is small compared to the well depth. The asymptotic limit $s_m = 1$ therefore is reached already not very far from the top of the well. Only in a thin shell (of order T_s) near the top s_m deviates from one and decreases with increasing energy to its asymptotic value zero at large energies.

We also mention without proof [2] that the quantity $N^a(T_s)$ occurring in (8) is given by

$$N^a(T_s) = \sum s_m \, N_m(T_s) \tag{13}$$

where $N_m(T_s)$ is the equilibrium (Boltzmann-) distribution of bound states in the potential well.

Since the dominant contributions to the r.h.s. of (13) come from the states m near the bottom of the well where s_m is close to one $N^a(T_s)$ is very well approximated by the equilibrium number of adsorbed particles, as mentioned already above.

Although the quantity r_d does not occur any more in the eq. (12) for s_m in the deep well limit it can be determined from the s_m via eqs. (8), (11) and (13).

III.   Microscopic Theory

We now turn to the calculation of transition probabilities from the Schrödinger equation. The strong coupling of atoms and molecules leads to a number of effects which can not be treated by the weak coupling theories used to treat the scattering of X-rays and neutrons from solids. Besides multiple scattering effects there are large energy transfers to the solid invalidating the first order distorted wave Born approximation (or single phonon approximation) which works so well for neutrons. (We are not going to consider energy transfer to the electrons. It can be expected to be at least an order of magnitude less important than to the phonons, may be even much smaller in many cases [10]).

A convenient measure of inelasticity is the average energy transfer $\Delta\epsilon$ after a single round trip to the solid. There are at least three more characteristic energies in the problem: The initial kinetic energy $\epsilon_p = p^2/2m$ of the scattering particle, the well depth D of the attractive well and the maximum energy $\hbar\omega_D$ of the phonon spectrum of the solid. Depending on the relative ratios of these energies one may distinguish different regimes.

A very useful simplification occurs if the energy transfer is small compared to the energy $\epsilon + D$ in the attractive well. It can therefore be neglected to first order. The scattering in lowest order thus occurs in a static potential. We shall see later on how the modification of the static potential (the socalled recoil effects) can be taken into account approximately.

A further simplification occurs if the energy transfer is small compared to $\hbar\omega_D$. This regime is well understood. It can be described by first order distorted wave Born approximation [11] and its various improvements [12]. We are now going to write down an expression for the transition rates interpolating between this approximation and the opposite limit $\Delta\epsilon \gg \hbar\omega_D$ where the lattice can be treated classically. For its derivation we again refer to the literature [13,14].

Apart from the restrictions on the energy transfer one usually introduces two further approximations:

1. One neglects momentum transfer parallel to the surface. This approximation is rather good [16,17] in particular for metal surfaces (which are quite flat).

2. One neglects collisions with more than one substrate atom at a time. The neglect of these so-called finite size effects is allowed for the collision of small atoms with large substrate atoms and can be corrected for larger adatoms to some extent at the end of the calculation by introducing effective masses [18].

In addition we treat the solid in the Debye model. The effect of Rayleigh surface phonons leads to some modifications [19]. The simplest way of correcting for all these approximations is by a "renormalization" of the two parameters (well depth and range) of the adatom-substrate potential which are fit parameters anyway: One fixes these parameters from one type of experiments (say on the accomodation coefficient) and predicts other experiments depending on the same type of transition probabilities, for instance the dependence of sticking coefficients on temperature and energy, the energy and angular distribution of desorbing particles.

The two ingredients of the theory are:
The spectrum of the force correlation function

$$S_i(\omega) = \langle p_i | f\delta(h-\epsilon_i-\omega)f | p_i \rangle m / |p_i| \tag{14}$$

with the vertical component of the force $f = \partial h/\partial z$, h the adatom Hamiltonian including the interaction with the substrate, and the spectral function of the vibrations of a substrate atom, which in the Debye approximation is given by

$$S(\omega) = 3\pi\omega\Theta(\omega_D-|\omega|)(1+n(\omega))/(2M\omega_D^3) . \tag{15}$$

Here Θ is the Heaviside step function and $n(\omega) = 1/(e^{\omega/T_s}-1)$ the average occupation number of phonons with frequency ω. If the parallel momentum is conserved one may introduce the energy or its "vertical part" $p_z^2/2m$ as the only relevant variable. Instead of the prompt transition probability p_{fi} one may then consider the quantity $p(\epsilon_f,\epsilon_i)$, or else, if one introduces the energy transfer $\epsilon = \epsilon_f-\epsilon_i$ and the initial energy as variables the function

$$p(\epsilon_f-\epsilon_i,\epsilon_i) = p_i(\epsilon) . \tag{16}$$

As a consequence of the approximations described so far this function can be expressed using "exponentiated perturbation theory" in terms of the two spectral functions mentioned (14,15) as [14]

$$p_i(\epsilon) = \int \exp[\int S_i(\omega)S(\omega)(e^{i\omega t}-1)d\omega + i\epsilon t]dt/(2\pi).\qquad(17)$$

A further approximation has turned out to be quite good, except at very low incident energies: the classical trajectory approximation [14,15]. In its simplest version the spectral function (14) is replaced by the absolute square of the Fourier transform $f_i(\omega)$ of the time dependence $f_i(t)$ of the classical force function of the scattering particle in the static potential of h (i.e. for fixed substrate atoms) and with an incident momentum p_i. This leads to the so-called "forced oscillator approximation":

$$S_i(\omega) = |f_i(\omega)|^2.\qquad(18)$$

This approximation violates the basic symmetry of the quantum mechanical matrix elements occurring in (14) between initial and final states. A more symmetric version would be

$$S_i(\omega) = |f_{\frac{p_i+p_f}{2}}(\omega)|^2\qquad(19)$$

or else

$$S_i(\omega) = |f_i(\omega) + f_f(\omega)|^2/4.\qquad(20)$$

Both versions are compared with the exact quantum results for He in fig. 5. As one can see the agreement with the exact result is surprisingly good and improved considerably as compared to the unsymmetric result.

For He the single phonon approximation is quite good. In this case p_f can be determined from p_i and ω from energy conservation

$$p_f{}^2/2m = p_i{}^2/2m + \hbar\omega.\qquad(21)$$

The classical trajectory approximation (18,19,20) now can be used to take into account the recoil effects discussed in connection with (17) in an

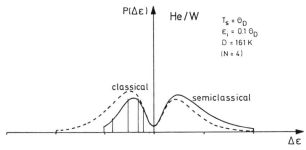

Fig. 5 Comparison of classical and quantum energy distributions in 1-phonon-approximation for He/W for incident energy $\epsilon_{i\perp} = 0.1\Theta_D$. --- is the classical result with unsymmetrized velocity (18). For $\Delta\epsilon = -0.1\Theta_D$ to $\Delta\epsilon = \Theta_D$. —— is the quantum mechanical result for free-free transitions (14) and also the classical analog with symmetrized velocity (19,20) both types of symmetrization falling together. For $\Delta\epsilon < -0.1\Theta_D$ —— is the symmetrized classical version for free-to-bound transitions and —·—·— indicates the corresponding intensity of the 4 discrete peaks in the quantum mechanical energy distribution.

approximate way: Instead of the static potential used in (18) one uses the correct classical trajectory including the response of the substrate during the collision. In a simplified version one uses again (19) or (20) with p_i and p_f taken from the actual classical trajectory including recoil, instead of (21). (21) may be called first order "quantum recoil"-correction. It occurs already in the static approximation just because in quantum theory initial and final state are different in a distorted wave approximation. There are of course higher order quantum recoil corrections which have to do with deviations from the first order DWBA [21].

A further quantum effect which is not contained in (18) is the total reflection of the wave function $\langle z|p_i\rangle$ occurring in (14) inside the attractive well for very low incident energy leading to a vanishing sticking coefficient. This effect, however, is not very relevant from an experimental point of view since it occurs only at extremely low energies [21].

Another quantum effect which is more important is the quantisation of substrate phonons. This is still contained in (17) even if the classical trajectory approximation (18,19,20) for the scattering particle is used. Phonon quantisation is particularly important for light particles such as He, Ne and H_2, or generally speaking for low average energy transfers $\Delta\epsilon$. It leads to large fractions of exactly elastically scattered particles as measured by the Debye Waller factor and a large broadening of the energy distribution of inelastically scattered particles (see the example of Ne in fig.6). For heavier particles the Debye-Waller factor becomes unobservably small but the "quantum broadening" of the energy distribution remains large. For Ar, Kr, Xe and N_2 for instance even at room temperature still about fifty per cent of the width of the energy distribution is due to the zero point vibrations of the substrate lattice (see fig.7,8).

Table 2 contains a list of quantum effects. For the calculation of sticking coefficients (and desorption rates) the most important effect is the quantum broadening, and for light particles such as He, Ne and H_2 the nonzero Debye-Waller factor. Since the sticking coefficient involves summation over the final phonon states and averaging over the initial energy distribution the most obvious quantum effect, namely the quantisation of adsorbate vibrations is of surprisingly little importance.

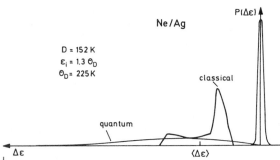

Fig. 6   Comparison of classical trajectory calculations for Ne/Ag [15] with the quantum energy distribution for $T_s=0$.   The classical curve is taken from Fig.10 [15].   The quantum curve is calculated for the same scattering condition and well depth. For the Morse length parameter we estimated $\ell^{-1}=1.5\text{\AA}^{-1}$.   The average energy loss is approximately equal in both calculations. Here the intensity of the elastic peak P_0 is illustrated by a Gaussian of finite width.

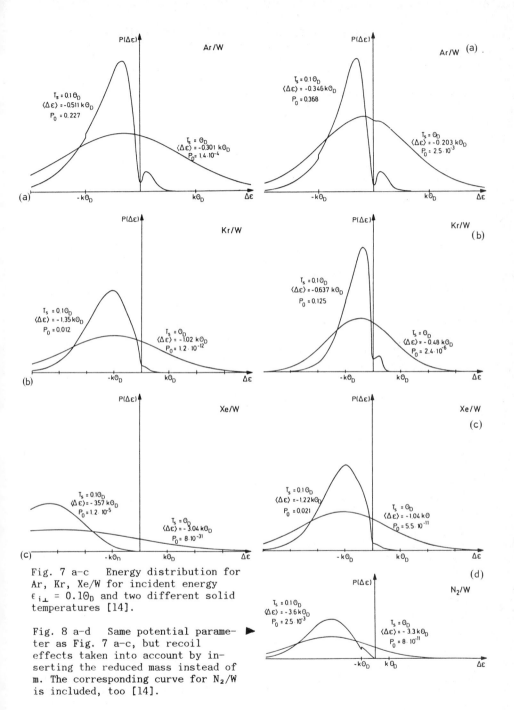

Fig. 7 a–c   Energy distribution for
Ar, Kr, Xe/W for incident energy
$\epsilon_{i\perp} = 0.1\Theta_D$ and two different solid
temperatures [14].

Fig. 8 a–d   Same potential parameter as Fig. 7 a–c, but recoil
effects taken into account by in-
serting the reduced mass instead of
m. The corresponding curve for N_2/W
is included, too [14].

Table 2: Survey of quantum effects in molecule-surface scattering

Quantum effect	Scattering property
"Quantum recoil" of scattered particle 1st order	heating of cold beam
Quantum recoil 2nd order	increase of inelasticity (below 1K for He)
"Transmission" problem	decrease of inelasticity and resonance behaviour (below 10^{-4}K for V ~ $1/Z^3$)
Diffraction effects	a) selective adsorption (below 40K for He, Ne) b) diffraction peaks
Vibrational level quantization	discrete final state contributions in $P(\epsilon)$
Phonon quantization of <u>substrate</u>	a) Debye-Waller factor b) broadening of energy and angular distribution

For heavier particles usually many phonons are excited (at least those of the lower part of the Debye spectrum) and the final energy distribution approaches a Gaussian, in particular at higher surface temperatures.

Rather than going into the evaluation of higher order corrections [20] of the approximation (17) we have checked it by comparing its results with exact numerical calculations for an Einstein model of the substrate phonons [22,23]. Fig. 9 shows the energy distribution of NO molecules scattered off Ag [22] and fig. 10 the temperature dependence of the Debye-Waller factor for He scattered off Xe/Graphite [23]. In both cases the many phonon and recoil effects are rather large, nevertheless the approximation (17) works quite well.

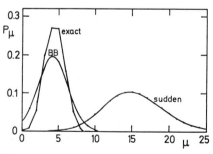

Fig. 9 Vibrational intensities from coupled channel calculations for NO/Ag compared to (17) denoted as BB and the so-called sudden approximation (see [22]).

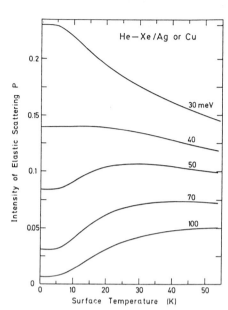

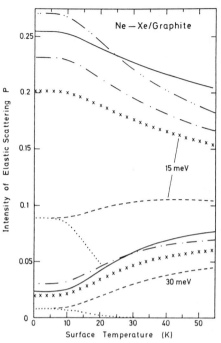

Fig. 10a   Exact elastic scattering intensity of He atoms from Xe adsorbed on Ag or Cu surface for various incident energies E_c=39.4 meV [23].

Fig. 10b   The elastic scattering probability of Ne atom from Xe adsorbed on graphite surface as a function of surface temperature without attractive part of scattering potential. E_c=12.4meV. —— is the exact calculation, – – – the sudden approximation, ··· the conventional Debye–Waller factor, x x x the strong coupling theory without and –·–·– with recoil effects in terms of the reduced mass. –··–··– is the strong coupling theory with recoil effects which is estimated by the use of classical trajectory calculation for coupled oscillator scattering particle system: The zero point motion of the oscillator has been taken into account and an average over initial phases of the oscillator has been taken [23].

At low incident energy $\epsilon \ll D$ of the scattered particle another simplification occurs: The scattering which is dominated by the energy $\epsilon + D \approx D$ inside the well becomes approximately independent of the incident energy and thus $p_i(\epsilon)$ in (16) becomes independent of i. In the Gaussian regime then the set of kinetic coefficients can be described in terms of two parameters only: The average energy transfer $\Delta\epsilon$ (the center of the

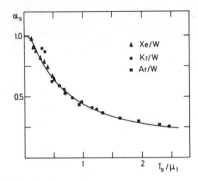

Fig. 11 Calculated isothermal accommodation coefficients α_s compared to experiment. The parameter μ_1 is the first moment of the Gaussian energy distribution [25].

Gaussian) and the width σ of the Gaussian. Both parameters, however, are not independent but related by reciprocity. Applying (3) to a Gaussian, one finds

$$\sigma^2 = \Delta\epsilon \, T_s . \tag{22}$$

Since the average energy transfer depends only little on surface temperature (see fig. 7,8), we have neglected this dependence in our kinetic calculations.

For a given substrate one then has only one independent energy scale in the problem. Experimental results confirm this prediction. Fig. 11 exhibits the dependence of the so-called accommodation coefficient

$$\alpha(T_s, T_g) = \sum P_{fi}(T_s)(\epsilon_f - \epsilon_i) J_i(T_g) / [2(T_s - T_g) \sum J_j(T_g)] \tag{23}$$

in the isothermal case $T_s = T_g$ plotted versus $T_s/\Delta\epsilon$ ($\Delta\epsilon$ called μ_1 in the figure). If $p_i(\epsilon) = p(\epsilon)$ independent of i a kind of "Wiedemann-Franz"-law holds: Sticking coefficient and accommodation coefficient are equal [2,25].

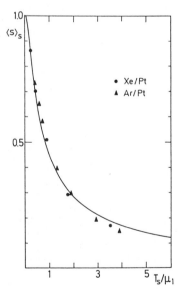

Fig. 12 Theoretical isothermal sticking coefficient in comparison with the computer simulated $\langle s \rangle_s$ as a function of the reduced temperature T_s/μ_1. ($\blacktriangle$)Ar/Pt:μ_1=130K, ($\bullet$) Xe/Pt:μ_1=575K [26].

The result of a calculation of the sticking coefficient using the simplified theory described above is compared to computer experiments [26] in fig. 12 [25]. The agreement is quite good demonstrating again, that the key quantity is the average energy transfer $\Delta\epsilon = \mu_1$. The same is true for the energy dependence on initial incident energy though not so obvious from fig. 13 since the two curves were obtained for the same surface temperature T_s [25].

Once the energy dependence of the sticking coefficient is known, the energy and angular dependence of desorbing particles can be calculated using reciprocity (3), (9): The intensity of the outgoing Maxwellian beam with temperature T_s has to be multiplied by the sticking coefficient. Usually this leads to a "cooling" of the outgoing beam: The "effective temperature" of the desorbing particles is lower than the surface temperature. A typical example is shown in fig. 14 giving the average energy of desorbing particles divided by that of a Maxwellian beam with temperature T_s.

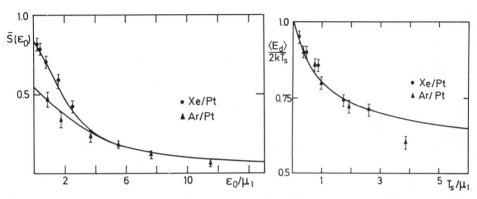

Fig. 13   Averaged sticking coefficients $\bar{s}(\epsilon_0) = \dfrac{1}{\epsilon_0} \int_0^{\epsilon_0} s(\epsilon_n)d\epsilon_n$ as a function of energy, compared with computer simulations [27]. The parameter μ_1 is the first moment of the Gaussian energy distribution. (▲) Ar/Pt: μ_1=130K, (●) Xe/Pt: μ_1=575K. Surface temperature is 250K. For $\epsilon_0 \gg \mu_1$ the sticking coefficient is given by $\bar{s}(\epsilon_0) \sim \mu_1/\epsilon_0$ independent of surface temperature T_s [26].

Fig. 14   Mean desorption energy, in units of $2kT_s$, as a function of T_s/μ_1 compared with [26], see [25].

Similar results can be found regarding the dependence on internal degrees of freedom, for instance rotational degrees for desorbing molecules. Since molecules with a high rotational energy have a tendency to transfer this energy into translational energy they have a low sticking coefficient. The decrease of sticking coefficient with rotational energy leads to "rotational cooling": The average rotational energy of desorbing molecules is lower than the average energy at the surface temperature (see fig. 14).

A simple semiquantitative explanation of this effect can again be given, using detailed balance. The energy distribution P_j of desorbing particles is

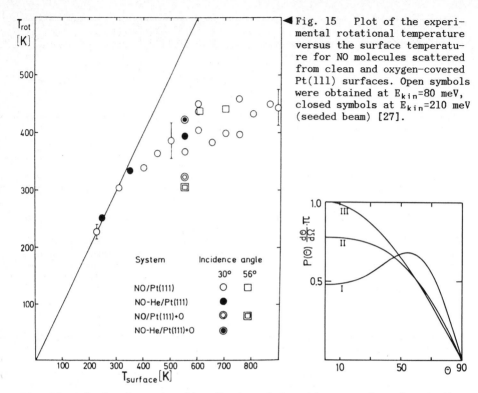

Fig. 16 Calculated angular distribution of desorbing Ar and Xe from Pt in polar angle θ, measured with respect to the surface normal. Surface temperature is 250 K. Curve I: Ar/Pt, curve II: Xe/Pt. The cosine distribution is shown for comparison (curve III) [25].

given by a Boltzmann distribution multiplied by the sticking coefficient $s(E_j)$. The linear dependence of ℓnP_j on E_j indicates an exponential dependence $s(E_j) = \exp(- E_j/E_r)$ of s leading to an effective temperature of desorbing molecules $T_{eff} = T_s E_r/(kT_s + E_r)$. E_r can be related to the average rotational energy transfer [28].

Angular distributions of desorbing particles have also been calculated using (9) [25]. One can find strong deviations from the "cos θ" behaviour of complete accomodation (see fig. 16).

Finally we mention that the validity of reciprocity (or detailed balance) has also been checked experimentally [6]. A typical example of angular distributions is shown in fig. 17.

IV. Discussion and Conclusion

A rather detailed description of sticking, desorption and inelastic scattering is possible if a clear separation of three time scales exists: Molecular vibrational periods, life times of vibrational states of adsorbed particles and residence times of adsorbed particles. In this case the theoretical problem can be reduced to the solution of the Schrödinger equation for the short time behavior and a kinetic equation for the intermediate and long time behavior.

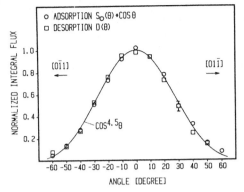

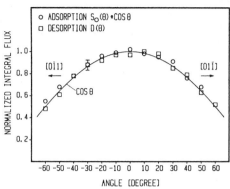

Fig. 17a   The expression $S(\theta)\cos\theta$ and the probability of desorption $D(\theta)$ for $H_2/Ni(111)$ show identical angular variation. Both follow a $\cos^{4.5}\theta$ function. Adsorption for a Maxwellian beam with $T_g=300K$ on a surface at $T_s=190K$ [6].

Fig. 17b   The expression $S(\theta)\cos\theta$ and the probability of desorption $D(\theta)$ for $CO/Ni(111)$ show identical angular variation. Both follow a cosine function. Adsorption for a Maxwellian beam with $T_g=300K$ on a surface at $T_s=300K$ [6].

A full quantum theory of kinetic processes in the more general case if the life time of vibrational states becomes short (so that the energy of vibrational levels is no longer well defined) does not exist so far.

The principle of detailed balance serves as an important check of the calculations and as a guide to reduce the number of independent parameters on all levels (microscopic, kinetic and macroscopic) of the calculation as well as a check of experimental results.

Since detailed balance is so important one might wonder about its validity. As pointed out in section II of this article the two ingredients of (3) are: (i) Time reversal invariance, (ii) initial equilibrium of the substrate. Since these two conditions are essentially always satisfied one might get the idea that detailed balance is always valid. In principle this is true. The problem in practice is the choice of the appropriate number of slow variables in the kinetic theory. If, for instance, there are other variables besides the number of adsorbed particles, varying on the time scale of the residence time, they have to be taken into account explicitly in the macroscopic kinetic treatment as well as in the detailed balance relation.

A typical example is the diffusion of reaction partners before the desorption of a reaction product. In this case something like the average distance between reaction partners on the surface may serve as a second slow variable. The macroscopic kinetics then can not be described simply in terms of a single sticking and desorption coefficient.

Similarly if precursors (intrinsic as well as extrinsic) are involved the macroscopic kinetics has to take into account the appropriate number of variables and kinetic coefficients.

In conclusion one may say, that the identification of a proper set of slow variables is the first important step in setting up a kinetic model. Detailed balance then always serves as an important principle to reduce the number of independent parameters of the model.

35

References

1. Kuscer, I.: Proc. of the 9th interntl. symp. on rarif. gas dyn. (1974), E.1-1
2. Brenig, W.: Z. Phys. B,48, 127 (1982)
3. Boato, G., Cantini, P., Tatarek, R.: J.Phys. F6, L 237 (1976)
4. Engel, T.: J. Chem. Phys. 69, 373 (1978)
5. Hurst, J.E., Becker, C.A., Cowin, J.P., Janda, K.C., Wharton, L., Auerbach, D.J.: Phys.Rev. Lett.43,1175(1979)
6. Steinrück, H.P., Rendulic, K.D., Winkler, A.: Surface Sci.154,99(1985)
7. Müller, H., Brenig, W.: Z. Phys. B 34, 165 (1979)
8. Kreuzer, H.J., Gortel, Z.W.: Physisorption Kinetics, Springer Series in Surface Sci.1, pge. 240
9. Iche, G., Nozieres, P.: J. Phys. (Paris) 37, 1313 (1976)
10. Schönhammer, K., Gunnarsson, O.: Phys. Rev. B 24, 7084 (1981) and Surf. Sci. 117, 53 (1982)
11. Lennard-Jones, J.E., Strachan, C.: Proc. Roy. Soc., A 150, 442 (1935)
12. Goodmann, F.O., Wachmann, Y.: Dynamics of Gas-Surface Scattering (Academic Press, N.Y. (1976))
13. Brenig, W.: Z. Phys. B 36, 81 (1979)
14. Böheim, J., Brenig, W.: Z. Phys. B 41, 243 (1981)
15. Barker, J.A., Dion, D.R., Merrill, R.P.: Surf. Sci. 95, 15 (1980) Compare also: Levi, A.C.: Nuovo Cim. 54 B, 357 (1979) and Kaplan,J.I., Drauglis, E.: Surf. Sci. 36, 1 (1973)
16. Brako, R., Newns, D.M.: Surface Sci. 117, 42 (1982) Brako, R.: Surface Sci. 123, 439 (1982)
17. Leuthäuser, U.: Surface. Sci. 145, 48 (1984)
18. Hoinkes, H., Nahr, H., Wilsch, H.:Surface Sci. 33, 516 (1972)
19. Stutzki, J., Brenig, W.: Z.Phys. B 45, 49 (1981)
20. Newns, D.M.: Springer Ser. in Sol. St. Sci. 59, 26 (1985)
21. Böheim, J., Brenig, W., Stutzki, J.: Z.Phys. B 48, 43 (1982)
22. Brenig, W., Kasai, H., Müller, H.: Springer Ser. in Sol. St. Sci. 59, 2 (1985)
23. Kasai, H., Brenig, W.: Z.Phys. B 59, 429 (1985)
24. Leuthäuser, U.: Z.Phys. B 44, 101 (1980)
25. Leuthäuser, U.: Z.Phys. B 50, 65 (1983)
26. Tully, J.C.: Surface Sci. 111, 461 (1981)
27. Segner, J., Robota, H., Vielhaber, W., Ertl, G., Frenkel, F., Häger, J., Krieger W., Walther, H.: Surface Sci. 131, 273 (1983)
28. Brenig, W., Kasai, H., Müller, H.: Surface Sci. 161, 608 (1985)

Energy Distributions of Thermally Desorbed Molecules: NO on AG(111) and Pt(111)

J.C. Tully

AT&T Bell Laboratories, 600 Mountain Ave., Murray Hill, NJ 07974, USA

1. Introduction

There has been considerable recent experimental and theoretical interest in the translational and internal energies of molecules desorbed from surfaces [1-13]. Deviations from equilibrium populations have been both predicted and observed. Such deviations can provide clues to the desorption mechanism and to the underlying gas-surface interactions.

In this paper we present calculations of the translational and rotational energies of NO molecules thermally desorbed from Ag(111) and Pt(111) for surface temperatures ranging from 100 K to 1400 K. Similar calculations have been reported previously [13]. However, the NO-Ag gas-surface interaction potential employed in the previous work has been shown to be inadequate to describe recent measurements of joint angular-rotational-translational distributions in hyperthermal scattering of NO from silver [14]. New interaction potentials have been developed which remove the shortcoming of the old potentials, and which accurately describe the recent measurements in full detail [14]. We employ these new potentials in the present work to investigate sticking probabilities and thermal desorption.

2. Calculations

The stochastic classical trajectory method has been described elsewhere [15,16]. Briefly, trajectories are calculated by integration of the three-dimensional classical mechanical equations of motion for the NO molecule and, in this case, the four nearest surface atoms. Energy flow between the four active surface atoms and the rest of the lattice is accounted for by the introduction of generalized friction and random force terms in the equations of motion, with memory chosen to reproduce the phonon spectrum of the metal surface [15]. With this procedure, any desired temperature of the local four-atom zone can be selected. Repeated redefinition of the active surface zone is required to follow the mobile adspecies laterally across the surface [15]. For the computations reported here, the friction and random forces employed for the Ag(111) and Pt(111) surfaces were identical to those reported in Ref. 13. The NO molecule was taken to be a rigid rotor in these calculations; i.e., vibrational effects were not included, as discussed below.

The interaction potentials employed for this study are of the same form as those reported previously [12], but with differences in choices of certain parameters. The potential V is given by

$$V = \sum_i V_i(r_N, r_O, r_i) \tag{1}$$

$$+ C(z-z_o)^{-9} - (D + E\cos^2\theta_s)(z-z_o)^{-3} ,$$

where r_N and r_O are the coordinates of the nitrogen and oxygen atoms, respectively, r_i is the coordinate of the ith substrate atom, z is the perpendicular distance of the center of mass of the NO molecule from the surface plane, and θ_s is the angle of the diatomic axis with respect to the surface normal ($\theta_s=0$ corresponds to the nitrogen end pointing to the surface).

The second and third terms of Eq. (1) describe a weak van der Waals type interaction, with the parameters C, D, E, and z_o chosen by the procedure employed in Ref. 13.

The directional chemical interaction between NO and the surface is represented by the first term of Eq. (1). The sum is carried out over the 14 nearest surface atoms, 4 moving and 10 fixed.

$$V_i(r_N, r_O, r_i) = A \exp(-\alpha|r_i - r_O|) \tag{2}$$
$$+ B\{\exp[-2\beta(|r_i - r_N| - r_e)]$$
$$- 2\cos^2\eta_i \exp[-\beta(|r_i - r_N|)]\}.$$

The first term in Eq. (2) is a simple exponential repulsion between the oxygen atom and surface atom i. The second term describes the attractive interaction between the nitrogen end of the molecule and the surface atom. It is a Morse potential, modified by the orientation dependent $\cos^2\eta_i$ term which insures that maximum binding is obtained only if the molecule is oriented normal to the surface.

The parameters are listed in Table I. They were chosen to be the same as those employed in Ref. 13, with the following changes. The major change was setting $z_o = 0$. This parameter is the "image plane" in the 3-9 term of the potential. The previous value of 1.0 Å appeared reasonable, but it produced non-physical behavior in some situations. The 3-9 potential is a rigid addition to the pairwise potentials, with no energy dissipation. For molecules scattered at very high energies, the rigid z^{-9} repulsion will dominate, resulting in an underestimate of the energy exchange. For energies in the 1 eV range this effect is not large, but is nevertheless significant. Moving the image plane back to $z_o = 0$ eliminates this problem.

Making this change, however, alters the corrugation of the potential. With $z_o = 1$ Å, the z^{-9} repulsion fills in somewhat between surface atoms, reducing the surface corrugation. With $z_o = 0$, the surface thus becomes too corrugated. In order to preserve the correct corrugation, as measured by the experimental angular

Table I. Interaction potential parameters

A	82170	kJ/mole
B (Ag)	13.0	kJ/mole
B (Pt)	59.0	kJ/mole
C	842.5	Å^9 - kJ/mole
D	106.6	Å^3 - kJ/mole
E	7.19	Å^3 - kJ/mole
z_o	0.	Å
α	3.366	Å^{-1}
β	1.683	Å^{-1}
r_e	2.60	Å

scattering distributions of Asada [17], the size parameters of the O and N atoms had to be altered: A from 6.1×10^4 to 8.2×10^4 kJ/mole and r_e from 1.5 to 2.6 Å. Finally, the new parameter values resulted in an increased energy exchange in NO-Ag(111) collisions. In order to reproduce the experimental translational energy distributions of Asada and Matsui [18], the parameter B was decreased to 13 kJ/mole for the silver surface. This produces a binding energy for NO on the Ag(111) surface of 23.8 kJ/mole, in relatively good accord with experimental estimates [19]. For NO on Pt(111), all parameters were chosen to be identical to those for NO on Ag(111), except the Morse energy parameter B which was chosen to be 59 kJ/mole in order to reproduce the experimental binding energy for NO on Pt(111) of 105 kJ/mole [20].

With the parameters listed in Table I, stochastic trajectory simulations for scattering of NO from Ag(111) reproduce accurately the detailed angle-translational energy-rotational energy distributions reported in Ref. 14. There are no experiments of comparable detail against which to test the NO-Pt(111) interactions.

3. Results and Discussion

Calculations of the rotational and translational energy distributions of thermally desorbed molecules were carried out as in previous studies, using detailed balance to circumvent problems associated with long residence times on the surface [2,13,21]. Molecules were directed at the surface with random initial molecular orientation and impact position, with random $(\cos\theta)$ incident angle of approach, with Maxwellian velocity distribution, and with Boltzmann initial rotational distribution. The temperatures of the latter two distributions were taken to be identical and equal to the surface temperature, T. A sample of 4000 trajectories was employed for each run.

Molecules were defined to be "stuck" if their total energy achieved a value below $-2kT$, where k is Boltzmann's constant. This criterion was shown to be essentially equivalent to requiring complete equilibration [2,21]. The mean translational and vibrational energies of that fraction of the 4000 trajectories which were stuck by this criterion were identified, via detailed balance, as the mean energies of molecules which were initially thermalized on the surface and subsequently desorbed thermally.

Fig. 1 shows calculated mean translational and rotational energies, expressed in degrees Kelvin, vs. surface temperature. The expected equilibrium values would fall along the diagonal lines. As shown in the figure, calculated results deviate significantly from the diagonal, especially at the higher temperatures.

Calculated mean rotational energies compare quite well with experimental determinations for NO desorbing from both Ag [22] and Pt [12]. There are no direct experimental comparisons for translational energy. However, the present results are qualitatively similar to those calculated for rare-gas desorption [2], which in turn are in quantitative agreement with experiment [1].

Thus both the translational and rotational energies of thermally desorbed NO are found to be "cooler" than the surface temperature, even though molecules have thoroughly equilibrated on the surface prior to desorption. The non-equilibrium behavior arises as the molecule escapes from the surface, and is due to the relative inefficiency of energy transfer between the surface and the molecule. The picture is quite simple for translational energy. In order for the molecule to break away from the surface, it must acquire an energy greater than its binding energy to the surface. Excess energy will appear as final translational energy. In order to acquire the required energy, the molecule vibrates in the binding well, continually exchanging energy with the surface. If the average energy exchanged per vibrational period is less than kT, then when the molecule is finally energized above the top of the binding well, its

39

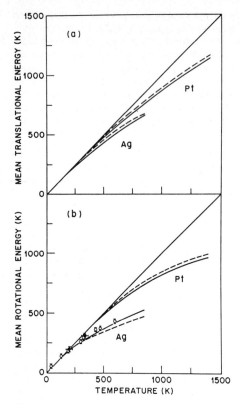

Fig. 1 - a) Calculated mean translational energy of NO thermally desorbed from Ag(111) and Pt(111), expressed in degrees Kelvin. Solid curves: energy associated with motion in the normal direction. Dashed curves: energy associated with parallel motion. b) Calculated mean rotational energy. Solid curves: angular momentum vector parallel to the surface normal. Dashed curves: angular momentum vector perpendicular to the surface normal. Open circles, Ag experiments from Ref. 22. Solid circles, Pt experiments from Ref. 12.

excess energy, on average, must be less than kT. Thus it will be translationally cold. Note that when the average energy exchanged per collision with the surface is less than kT, the sticking probability will be correspondingly less than unity, as required by detailed balance.

If the surface were perfectly flat, only translational energy in the normal direction would be depleted. Parallel translational motion would be completely thermal. As shown in Fig. 1, parallel motion is calculated to be more nearly thermal than normal motion, but is still significantly cooler than the surface temperature. Thus, even though the Ag(111) and Pt(111) surfaces are quite flat, corrugations are sufficiently large to quite effectively couple normal and parallel translational motion.

Depletion of rotational energy is somewhat more complex, with two effects operative. First, there is a strong coupling between rotation and translation induced by the surface. It has been shown [13] that transfer of translational to rotational energy upon impact of a molecule with the surface can make an important contribution to sticking. If so, the reverse must also be true. Transfer of (frustrated) rotational energy into translation must make an important contribution towards supplying the translational energy required to produce desorption. If rotational energy is not replenished sufficiently fast, this will result in depleted rotational energy [8,9]. Since rotational motion with angular momentum vector aligned perpendicular to the surface normal is most strongly coupled to translation, via the strong orientational forces in the interaction potential, it is this component that is most significantly depleted of energy by this mechanism.

The second mechanism is related to that proposed by Gadzuk, et al. [4] with the interaction potentials employed for this calculation, the NO molecule in its most stable binding configuration is oriented perpendicular to the surface. Thus, the bending vibrations (frustrated rotations) that are thermally populated prior to desorption will evolve into rotational motion with angular momentum vector perpendicular to the surface normal; i.e., tumbling motion. The adsorbed molecule can have essentially no component of angular momentum along the surface normal until it breaks out of the binding well that holds it perpendicular to the surface. Trajectories demonstrate that prior to desorption molecules do achieve a state of activation involving nearly free rotation and lateral translation. However, memory of the oriented configuration from which it originated may remain. This mechanism would deplete predominantly rotational motion associated with angular momentum vector along the surface normal (solid curve in Fig. 1b). The effects of the two mechanisms are comparable, in contrast to the conclusions of Ref. 13. In the case of NO on Ag(111), with relatively weak orientational forces, the first mechanism is predominant. However, the calculations predict that the second mechanism may actually be more important for the strong orientation case of NO on Pt(111). Measurements of the polarization of rotation of desorbing molecules could distinguish between these mechanisms and possibly provide a clue to the orientation of the bound molecule on the surface.

As reported in Ref. 13, the sticking probability of NO on Ag(111) and Pt(111) is calculated to be independent of incident vibrational energy. Coupling of vibration to translation or rotation is too weak to have a significant effect. Thus, the distribution of vibrational energies of NO molecules thermally desorbed from Ag(111) or Pt(111) are expected to be Boltzmann at the surface temperature for the entire range of temperatures considered here.

Since the calculations reported here were performed backwards, obtaining sticking distributions rather than desorption distributions, the overall sticking probabilities are also obtained. These are reported in Fig. 2. Note that these sticking probabilities refer to surface and gas at equilibrium at the same temperature. Sticking probabilities are higher for the more strongly bound NO-Pt(111) system, but the shapes of the two curves are quite similar. In fact, the curves nearly scale, with the sticking probability for NO on Ag(111) at temperature T nearly equal to that for NO on Pt(111) at temperature 3T. Note that the scaling factor of 3 is not equal to the ratio of binding energies of about 5. This could arise for two reasons. First, since Pt atoms are more massive than Ag atoms and the surface Debye frequency is somewhat higher for Pt(111), energy exchange is expected to be somewhat less efficient for NO collisions with Pt(111) than with Ag(111), at least at high collision energies. Secondly, because of the strong orientational dependence of the potentials, regions near the minimum energy configuration contribute only a small fraction of the phase space available for adsorption. Some weighted average over configurations would be more appropriate for comparison, and this would reduce the factor below 5. Comparing Figures 1 and 2,

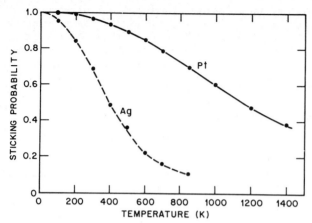

Fig. 2 - Calculated sticking probabilities for NO on Ag(111) (dashed curve) and Pt(111) (solid curve). Initial translational and rotational distributions were Boltzmann at the same temperature as the surface, and the initial angular distributions were cosθ. Trajectories were defined as "stuck" if the total energy of the molecule achieved a value less than −2kT.

translational and rotational energies are near thermal at temperatures for which the sticking probability is near unity, as required by detailed balance, and deviate significantly from thermal only when sticking probabilities become much less than unity.

4. REFERENCES

[1] J. E. Hurst, Jr., L. Wharton, K. C. Janda and D. J. Auerbach, J. Chem. Phys. *83*, 1376 (1985).

[2] J. C. Tully, Surf. Sci. *111*, 461 (1981).

[3] R. R. Cavanagh and D. S. King, Phys. Rev. Lett. *47*, 1829 (1981); D. S. King and R. R. Cavanagh, J. Chem. Phys. *76*, 5634 (1982).

[4] J. W. Gadzuk, U. Landman, E. J. Kuster, C. L. Cleveland and R. N. Barnett, Phys. Rev. Lett. *49*, 426 (1982).

[5] M. Asscher, W. L. Guthrie, T. H. Lin and G. A. Somorjai, J. Chem. Phys. *78*, 6992 (1983).

[6] D. Burgess, Jr., R. Viswanathan, J. Hussla, P. C. Stair and E. Weitz, J. Chem. Phys. *79*, 5200 (1983).

[7] J. Segner, H. Robota, W. Vielhaber, G. Ertl, F. Frankel, J. Häger, W. Krieger and H. Walther, Surf. Sci. *131*, 273 (1983).

[8] J. M. Bowman and J. L. Gossage, Chem. Phys. Lett. *96*, 481 (1983).

[9] J. E. Adams, Chem. Phys. Lett. *110* 155 (1984).

[10] R. R. Lucchese and J. C. Tully, J. Chem. Phys. *81*, 6313 (1984).

[11] J. C. Polanyi and R. J. Wolf, J. Chem. Phys. *82*, 1555 (1985).

[12] D. S. King, D. A. Mantell and R. R. Cavanagh, J. Chem. Phys. *82*, 1046 (1985).

[13] C. W. Muhlhausen, L. R. Williams and J. C. Tully, J. Chem. Phys. *83*, 2594 (1985).

[14] J. Kimman, C. T. Rettner, D. J. Auerbach, J. A. Barker and J. C. Tully, Phys. Rev. Lett. *57*, 2053 (1986).

[15] J. C. Tully, J. Chem. Phys. *73*, 1975 (1980).

[16] J. C. Tully, in Many-Body Phenomena at Surfaces, D. Langreth and H. Suhl, eds. (Academic Press, NY, 1984), p.377.

[17] H. Asada, Jpn. J. Appl. Phys. *20*, 527 (1981).

[18] H. Asada and T. Matsui, Jpn. J. Appl. Phys. *21*, 259 (1982).

[19] D. J. Auerbach, A. W. Kleyn and A. C. Luntz, to be published.

[20] J. A. Serri, J. C. Tully and M. J. Cardillo, J. Chem. Phys. *79*, 1530 (1983).

[21] E. K. Grimmelmann, J. C. Tully and E. Helfand, J. Chem. Phys. *74*, 5300 (1981).

[22] A. W. Kleyn, A. C. Luntz and D. J. Auerbach, unpublished. We thank these authors for their permission to reproduce their results prior to publication.

Nonequilibrium Desorption of Physisorbed Atoms

Z. W. Gortel[1] *and H. J. Kreuzer*[2]

[1]Department of Physics, The University of Alberta,
Edmonton T6G 2J1, Alberta, Canada
[2]Department of Physics, Dalhousie University,
Halifax B3H 3J5, Nova Scotia, Canada

1. Introduction

Thermodynamic arguments of detailed balancing are frequently used to correlate experimental data on adsorption, desorption and reflection of particles from solid surfaces /1,2,3/. These arguments are based on the statement that in equilibrium the flux of particles leaving a surface must balance the flux of particles arriving at it. PALMER et al. /2/ have shown experimentally that even if the whole system is not in equilibrium, the relationship between adsorption and desorption resulting from the application of detailed balance arguments still holds. For example, studying desorption and sticking of various isotopes of hydrogen on Ni(111) they find that while desorption varies with the angle to the normal to the surface as $\cos^n\theta$, sticking was found to behave like $\cos^{n-1}\theta$. Similar conclusions are reached by CARDILLO et al. /3/. According to GOODSTEIN /4/ the detailed balance arguments do not require that the entire solid – adsorbate – gas system is in thermal equilibrium, but only that the adsorbate itself be in internal equilibrium.

Obviously the most interesting cases are those in which detailed balance predictions fail. Such a failure may contain valuable information on the processes leading to desorption, sticking or scattering. This information may be extracted from the experimental data provided microscopic models and theories exist. This is the case for physisorption, in which a knowledge of interactions and completeness of existing microscopic theories allow one to approach the question of a relationship between adsorption, desorption and scattering from a microscopic perspective and to identify circumstances in which detailed balance arguments should fail. Here we consider adsorption and desorption only, and in the next section we relate one to the other, starting from a phonon cascade model of physisorption kinetics /5/. We consider both localized and mobile adsorbates and point out differences between flash and isothermal desorption. Section 3 is devoted to a detailed numerical study of a model system of a mobile adsorbate, aiming at a justification of some of the approximations made in Sec.2, and providing a numerical example of nonequilibrium desorption. In Sec. 4 final conclusions are drawn.

2. Time-of-Flight Spectra and Sticking Coefficient

The starting point for the studies of the physisorption kinetics at the microscopic level is a master equation from which the experimentally measured quantities like the desorption time, the sticking and the accommodation coefficients, the time-of-flight (TOF) spectra and others can be derived and related to the microscopic parameters of the system studied. In this section we want to concentrate on the velocity-dependent sticking coefficient and the TOF spectra and to identify the conditions necessary to relate one to the other. We note here that each of them is measured in a different kind of experiment: the TOF spectra contain the information about the dynamical parameters of desorbing particles, whereas the sticking coefficient characterizes adsorption.

To provide the starting point and to introduce the notation used we quote here the master equation appropriate for physisorption kinetics /6/

$$\frac{dn_{\vec{i}}}{dt} = \sum_{\vec{i}'} \widetilde{W}(\vec{i},\vec{i}') n_{\vec{i}'} + \sum_{\substack{\vec{i}' \\ \vec{k}}} W(\vec{i},\vec{k}) n_{\vec{k}} \qquad (1)$$

$$\frac{dn_{\vec{k}}}{dt} = \sum_{\vec{k}'} \widetilde{W}(\vec{k},\vec{k}') n_{\vec{k}'} + \sum_{\vec{i}} W(\vec{k},\vec{i}) n_{\vec{i}} \quad . \qquad (2)$$

The index $\vec{i}$ labels all bound states into which a particle can be trapped when it is adsorbed and $\vec{k}$ is the wavevector of the particle in the gas phase. $n_{\vec{i}}$ and $n_{\vec{k}}$ are the corresponding nonequilibrium occupation probabilities and $\widetilde{W}$'s are given by

$$\widetilde{W}(\vec{i},\vec{i}') = W(\vec{i},\vec{i}') - \delta_{\vec{i},\vec{i}'} \left(\sum_{\vec{i}''} W(\vec{i}'',\vec{i}) + \sum_{\vec{k}} W(\vec{k},\vec{i}) \right) \qquad (3)$$

$$\widetilde{W}(\vec{k},\vec{k}') = W(\vec{k},\vec{k}') - \delta_{\vec{k},\vec{k}'} \left(\sum_{\vec{k}''} W(\vec{k}'',\vec{k}) + \sum_{\vec{i}} W(\vec{i},\vec{k}) \right) \qquad (4)$$

with $W(\vec{i},\vec{i}')$ being the probability per unit time of transition from $\vec{i}'$ to $\vec{i}$. In physisorption these transitions are accompanied by the simultaneous emission/absorption of phonons into/from the solid. All W and $\widetilde{W}$ satisfy the condition of the microscopic detailed balance; for example

$$W(\vec{i},\vec{i}') n_{\vec{i}'}^{eq}(T_f) = W(\vec{i}',\vec{i}) n_{\vec{i}}^{eq}(T_f) \quad . \qquad (5)$$

Here $n_{\vec{i}}^{eq}(T_f)$ is the equilibrium Boltzmann probability distribution at the temperature of the solid T_f. The initial occupations at t=0 depend on the type of experiment the master equation is intended to describe.

The eigenvalues $\lambda_{\vec{k}}$ of the matrix $\widetilde{W}(\vec{i},\vec{i}')$ provide the time scales with which the occupation probabilities evolve as the functions of time. In most situations the eigenvalues group into a slow set, for which λ_0 is a characteristic one, and all other fast ones, for which $\lambda_{\vec{k}} \gg \lambda_0$. Such separation of the time scales allows one to identify the desorption time t_d as λ_0^{-1}, and is a necessary condition for the meaningful definition of the sticking coefficient $S_{\vec{k}}(T_f)$. For particles of mass m and momentum $\hbar\vec{k}$ striking solid surface kept at T_f, we get from (1) and (2) /7/

$$S_{\vec{k}}(T_f) = \frac{Vm}{A\hbar\vec{k}\cdot\vec{n}} \sum_{\vec{i}} W(\vec{i},\vec{k}) \sum_{\vec{k}}{}' \widetilde{e}_{\vec{i}}^{(\vec{k})} \sum_{\vec{j}} e_{\vec{j}}^{(\vec{k})} \qquad (6)$$

where V, A and $\vec{n}$ are the volume of the gas phase, the solid surface area and the unit vector normal to it, respectively. The prime indicates that the summation over $\vec{k}$ includes the slow set only and $\widetilde{e}_{\vec{i}}^{(\vec{k})}$ and $e_{\vec{i}}^{(\vec{k})}$ are the left and the right eigenvectors, properly normalized, of the matrix $\widetilde{W}(\vec{i},\vec{i}')$. Their general properties are listed in /6/.

The rate with which the desorbing particles are registered by the detector at the position $\vec{L}$ with respect to the solid surface (TOF spectrum) is given by /8/

$$\frac{dN_{reg}}{dt} = CONST \cdot t^{-4} \sum_{\vec{i}} W(\vec{k},\vec{i}) \int_0^t d\tau \, n_{\vec{i}}(t-\tau) \Big|_{\vec{k}=m\vec{L}/\hbar t} \tag{7}$$

where $n_{\vec{i}}(t)$, the solution of (1), is

$$n_{\vec{i}}(t) = e^{-\lambda_0 t} n_{\vec{i}}^{eq}(T_f) \sum_{\vec{k}}' \tilde{e}_{\vec{i}}^{(\vec{k})} \sum_{\vec{j}} e_{\vec{j}}^{(\vec{k})} n_{\vec{j}}^{eq}(T_i)/n_{\vec{j}}^{eq}(T_f) . \tag{8}$$

Here $n_{\vec{i}}^{eq}(T_i)$ stands for the initial occupation probabilities for the adsorbate in equilibrium with the solid at the initial temperature T_i. $T_i=T_f$ or $T_i<T_f$ for isothermal or flash desorption, respectively. We stress here that, because of $\tilde{e}_{\vec{i}}^{(\vec{k})}$ present in (8), the adsorbate is not necessarily in a state of internal equilibrium at T_f during desorption as the form of (8) might suggest.

Before discussing in detail the case of flash desorption TOF we observe that for isothermal desorption (8), (7) and (6) imply, for $t \gg \lambda_0^{-1}$ and $\vec{k} \cdot \vec{n} \propto t^{-1} \cos\theta$, that

$$\left(\frac{dN_{reg}}{dt}\right)_{iso} = CONST \cdot t^{-5} \cos\theta \, n_{\vec{k}}^{eq}(T_f) S_{\vec{k}}(T_f) \Big|_{\vec{k}=m\vec{L}/\hbar t} . \tag{9}$$

Here θ is the angle between $\vec{L}$ and the surface normal. Thus, the angular dependence of sticking is seen to be $\cos\theta$ times weaker than that of the TOF spectra for isothermal desorption. Such a relationship between adsorption and desorption, implied also by the principle of detailed balance /1/, was first experimentally confirmed by PALMER et al. /2/. We note here that (9) is valid even if the adsorbate is not in a state of internal equilibrium. Deviations from this state enter in the same way both the sticking coefficient (containing all information about the microscopic processes involved) and the isothermal TOF signal.

Turning now to the flash desorption for which (9) does not in general hold, we focus our attention first on the localized adsorbate. Experience gained in numerical studies of one-dimensional models of adsorption-desorption kinetics strongly suggests that for the localized adsorbate only one term ($\vec{k}=0$, say) of the sum over $\vec{k}$ contributes significantly to (6) and (8). In this case (9) is again recovered with CONST containing now both T_i and T_f. Again, fast thermalization of the adsorbate is not a necessary condition in the analysis. Indeed, only if the conditions of applicability of the perturbation theory /9/ to the master equation (1) are satisfied, i.e. if the term proportional to $\sum W(\vec{k},\vec{i})$ is negligible for all pairs $(\vec{i},\vec{i}')$ in (3), one has

$$e_{\vec{i}}^{(0)} = n_{\vec{i}}^{eq}(T_f) \tilde{e}_{\vec{i}}^{(0)} \approx n_{\vec{i}}^{eq}(T_f)/\left(\sum_{\vec{j}} n_{\vec{j}}^{eq}(T_f)\right)^{1/2} \tag{10}$$

and, consequently, (8) becomes

$$n_{\vec{i}}(t) = e^{-\lambda_0 t} n_{\vec{i}}^{eq}(T_f) \left(\sum_{\vec{j}} n_{\vec{j}}^{eq}(T_i)/\sum_{\vec{j}'} n_{\vec{j}'}^{eq}(T_f)\right) , \tag{11}$$

46

meaning that the adsorbate is thermalized (at T_f) before any significant desorption takes place.

In order to examine now mobile adsorbates we observe that a quantum state of motion of the adsorbed particle is characterized by a two-dimensional wavevector $\vec{K}$ parallel to the solid surface, and a discrete index i due to the quantized motion in the direction normal to it. Thus $\vec{i}=(i,\vec{K})$ and for one-phonon processes taking the adsorbed particle from $\vec{i}$ to $\vec{i}'$ one encounters momentum conservation, $\hbar\vec{K}=\hbar(\vec{K}+\vec{P})$, where $\vec{P}$ is the parallel (to the surface) component of the wavevector of the phonon participating in the process. We now note that, at a given energy $k_B T_f = \hbar c |\vec{p}| = \hbar^2 k^2/2m$, phonon and particle momenta $\hbar\vec{p}$ and $\hbar\vec{k}$ are very different,

$$|\vec{p}|/|\vec{k}| = (k_B T_f/2m)^{1/2}/c \ll 1 \tag{12}$$

where c is the average sound velocity. The ratio (12) is only 10^{-2} for He at 10K and of the same order for heavier particles at higher temperatures. This suggests that the phonon momentum plays no major role in the desorption process, and that a simplified momentum conservation condition, $\vec{K}=\vec{K}'$, can be used in the transition probabilities $W(\vec{i},\vec{i}')$. We then get

$$W(\vec{i},\vec{i}') = \delta_{\vec{K},\vec{K}'} \; W(i,i') \; . \tag{13}$$

The same factorization for the matrix $\tilde{W}(\vec{i},\vec{i})$ follows and consequently, the index $\vec{k}$ labelling its eigenvectors can be written as $\vec{k}=(\kappa,\vec{X})$. Each eigenvalue $\lambda_{\vec{k}}$ becomes degenerate and equal to λ_κ – the corresponding eigenvalue of $\tilde{W}(i,i')$. If $\tilde{e}_i^{(\kappa)}$ and $e_i^{(\kappa)}$ are, respectively, the left and the right normalized eigenvectors of this matrix, then the corresponding eigenvectors of the full matrix (13) are

$$e_{\vec{i}}^{(\vec{k})} = \delta_{\vec{K},\vec{X}} \; e_i^{(\kappa)} \, (n_{\vec{K}}^{eq}(T_f))^{1/2} \tag{14}$$

$$\tilde{e}_{\vec{i}}^{(\vec{k})} = \delta_{\vec{K},\vec{X}} \; \tilde{e}_i^{(\kappa)} \, (n_{\vec{K}}^{eq}(T_f))^{-1/2} \tag{15}$$

where $n_{\vec{K}}^{eq}(T_f)$ is the two-dimensional equilibrium probability distribution. $n_{\vec{i}}^{eq}(T_f)$ is a product of $n_{\vec{K}}^{eq}(T_f)$ times the one-dimensional distribution $n_i^{eq}(T_f)$. The primed sums over $\vec{k}$ in (6) and in (8) become now sums over $\vec{X}$ with $\kappa=0$. With (14) and (15) we get from (8)

$$n_{\vec{i}}(t) = n_{i,\vec{K}}(t) = e^{-\lambda_o t} \, n_{\vec{K}}^{eq}(T_f) n_i^{eq}(T_f) \tilde{e}_i^{(o)} \sum_j e_j^{(o)} n_j^{eq}(T_i)/n_j^{eq}(T_f) \tag{16}$$

implying, through the factor $n_{\vec{K}}^{eq}(T_i)$, that during desorption the motion along the surface remains in its initial equilibrium state whereas the probability distribution of the motion normal to the surface is modified by the interactions with the phonon system of the solid. Only when $\tilde{e}_i^{(o)}$ and $e_i^{(o)}$ can be approximated by the one-dimensional analogues of (10) (see comments there), the motion normal to the solid surface thermalizes at T_f before any significant desorption occurs.

Combining (16), (6) and (7), factorizing $W(\vec{K},\vec{i})$ as in (13) and applying the microscopic detailed balance, we arrive at the following generalization of (9) for the flash desorption TOF spectrum for the mobile adsorbate:

$$\left(\frac{dN_{reg}}{dt}\right)_{flash}^{mobile} = CONST \cdot t^{-5} \cos\theta \, (n_{\vec{K}}^{eq}(T_i) n_{\vec{k}}^{eq}(T_f)) s_{\vec{K}}(T_f) \Big|_{\vec{K}=L\vec{m}/\hbar t} \tag{17}$$

where $\vec{k}=(k,\vec{K})$. Deviations from equilibrium for the motion normal to the surface are contained entirely in the sticking coefficient in (17). The angular dependence of the factor in the round brackets depends on the ratio T_i/T_f, and can lead to strong forward peaking of the desorption signal, and to a significant rainbow effect, i.e. a lowering of the average kinetic energy of desorbing particles in directions away from the surface normal. Strong angular dependence of this factor makes flash desorption a rather poor candidate for experimental determination of the angle-dependent sticking coefficient. Simultaneously, the strong forward peaking and the rainbow effect observed in flash desorption of physisorbed atoms can be an indication of a mobile adsorbate.

In deriving (17) a number of approximations were made. In particular, the factorization (13) is a key simplification leading directly to the lack of equilibrium between the two modes of the translational motion of adsorbed atoms. In the next section we will solve the master equation (1) numerically for a model system of mobile adsorbate without employing (13), and will compare the obtained solutions for $n_{\vec{K}}(t)$ with (16). We will also analyze the isothermal and the flash desorption TOF spectra computed for this system and will compare them with the existing experimental data.

3. Numerical Analysis

As our numerical example we choose a system of helium physisorbed on nichrome for which angle-dependent TOF spectra for flash desorption /10/, exhibiting both strong forward peaking and the rainbow effect, were measured.

Assuming that He atoms can move freely along the surface, we represent their static interaction with the solid by a Morse surface potential

$$V_s(\vec{r}) = V_s(z) = V_o (e^{-2\gamma(z-z_o)} - 2e^{-\gamma(z-z_o)}) \tag{18}$$

which depends only on the distance z from the surface, and develops i_o+1 bound states with energies ϵ_i. The measured /10/ heat of adsorption is about 30K. This fixes the depth of (18) at $V_o/k_B=45K$, if we rather arbitrarily choose $\gamma^{-1}=1A$. The surface potential then develops three bound states.

To describe the desorption experiment we set $n_{\vec{K}}(t)\equiv 0$ in (1) thus eliminating readsorption processes. The transition probabilities in (1) are calculated using $-\vec{u}(t)\cdot W_s(\vec{r})$ as a coupling of the adsorbate to the thermal vibrations of the solid. Phonon momentum is included in the momentum conservation entering W's, i.e. the factorization (13) is not used. The resulting transition probabilities are listed in /6/, and in /11/, together with all others used below. To keep (1) as a discrete matrix equation, we consider a finite surface area L^2 with periodic boundary conditions, so that $\vec{K}=(n_x,n_y)\pi/L$ for integer n_x and n_y. Because the initial occupation probabilities depend only on $E_\parallel=\hbar^2K^2/2m$, and not on the direction of $\vec{K}$, and because on an isotropic surface the W's do not favor any given direction, the time-dependent probabilities remain isotropic: $n_{\vec{K}}(t)=n_i(E_\parallel,t)$. To reduce the size of the matrices involved we coarse grain the important range of lateral energies, $0<E_\parallel<E_{max}$, into N discrete values $\epsilon_\parallel^l$ (l=1,....N) with E_{max} of the order of several k_BT and N less than 100. The resulting master equation can be written

$$dn_i(\epsilon_\parallel^\ell,t)/dt = \sum_{i'=0}^{i_0} \sum_{\ell=1}^{N} T_{ii'}(\epsilon_\parallel^\ell,\epsilon_\parallel^{\ell'})n_{i'}(\epsilon_\parallel^{\ell'},t) \qquad (19)$$

with T directly related to $\widetilde{W}$ in (3). Equation (19) can be solved by dia-
gonalizing T numerically, and with the initial condition

$$n_i(\epsilon_\parallel^\ell,0) = n_{i,\ell}^{eq}(T_i) \propto \exp[-(\epsilon_i + \epsilon_\parallel^\ell)/k_B T_i] \qquad (20)$$

we get

$$n_i(\epsilon_\parallel^\ell,t) = n_{i,\ell}^{eq}(T_f) \sum_\kappa e^{-\lambda_\kappa t} \tilde{e}_{i,\ell}^{(\kappa)} \sum_{i',\ell'} e_{i',\ell'}^{(\kappa)} n_{i',\ell'}^{eq}(T_i)/n_{i',\ell'}^{eq}(T_f) \qquad (21)$$

where λ_κ, $\tilde{e}^{(\kappa)}$ and $e^{(\kappa)}$ are, respectively, eigenvalues and eigenvectors
of T and $\kappa=0,\ldots(i+1)N-1$. Note that (21) is the exact solution of (19)
i.e. all time scales λ_κ are included, in contrast to the approximate form
(8) used before. The time-dependent coverage $\theta(t)$ is proportional to
$\Sigma_{i,\ell} n_i(\epsilon_\parallel^\ell,t)$.

Let us now discuss helium desorption from a nichrome substrate which
was instantaneously heated from $T_i=2K$ to $T_f=8K$. We typically choose the
cutoff energy E_{max} of the order 10-20K and pick N between 50 and 100 to
ensure that our coarse graining does not affect the numerical results. We
note first that $\ln \theta(t)$ versus t is very linear, with a slope corresponding to a desorption time $t_d=6.25\times10^{-7}s$, with the linearity pertaining for
t up to $5t_d$. This implies that, out of many eigenvalues λ_κ of T, only the
slowest ones contribute significantly in the sum (21). We also note that
if the factorization (13) is employed, then $t_d=\lambda_0^{-1}=6\times10^{-7}s$ is obtained,
λ_0 here being the smallest eigenvalue of $\widetilde{W}(i,i')$.

To check how close (16) approximates the individual occupation probabilities (21), we plot in Fig. 1

$$n_i(\epsilon_\parallel^\ell,t)e^{\lambda_0 t} \exp[(\epsilon_i/T_f + \epsilon_\parallel^\ell/T_i)/k_B] \qquad (22)$$

versus $\epsilon_\parallel^\ell$ for various $\lambda_0 t$. We can infer the following: (i) from the fact
that the curves for $i=0,1,2$ are more or less identical we conclude that
the motion perpendicular to the surface thermalizes to T_f as early as
$0.2/\lambda_0$; i.e. with more than 80% of the adsorbate still present; (ii)
because all the curves are nearly horizontal, the motion parallel to the
surface remains characterized by the initial temperature T_i, i.e. does not
thermalize, even after two desorption times; (iii) the rise of all curves
near E_{max} is caused by the numerical procedure used: increasing E_{max} from
10K to 15K also shifts the region of upward bending to higher $\epsilon_\parallel^\ell$; (iv) the
fact that for later times the slope of the curves slightly increases indicates a slow increase in lateral temperature. This is more apparent if we
plot $\ln(n_0(\epsilon_\parallel^\ell,t)\exp(\lambda_0 t))$ versus $\epsilon_\parallel^\ell$ for various $\lambda_0 t$ in Fig. 2. A decreasing slope signals an increasing temperature. The rate of the lateral ther-
malization can be estimated as $5\times10^4 K/s$; i.e. it takes about $10^{-4}s \approx 200t_d$
to thermalize the lateral motion completely. Similar conclusions follow
from the calculations performed for various combinations of T_i and T_f.

From the numerical analysis presented thus far we conclude that (16) is
a faithful representation of nonequilibrium occupation probabilities for a
mobile adsorbate suddenly flashed from T_i to T_f. Consequently, the factorization (13), reducing three-dimensional theories of physisorption to
one-dimensional ones, is a very good approximation for mobile adsorbates.
Not only does the lateral motion thermalize so slowly that during desorption it is still characterized by the initial temperature of the solid,

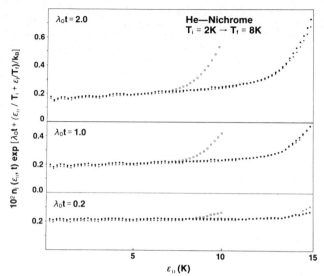

Fig. 1. The nonequilibrium occupation probabilities (22) for a desorption flash from T_i=2K to T_f=8K as a function of lateral energy $\epsilon_{\parallel}$ for He on a nichrome surface. Points are for the ground state i=0, crosses for i=1, both for E_{max}/k_B=15K. Circles are for i=0 and E_{max}/k_B=10K. The wiggles are indicative of the numerical errors due to the coarse graining of $E_{\parallel}$ to 75 and 50 points, respectively. λ_0^{-1} =6.25×10^{-7}s is the desorption time (from /11/).

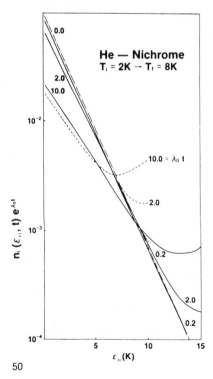

Fig. 2. Nonequilibrium occupation probabilities (21) as a function of lateral energy $\epsilon_{\parallel}$ for various times $\lambda_0 t$. The decrease in slope reflects the heating of the lateral degrees of freedom from an initial temperature T_i=2K (dashed-dotted line). The deviations from straight lines close to the upper cutoff energy (E_{max}/k_B=10K for dashed lines) reflect numerical errors and nonequilibrium effects (from /11/).

but also, at least for weakly bound light adsorbed atoms, the motion perpendicular to the surface evolves towards the final equilibrium at T_f during initial transient stage of desorption. This fact justifies use of the lowest order perturbation theory (see comments above (10) and below (16)) to the master equation (1). For heavier and more strongly bound adsorbates, the transient evolution leads to a quasi-stationary state, which differs from equilibrium at energies close to the desorption continuum /9/.

The important result of our analysis is the fact that the relations between desorption and sticking, (9) and (17) are valid even if the adsorbate does not reach equilibrium before desorbing. The most important condition for their validity is also the necessary condition which allows one to define the desorption time and the sticking coefficient at all: there must be a vast separation between slow time scales, responsible for desorption and/or sticking, and fast transient time scales in the system. In a sense (9) is more universal than thermodynamic arguments of detailed balance allow and (17) is its further generalization.

As final numerical example we present in Figs. 3 and 4 isothermal and flash desorption TOF spectra, respectively, for He desorbing from a nichrome surface. Both are obtained using (7) and the perturbation theory version of (16):

$$ n_{\vec{k}}(t) = CONST \cdot n_{\vec{k}}^{eq}(T_i) n_i^{eq}(T_f) e^{-\lambda_0 t} . \qquad (23) $$

The spectra are the same within the accuracy of the plots whether they are calculated (i) with the factorization (13) applied to $W(\vec{k},\vec{i})$ in (7) – a one-dimensional theory, or (ii) without employing (13) – a three-dimensional theory.

The isothermal desorption spectra in Fig. 3 are slightly narrower than a Maxwellian at 8K. Their maxima decrease as a function of θ roughly like $\cos^{1.2}\theta$. Both facts indicate an almost velocity-independent isotropic sticking coefficient. The flash desorption TOF spectra in Fig. 4 exhibit both strong θ dependence and the rainbow effect, and they can be quite well reproduced by the ad hoc formula

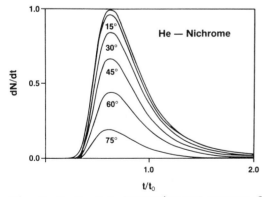

Fig. 3. Isothermal desorption TOF spectra for helium desorbing from nichrome at T_f=8K as a function of angle θ to the surface normal. t_0 is the time needed by a thermal atom to reach the detector, $t_0 = L(m/2k_B T_f)^{1/2}$ (from /11/).

51

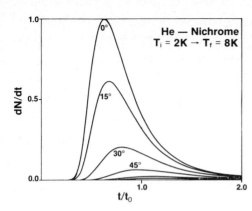

Fig. 4. Flash desorption TOF spectra for T=2K and T=8K (from /11/).

$$\frac{dN_{reg}}{dt} = \text{CONST} \cdot t^{-5} \cos\theta \, \exp\left[-L^2 m(\cos^2\theta/T_f + \sin^2\theta/T_i)/2k_B\right] \tag{24}$$

obtained from (17) if the $\vec{k}$ dependence of the sticking coefficient is entirely neglected.

4. Conclusion

We have shown that starting from a master equation for a physisorption kinetics, we can derive a simple relation between desorption and sticking provided certain plausible conditions are met. For an isothermal desorption and/or for a localized adsorbate this relation, (9), agrees with the one which is obtained using thermodynamic arguments of detailed balancing, even if not all criteria of their applicability are met. For a flash desorption from mobile adsorbates these arguments cannot be applied, and the more general relation (17) was obtained.

The only experimental TOF spectra existing at a present time, for desorption of He from a nichrome surface flashed from 2 to 8K, are consistent with our findings. Considering, however, the high coverages at which the measurements are made, one cannot exclude other possible sources of the observed strong forward peaking of a desorption flux, i.e. scattering in a cloud of desorbed particles /12/. Collisions between particles in the adsorbate do not promote thermalization because no energy transfer; neither from the solid, nor between the normal (to the surface) and the lateral components of the motion would be involved.

This work was supported in part by a grant from the Natural Sciences and Engineering Research Council of Canada. We would like to thank R. Teshima for writing the computer codes.

1. G. Comsa, R. David: Surf. Sci. Rep. 5, 145 (1985).
2. R.L. Palmer, J.N. Smith, H.Saltsburg, D.R. O'Keefe: J. Chem. Phys. 53, 1666 (1970).
3. M.J. Cardillo, M. Balooch, R.E. Stickney: Surf. Sci. 50, 263 (1975).
4. D. Goodstein: In Many-Body Phenomena at Surfaces, ed. by D. Langreth and H. Suhl (Academic Press, New York, 1984) p.277.

5. Z.W. Gortel, H.J. Kreuzer, R. Teshima: Phys. Rev. B22, 5655 (1980).
6. H.J. Kreuzer, Z.W. Gortel: Physisorption Kinetics, Springer Ser. Surf. Sci., Vol. 1 (Springer, Berlin, Heidelberg, 1986).
7. W. Brenig: Z. Phys. B48, 127 (1982); ref. 6, §7.2.
8. Z.W. Gortel, H.J. Kreuzer, M. Schaff, G. Wedler: Surf. Sci. 134, 577 (1983); ref. 6, §6.2.
9. H.J. Kreuzer, R. Teshima: Phys. Rev. B24, 4470 (1981); ref. 6, §5.3.
10. P. Taborek: Phys. Rev. Letters 48, 1737 (1982).
11. Z.W. Gortel, H.J. Kreuzer: Phys. Rev. B31, 3330 (1985).
12. J.P. Cowin, D.J. Auerbach, C. Becker, L. Wharton: Surf. Sci. 78, 545 (1978).

Studies of Surface Kinetics
Using Second-Harmonic Generation

H.W.K. Tom

AT&T Bell Laboratories, Holmdel, NJ 07733, USA

Recently, optical second-harmonic generation (SHG) has been shown to be a sensitive and versatile surface probe [1]. SHG is sensitive to adsorbate coverage, composition, binding site [2], and molecular orientation on surfaces [3,4]. In addition SHG has been shown to reveal the symmetry of atomic order on surfaces [5]. It is thus able to monitor symmetry-changing surface phase transitions [6]. Here, the features of SHG that make it a unique tool for monitoring surface kinetics are discussed. Results from SHG measurements with time resolutions of 100 msec and 100 fsec are presented.

1. Time-Resolved Surface Studies

In this workshop, it has been noted that in kinetic studies of reaction rates, measurements must be performed over many orders of magnitude of rates to establish the importance of non-equilibrium effects [MENZEL, these proceedings]. For the most part, surface science techniques are limited to measuring changes on the surface on the time scale of seconds. Reaction rates are adjusted to this time regime by reducing temperature or reactant concentrations. However, since statistical equilibrium between modes of a single degree of freedom occurs in picoseconds and between degrees of freedom occurs in less than nanoseconds, conventional surface techniques cannot probe non-equilibrium effects directly. Furthermore, Menzel pointed out that in identifying reaction intermediates such as adsorption precursors, measurements must be performed at "normal" reaction rates and temperatures to ensure proposed intermediates are not side reactions. The theoretically proposed time scales of such intermediates is also less than 10^{-9} sec (10^{-12} sec for adsorption precursors [HOOD, these proceedings]). Finally, in discussions of surface phase transitions, while domain walls move collectively at rather slow rates resolvable with conventional LEED, the primary events of atomic movements, diffusion, and rearrangements occur rapidly again on time scales of 10^{-9} sec or less.

Several techniques have been developed recently to access these faster time scales. Time-resolved electron energy loss has been able to probe the millisecond time-scale [ELLIS, these proceedings]. Molecular beam experiments can limit the time of particle-surface interaction to the particle's residence time [AUERBACH, these proceedings], which may be on the order of 10^{-9} or less seconds. However, direct measurement of the residence time is a formidable problem even if the incident particles are prepared with a short laser pulse due to the spread in particle velocities after the scattering event.

Surface techniques that incorporate lasers to initiate a reaction or surface response can obtain time-resolution under 10^{-9} sec. By using a laser pulse to heat the sample, temperature ramps in excess of 10^5 K/sec can be obtained and thermal desorption spectroscopy and reaction kinetics have been studied in this way [7]. In such experiments however, the time or temperature at which the particles desorb or react cannot be measured directly. Low-Energy Electron Diffraction following an intense laser pulse that initiates structural change has been reported with 10 nsec time-resolution [8]. Photoemission induced by a laser pulse following excitation of the electronic surface states by another laser pulse has been reported with 70 psec resolution [9]. Better time-resolution is possible with these pulsed electron spectroscopy techniques, however, there is a tradeoff between time-resolution and signal due to space charge effects.

Second-harmonic generation (SHG) is an *insitu* probe so molecular scattering or desorption events can be time-resolved directly. In addition, intermediate states that may not be stable or distinguishable in the gas phase may be probed with SHG. SHG is also an all optical technique. It is therefore not as drastically effected by space charge limitations as are electron spectroscopies. By using high repetition rate pulse lasers like the 8KHz Cu vapor laser, 0.12 msec resolution is easily obtainable for kinetic measurements. By using pump-probe techniques, the ultimate time resolution is limited only by the time-duration of a single pulse. With time-resolution as low as a few fsec, SHG makes possible direct measurement of non-equilibrium effects of surface reactions.

2. Second-Harmonic Generation: Experimental Considerations

In SHG experiments, a laser pulse at the excitation frequency induces a polarization response at twice that frequency in the surface and near-surface bulk. This polarization radiates at the SH frequency and is detected in reflection with gated photon-detecting electronics. From clean metal or semiconductor surfaces, several 100's of SH photons are generated for an input pulse of 10 mJ focused to 1 cm^2 with 10 nsec duration. These photons are detected with typically 5 to 8% efficiency. The SH yield scales as the square of the input fluence and inversely with the area and pulse time duration.

The extreme sensitivity to surface (vs. bulk) electronic states obtained in SHG comes from the fact that SHG is dipole-forbidden in the bulk of centrosymmetric media but always dipole-allowed at the surface where centrosymmetry is broken normal to the surface. In addition, certain surface states may be intrinsically more highly polarizable than corresponding bulk states. Finally, the large gradient in the electric field at the surface may make the radiation from even surface quadrupole-allowed SH as large as that from bulk quadrupole-allowed SH [10].

The polarization for the surface and bulk regions may be written as [5,10,11]:

$$P_i^s(2\omega) = X_{ijk}^s(2\omega)E_j(\omega)E_k(\omega) \qquad (1)$$

$$P_\ell^b(2\omega) = X_{ijk\ell}^b(2\omega)E_j(\omega)\nabla_k E_\ell(\omega) \qquad (2)$$

where the superscripts s and b denote surface and bulk, subscripts i,j,k and ℓ denote vector components, the sum over repeated subindices is implied, $\breve{E}(\omega)$ is the fundamental field amplitude and $\breve{X}(2\omega)$ is the SH-nonlinear susceptibility. All the material specific information is in the tensor elements of $\breve{X}^s(2\omega)$ and $\breve{X}^b(2\omega)$ tensors. The second-harmonic field $\breve{E}(2\omega)$ is calculated from vector $\breve{P}(2\omega)$ by Maxwell's Equations. For a given excitation geometry (angle of incidence and field polarizations) the SH signal is proportional to the square modulus of some linear combination of the elements of $\breve{X}(2\omega)$ which we will denote as $|X_{eff}|^2$ in this text. We note here that the surface dipole and quadrupole contribution to SH are incorporated into $\breve{X}^s$ in (1).

3. SH Studies of Adsorption Kinetics

The first figure demonstrates the sensitivity and time-resolution available with SHG in a typical kinetic rate measurement of the adsorption of O_2 on Rh(111) [2]. Starting with

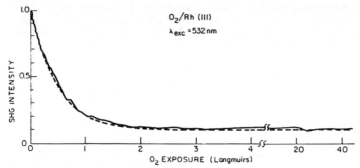

Fig. 1. Adsorption of O_2 on Rh(111). SH intensity normalized to signal for clean Rh(111). Data (solid line) obtained as a continuous function of O_2 exposure (1 L in 20 sec). Theory (dashed line).

an atomically clean and well-ordered Rh(111) surface at 315K, the SHG signal was monitored as the UHV chamber was backfilled with 5×10^{-8} torr O_2. The signal, shown in Fig. 1, decreases smoothly to a saturation value of 0.12 times the bare metal value at about 1.8 Langmuirs (1 Langmuir = 10^{-6} torr-sec corresponds to 20 seconds). A sharp 2X2 LEED pattern was not obtained until about 20 L. The time resolution is limited here by the pulse repetition rate of the laser which was 0.1 sec.

One sees immediately that SHG is extremely sensitive to the oxygen overlayer. The dashed line shows a fit to the data that is consistent with a model of Langmuir adsorption kinetics (sticking coefficient is coverage independent). The kinetics were previously established with Auger by YATES, et al [12]. Our model is that the change in surface nonlinear response per adsorbate is constant with coverage (consistent with all empty sites being equivalent) and that therefore the the nonlinear susceptibility of the surface may be given by the expression:

$$X_{eff} = A + BD/D_s ,$$ (3)

where D and D_s are the fractional and saturation surface coverage with respect to surface Rh atoms, and A and B are coefficients associated with the nonlinearity of the bare metal and the induced change in surface nonlinearity per adsorbate.

The coverage as a function of exposure is just given by Langmuir adsorption kinetics in the limit of negligible desorption: $D(t)/D_s = 1 - \exp(-Kpt)$ where K is a constant accounting for the sticking coefficient and p is the oxygen pressure. The SH signal is proportional to $|X_{eff}|^2$. The fit gives $B/A = 1.03 \exp(i160°)$ and $K/D_s = 0.93$/layer. The latter is in close agreement with the value 0.78/layer obtained by YATES, et al [12] considering possible differences in pressure calibration.

SHG from Rh(111) shows similar high sensitivity to other adsorbates, among them CO, benzene and pyridine and their hydrocarbon fragments upon dehydrogenation [13], and the alkali metals [2,14]. Similar results have been obtained on Cu(111) [15], Ni(111) [15,16], Pt(111) [16], Ag(110) [17], and Si(111)-7X7 [18,19,20]. SHG then promises to be generally applicable to metal and semiconductor-adsorbate systems. Readers should also be aware of the great sensitivity of SH to adsorbates on surfaces in electrochemical environments [21,22].

For O_2/Rh(111), the sticking coefficient was independent of coverage and the SH took a particularly convenient form. By calibrating the SH signal to coverage by an independent technique such as thermal desorption yield or Auger it would be possible to extract coverage as a function of time or sticking coefficient as a function of coverage. Such studies have been performed for CO on Rh(111) [2], Cu(111) and Ni(111) [15], and for CO, O_2 and H_2 on Pt(111) and Ni(111) [16] and departure from Langmuir kinetics has been distinguished.

4. SH Studies of Desorption and Reaction Kinetics

While the details of the SHG function of time during a reaction may be difficult to interpret without extensive calibration of surface coverage or composition at intermediate points, relative reaction rates can be compared easily. In Fig. 2, the SHG signal is shown during the thermal decomposition of a thin thermally grown oxide layer on Si(111) at 900° C [18]. The time origin in Fig. 2 begins immediately after dosing the 900° C sample with 40 L of O_2 at 10^{-6} torr. Immediately the SHG signal begins to increase toward the clean Si(111) level, indicating the oxide is decomposing. After approximately 2 minutes, Auger showed that the oxygen coverage was equivalent to that of a saturated room temperature chemisorbed layer. The fit to the data shown as the dashed curve is obtained with the same model of the surface SH susceptibility described in (3), except the fit was made with $B/A = -0.412$ and $D(t)/D_s = \exp(-kt)$ where $k = 1.7 \times 10^{-3}$/s. The same experiment was performed at 1100° C and the shape of the SH data was the same except it was fit with a faster rate constant, $k = 5.3 \times 10^{-2}$/s. If we assume there is only a single activation energy such that $k = \nu_0 \exp(-E_D/k_B T)$ where ν_0 is the pre-exponential factor, E_d is the energy of desorption and k_B is the Boltzmann constant, then E_D, is 2.4 eV and $\nu_0 = 3.1 \times 10^7$/s. Our value of E_D may be compared to the value of 3.1 eV obtained for desorption from Si(100) between 700 and 790°C [D'EVELYN, et al, these proceedings]. A similar SHG study of the effect of coadsorbed sulfur on methoxy decomposition on Ni(111) has been performed [23].

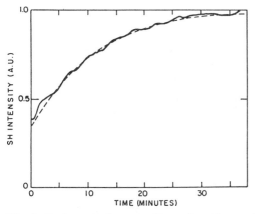

Fig. 2. Isothermal decomposition of oxide on Si(111) surface formed at 900°C by exposure to 40 Langmuirs of O_2. Time axis begins immediately after O_2 exposure. SH data (solid line) normalized to signal for clean Si(111) at 900° C. Theory (dashed line).

It is easy to imagine extending this kind of measurement to much shorter time scales by using a laser pulse to rapidly heat the sample and then using a time-delayed probe pulse to time-resolve the evolution of the reaction. Recently, the laser-induced desorption of Rhodamine 610 dye molecules from a fused silica substrate was studied with a time-delayed SH probe with 6 psec time-resolution [24]. For a pump fluence of 0.5 J/cm^2, the change in the SH probe was so rapid that it could not be resolved in 6 ps. The results indicate that either the molecules photofragmented or that they were desorbed in a few psec with a thermal velocity distribution in excess of 5000K.

5. SH Studies of Phase Transitions

SHG is also highly sensitive to surface order as manifested in the symmetry of the surface electronic susceptibility. As shown in Fig. 3, the SH intensity for a given excitation geometry depends on the orientation of the sample as it is rotated about the surface normal [25]. In the case shown, the SH from a Cu(111) surface shows the 3m symmetry of the surface. For 3m symmetric surfaces, the SH susceptibility is generally given by:

$$X_{eff} = L + Mf(3R),\qquad (4)$$

where R is the rotation angle and coefficients L and M are the angle-independent and angle-dependent contributions to SH, respectively. It can be shown that for p̂-polarized input and ŝ-polarized SH, the angular function f(3R) is simply sin(3R) and the angle-independent coefficient L is zero. It is important to realize that the value of M may be used as a monitor of surface order because M is only non-zero if the surface and near-surface bulk have 3m symmetry.

Rotational anisotropy is predicted from the form of the tensors $\tilde{X}^s(2\omega)$ and $\tilde{X}^b(2\omega)$ in (1) and (2), for all surface symmetries 4m and lower [5]. The dependence appropriate for the 4m symmetry (Si(100)) [5] and 2m symmetry (Si(111)-2X1) [26] have also been observed.

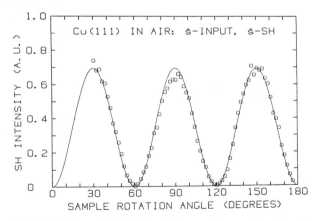

Fig. 3. SH from Cu(111) in arbitrary units as a function of sample rotation angle about its normal. The sample was cleaned and atomically-ordered in UHV before exposing to air. The rotation angle is measured between the surface mirror plane and the optical plane of incidence. Data (points). Theory (solid line) is fit with $X_{eff} = Msin3R$.

Phase-transitions that change symmetry may then be studied with SHG. The transformation of a Si(111)-2X1 to Si(111)-7X7 (2m to 3m symmetry change) as the sample temperature was ramped at 3K/s was studied with SHG [26]. The transition occurs rapidly in 10's of seconds around 285° C. Those authors also studied ion-beam-induced disordering [19], and annealing of an amorphous Si overlayer on Si(111)-7X7 [27], all as continuous functions of time on the scale of seconds which is suitable for many processing applications.

The time-evolution of surface order of a Si(111) surface irradiated with a laser pulse intense enough to induce melting was recently measured with 100 femtosecond time-resolution [6]. The rotational anisotropy of the SH signal was measured as a function of delay time with respect to the intense melting pulse. The rotational anisotropy decreased rapidly within the first 150 fs but had a slow decay component lasting 1 to 3 psec. The authors interpreted the SH signal decrease as consistent with the transformation from crystalline solid to a disordered liquid. Similar studies [28,29] of optical melting of GaAs surfaces indicated that disorder was induced in less than 2 psec.

6. SH Study of Optically-Induced Disorder of Si(111)

In Fig. 4 we present results from a more recent study of the order-disorder transition of Si(111) (held in air) after optical excitation. In this experiment, the probe SH is excited with $\hat{p}$-polarized light and the sample is rotated to a position such that the $\hat{s}$-polarized signal is proportional to $|M|^2$ and the $\hat{p}$-polarized signal is proportional to $|L|^2$ where L and M were defined in (4). Pulses were 100 fsec in duration derived from an amplified CPM laser at 610 nm and had an Airy function beam profile after a spatial filter. The weaker probe pulse was split off from the laser beam, passed through a variable delay line, and focused to a spot size of 25 micron diameter at the center of the pump spot on the sample surface. The pump spot was 80 microns in diameter to insure even excitation over the probe area. A new spot on the sample was used for each shot of the laser. The pump and probe were incident at 20° and −25° with respect to the surface normal in order to reduce the detection of SH induced by the intense pump.

The top register of Fig. 4 shows the cross-correlation of the pump and probe on the surface obtained from the second-harmonic generated by the product of the pump and

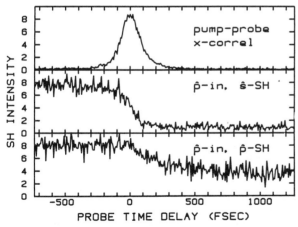

Fig. 4. SH from Si(111) as a function of probe delay with respect to pump arrival at surface. Pump intensity is 2 times the threshold for creating a melted spot in the Si(111) surface.

probe. This signal marks the arrival time of the probe with respect to the pump and the time-resolution of the measurement. The lower two registers of Fig. 4 show the ŝ- and p̂-polarized SH probe signals. All signals are shown as a function of probe delay with respect to the pump.

We see in the figure that the ŝ-SH signal that depends on order is reduced within 150 fsec or the resolution of the experiment. The p̂-SH changes much more slowly, on the time-scale of 500 fsec. Both changes in SH are threshold dependent. At 0.5 X threshold no change in M is observed with delay time. This indicates that the rapid decrease in M depends critically on the density of excited electronic states and is not due to bleaching of the nonlinear susceptibility which should not depend on threshold. The SH signal does not recover from its value at 250 fsec indicating that whatever changes occur do so permanently. We note that the threshold (around 0.1 J/cm^2) for the rapid SH change corresponds to the melting threshold of the bulk as well: melt spots in the Si wafer are observed with an optical microscope only for pump intensities exceeding threshold. The slow decrease of the p̂-SH is still under investigation, but is thought to arise from melting of the bulk. Quadrupole-allowed SH may arise from the topmost 100Å of the bulk and may be contributing to this p̂-SH signal. Previous studies using linear reflection have shown that the liquid melt front propagates into the bulk at rates in excess of 10^6 cm/sec [31] which corresponds to 100Å in 1 psec.

The less than 150 fsec change in surface order is most surprising. These new results are consistent with results published earlier[6] if one ignores the longer decay feature, takes into account a possible shift in the origin of time, and the fact that the previous study measured a signal proportional to $|L + M|^2$ where L/M was roughly 1 whereas we measure $|L|^2$ and $|M|^2$ directly. However, in contrast to the previous work which deduced a disorder time around 1 psec which would be consistent with thermal melting, this work is consistent with a complete loss of order within 1 or 2 electron-phonon relaxation times, a time much too short to have phonon equilibrium or thermal melting. In this case, disorder would be electronically-induced.

Such rapid disordering of the surface Si atoms could occur in an optical process analogous to photodissociation for molecules. Roughly speaking, the bonding orbitals that hold the Si atoms in their lattice positions are depopulated so efficiently that the bonds are temporarily broken and the atoms are pushed to new positions by the remaining lattice electric fields and new local fields due to electrons in highly excited "anti-bonding" states. This may be more likely to occur at the Si−SiO$_2$ interface rather than in the bulk because the amorphous oxide provides less steric resistance to movement from the crystalline lattice positions. At this point, the role of the oxide or defects at the surface is unclear, but repeating this experiment on a clean sample in UHV should clarify such effects.

The results do not necessarily imply atomic disorder occurs in less than 150 fsec: SH measures the instantaneous electronic susceptibility and not the real atomic periodicity directly. For example, the bonding state susceptibility that dominates M might be reduced immediately upon laser excitation and the surface could melt before M could recover. This could occur through bleaching if a significant amount of the surface conduction band were filled and the relaxation time of such states were long compared to the melting process. This appears inconsistent with the lack of bleaching effects even near threshold. Alternatively, because M is a measure of 3m symmetry, M could be reduced by transforming the 3m symmetry to 6m order. This might be possible if sufficient electronic screening were established to isolate the single outermost layer of bulk terminated Si (which has 6m symmetry) from the influence of lower layers in registry with the 3m symmetric bulk (which normally makes the surface 3m symmetric). Electron screening might also weaken the bonding state susceptibility directly. However, if one extrapolates from the carrier density induced by excitation at 0.2X

60

threshold, e.g., assuming no saturation, the maximum number of excited states at threshold is $3.2 \times 10^{21}/cm^3$ or 6% of the Si atoms[31]. Even at such density it is unlikely that screening alone could dramatically reduce the nearest neighbor bond susceptibility. We can also dismiss the possibility that the Si surface and native oxide could be evaporated in 100 fsec considering that even if the atoms instantaneously acquired a velocity of 10^5 cm/s, they could only move 1Å or a half of the interatomic distance in 100 fsec.

If the Si atoms are indeed disordered by a photodissociation-like process, it would be the first such observation for a solid surface. This fast surface structural change may prove to be an essential step (in analogy to nucleation) in the slower bulk structural change and related advance of the liquid melt front away from the surface into the bulk. Similar experiments on metal surfaces are in progress.

7. Conclusion

Second-harmonic generation has the sensitivity to adsorbates and surface order and also the time-resolution to be a useful probe of surface reaction kinetics. Certainly, several problems remain with SHG as a surface probe, among them how to relate the changes in electronic susceptibility as measured by SHG to the changes in adsorbate composition in studies of reaction intermediates or to changes in atomic structure in structural phase transition studies. However, the theory of the electronic structure of surfaces is in itself a growing field and there is no reason that the SHG experiments should not contribute to that understanding as well as benefit by its development. Even under the present lack of microscopic theory, SHG promises to be a most useful tool for investigating a host of surface science questions in a time-regime hard to access by any other means. At the very shortest time-scales, it may be possible to observe reaction intermediates, precursors to adsorption, and non-equilibrium distributions of states. In addition, new effects due to high densities of highly excited surface electronic states may reveal new and interesting physics and photochemistry.

References

1. See review by Y. R. Shen, *J. Vac. Sci. Technol.* **3**, 1464 (1985).
2. H. W. K. Tom, C. M. Mate, X. D. Zhu, J. E. Crowell, T. F. Heinz, G. A. Somorjai and Y. R. Shen, *Phys. Rev. Lett.* **52**, 348 (1984).
3. T. F. Heinz, H. W. K. Tom, and Y. R. Shen, *Phys. Rev.* **A28**, 1883 (1983).
4. Th. Rasing, Y. R. Shen, M. W. Kim, and S. G. Grubb, *Phys. Rev. Lett.* **55**, 2903 (1985).
5. H. W. K. Tom, T. F. Heinz, and Y. R. Shen, *Phys. Rev. Lett.* **51**, 1983 (1983).
6. C. V. Shank, R. Yen, and C. Hirlimann, *Phys. Rev. Lett.* **51**, 900 (1983).
7. R. B. Hall and A. M. DeSantolo, *Surf. Sci.* **37**, 421 (1984).
8. R. S. Becker, G. S. Higashi, and J. A. Golovchenko, *Phys. Rev. Lett.* **52**, 307 (1984).
9. R. Haight, J. Bokor. J. Stark, R. H. Storz, R. R. Freeman, and P. H. Bucksbaum, *Phys. Rev. Lett.* **54**, 1302 (1985); R. Haight and J. Bokor, *Phys. Rev. Lett.* **56**, 2846.
10. P. Guyot-Sionnest and Y. R. Shen, *Phys. Rev.* **B33**, 8254 (1986); and submitted to *Phys. Rev. B.*
11. N. Bloembergen, R. K. Chang, S. S. Jha, and C. H. Lee, *Phys. Rev.* **174**, 813 (1968); **178**, 1528(E) (1969).
12. J. T. Yates, P. A. Thiel, and W. H. Weinberg, *Surf. Sci.* **82**, 45 (1979).
13. H. W. K. Tom, Ph. D. Thesis, Univ. Cal. Berkeley, (1984).
14. H. W. K. Tom, C. M. Mate, X. D. Zhu, J. E. Crowell, Y. R. Shen, and G. A. Somorjai, *Surf. Sci.* **172**, 466 (1986).

15. X. D. Zhu, Y. R. Shen, and R. Carr, *Surf. Sci.* **163**, 114 (1985).
16. S. G. Grubb, A. M. DeSantolo, and R. B. Hall, submitted to *J. Phys. Chem.*
17. A. Burns, H. -L. Dai, D. Heskett, E. W. Plummer, and K. -J. Song, to be published in *J. Chem. Phys.*
18. H. W. K. Tom, X. D. Zhu, Y. R. Shen, and G. A. Somorjai, *Surf. Sci.* **167**, 167 (1986).
19. T. F. Heinz, M. M. T. Loy, and W. A. Thompson, *J. Vac. Sci. Technol.* **B 3**, 1467 (1985).
20. H. W. K. Tom and G. D. Aumiller, submitted to *Phys. Rev. B.*
21. C. K. Chen, T. F. Heinz, D. Ricard, and Y. R. Shen, *Phys. Rev. Lett.* **46**, 1010 (1981).
22. G. L. Richmond, *Chem. Phys. Lett.* **110**, 571 (1984).
23. R. B. Hall, A. M. DeSantolo, and S. G. Grubb, submitted to *J. Vac. Sci. Technol.*
24. G. Arjavalingam, T. F. Heinz, J. H. Glownia, in *Ultrafast Phenomena*, **V**, eds. G. R. Fleming and A. E. Siegman, (Springer-Verlag, 1986).
25. H. W. K. Tom and G. D. Aumiller, *Phys. Rev. B* **33**, 8818 (1986).
26. T. F. Heinz, M. M. T. Loy, and W. A. Thompson, *Phys. Rev. Lett.* **54**, 63 (1985).
27. T. F. Heinz, G. Arjavalingam, M. M. T. Loy, J. H. Glownia, paper THII1, International Quantum Electronics Conference, June 9-13, 1986, San Francisco, CA.
28. S. A. Akhmanov, N. I. Koroteev, G. A. Paition, I. L. Shumay, M. F. Guljaudinov, and E. I. Shtyrkov, *Opt. Commun.* **47**, 202 (1983).
29. A. M. Malvezzi, J. M. Liu, and N. Bloembergen, *Appl. Phys. Lett.* **45**, 1019 (1984).
30. C. V. Shank, R. Yen, and C. Hirlimann, *Phys. Rev. Lett.* **50**, 454 (1983).
31. M. C. Downer and C. V. Shank, *Phys. Rev. Lett.* **56**, 761 (1986).

Direct Measurements of Surface Kinetics by Time-Resolved Electron Energy Loss Spectroscopy

T.H. Ellis[1], M. Morin[1], L.H. Dubois[2], M.J. Cardillo[2], and S.D. Kevan[3]

[1]Département de chimie, Université de Montréal,
  C.P. 6128, Succ. A, Montréal, Québec H3C 3J7, Canada
[2]AT&T Bell Laboratories, 600 Mountain Ave., Murray Hill, NJ 07974, USA
[3]Department of Physics, University of Oregon, Eugene, OR 97403, USA

1. Introduction

The technique of time-resolved electron energy loss spectroscopy (TREELS) has progressed to the point where it is now appropriate to assess its present capabilities and future impact. In this article we will illustrate the way in which TREELS can contribute to our understanding of surface rate processes, specifically chemical kinetics. As part of this assessment, it is useful to consider the basis for TREELS. In particular, we will highlight two key experimental developments, the incorporation of dispersion compensation and dispersive detection. These have resulted in greatly enhanced signal levels compared to conventional EELS ($\sim 10^2$–10^3) and allowed the resolution of EELS measurements into small time units, of the order of 1 millisecond. With this result now experimentally confirmed, we will informally discuss selected traditional techniques used to obtain surface kinetic parameters and compare them to TREELS. A specific example of a surface rate process, the adsorption and desorption of CO from Cu(100), as studied by TREELS, is presented as a demonstration of present day capabilities.

2. Background

The two new design improvements to EELS spectrometers are dispersion compensation [1] and dispersive detection [2]. Both of these ideas are in fact not new, but are adapted from other electron spectroscopies [3-6]. The design of these spectrometers has been discussed thoroughly elsewhere [1,2,7] and will not be repeated here. In practice, the prototype spectrometers of both designs have achieved quite similar performance, namely it is routinely possible to record a complete vibrational spectrum in 1 second with very good signal to noise (S/N) and an energy resolution of better than 15 millivolts. A single vibrational peak can be measured, in position and intensity in 10 milliseconds or less [8,9]. With some design improvements it will be possible to gain an order of magnitude improvement in signal, and the energy resolution should improve to the level of conventional EELS (< 5 meV). In addition, a single spectrometer can be designed to incorporate both dispersion compensation and dispersive detection, so that sub-millisecond time resolution may be possible.

This improvement in equipment has brought about a renewed interest in time-resolved studies of surface species. It has been recognized for some time that such a technique was needed to complement existing time-resolved measurements of incoming and desorbing fluxes, i.e. temperature-programmed desorption and molecular beam scattering. The motivation is clear – a direct measurement of the surface species will provide a very detailed description of most kinetic processes, particularly when combined with measurements of the molecules arriving and leaving. There have been in the past a limited number of time-resolved measurements with other techniques (as discussed by HO [2]), and together with TREELS this has recently sparked much more experimental effort towards this goal. TREELS is perhaps the most versatile of these techniques, because a vibrational spectroscopy in general provides more information with regard to chemical kinetics. It not only measures coverages, but also the detailed nature of the bonding. It is a generally useful technique for a wide variety of systems, and it is the method of choice to measure hydrogen.

There has recently been progress in time-resolved infrared spectroscopy (TRIRS?) brought about largely due to improvements in Fourier Transform spectrometers [10], but also with conventional grating instruments [11]. Other surface spectroscopies are continuing to show progress as well [12,13], and this will be important since as always in surface science complementary measurements are vitally important. The purpose of this paper is to address a problem facing all of these techniques. The time-resolved spectrometer is only the beginning - the next step is to create a time-resolved event to study.

The creation of time-resolved events on the minute time scale is done routinely in surface science experiments, while heating, cooling and dosing. Indeed, even on this time scale TREELS is extremely valuable. For example, using conventional EELS, SEXTON [14] observed that on Cu(100) a surface formate underwent a transformation from perpendicular to tilted bonding upon cooling, occurring over a period of about twenty minutes. This was recently studied in more detail with TREELS [15], by recording complete EELS spectra every 30 seconds while dosing, heating and cooling. It was shown that the tilting was brought about by readsorption of formic acid from the background. This result also illustrates a simple yet important reason for pushing TREELS into the millisecond time scale. One always wants to study kinetics over a large range of rates. The minimum time scale is determined by the signal to noise of the spectrometer, but there is also an upper time limit, which is the stability of the apparatus over a long period of time, with contamination being the most serious problem.

The most important new application of TREELS is on the fastest possible time scale. To measure kinetics, one must change the temperature or pressure (or coverage) on an equally fast time scale. There are several possible experimental configurations, as discussed below. For each case there is an analogous existing experimental technique which does not involve a direct measurement of the adsorbed surface species. With the exception of laser heating rates, experimental constraints often limit events from occurring faster than 0.1 milliseconds. Therefore, TREELS can be added to any of these experiments on about the same time scale as is presently used. This is where TREELS will have its biggest impact in the near future, since the experimental and theoretical foundation has already been laid. The problem now is to combine these existing techniques with TREELS without making too many experimental compromises.

3. Experimental Techniques

A general kinetic experiment involves a perturbation of temperature or pressure and subsequent measurement of the system as it returns to equilibrium. The temperature or pressure can be changed in the form of a ramp, jump or waveform, as shown in Fig.1. Shown also is the name of the related technique which measures the desorbed flux, in the case of

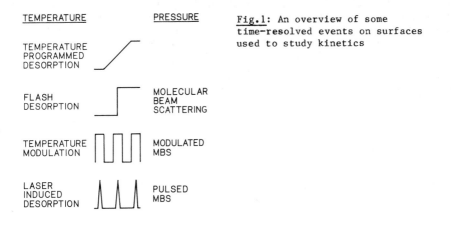

TEMPERATURE	PRESSURE

TEMPERATURE
PROGRAMMED
DESORPTION

FLASH
DESORPTION   MOLECULAR
                          BEAM
                          SCATTERING

TEMPERATURE   MODULATED
MODULATION   MBS

LASER
INDUCED   PULSED
DESORPTION   MBS

Fig.1: An overview of some
time-resolved events on surfaces
used to study kinetics

adsorption, desorption or diffusion kinetics. The extension to reaction
kinetics will not be discussed here, but the same description applies.

1) **Temperature.** Temperature-programmed desorption (TPD) [16,17] is a
widely used technique, and usually involves linear heating rates of
1-100K/sec. One limitation is that peaks in the desorption spectra can
originate from several sources, which creates ambiguity. Temperature-
programmed EELS (TPEELS) is the ideal technique for resolving this
problem. Time-resolved LEED would be useful as well when structural
phase transitions are involved. The data analysis of TPEELS is straightforward. For simple desorption processes, the time derivative of the
TPEELS signal should be the same as a TPD peak. HO and co-workers
[2,9,18] have performed many experiments of this type, and BENZIGER and
SCHOOFS have done TPIRS [11]. A related technique, where the temperature
is quickly increased and held constant, has been called flash desorption
by EHRLICH [19]. Here, the integrated desorbed flux (i.e. pressure rise)
is measured in real time, and this is the complement of a measurement of
surface coverage. The technique is kinetically simpler than TPD since
the desorption occurs under isothermal conditions. Conventional flash
desorption is limited because it must be done with zero pumping speed,
but TREELS does not have this same constraint.

Recently, ENGSTROM and WEINBERG [20] have shown the usefulness of temperature modulation techniques, using 1K modulation at 0.01 to 1.0 Hz.
This would be ideal for coupling to TREELS. Another method in the same
category is laser-induced desorption (LID). One example is the study of
surface diffusion kinetics recently probed on the ten second time scale
by monitoring the rate at which a laser desorption spot is filled in from
the edges [21]. In general, temperature pulses are used to create a
sudden change in coverage, and the system is studied as it returns back
to equilibrium. Lasers are not needed, and in fact their rapid heating
rates can impede desorption. Flash lamps or electrical discharge heating
may also be used.

2) **Pressure** The most straightforward form of molecular beam scattering experiments [22,23] is a pressure jump. It is a convenient way to
measure adsorption or desorption kinetics under isothermal conditions.
An upward pressure step is most commonly used to measure sticking coefficients as a function of both surface and beam conditions. TREELS will be

extremely useful for the study of reactive sticking probabilities [24].
CEYER and co-workers [25] have used conventional EELS to study adsorption. A downward pressure step can also be used to measure desorption
kinetics, although generally on a fairly slow time scale.

For faster times, the pressure waveform is used. This is generally
known as modulated molecular beam scattering [26,27]. It is an extremely
powerful technique carried out under very well defined conditions.
Normal molecular beam sources create a small pressure modulation at the
surface ($< 10^{-4}$ Torr). The resulting coverage change can be detected by
TREELS. The time resolution can remain at one millisecond even for small
coverage changes since the waveform is repeated and signal averaged. The
kinetics can easily be measured as a function of temperature and
coverage, the latter made possible by using a second, effusive beam to
establish an equilibrium coverage. Since the coverage modulation is
small, tuning problems associated with work function changes are
minimized. An extensive body of literature exists for data analysis,
including powerful Fourier transform methods. For this purpose square
wave chopping has some advantages, although any waveform can be modeled.
In addition to the measurements performed in our laboratory [8,24],
FROITZHEIM et al. [28] have performed modulated molecular beam studies
with a conventional EELS spectrometer.

One general comment on all of the above. TREELS must be pushed to its
present limit to work on the time scale attainable with most of the
techniques which measure desorbed fluxes. However, it is possible to
measure continously from this time scale to as slow as possible, which is
accompanied by an increase in the S/N. Such is not the case for measurements of the desorbed flux, where the S/N generally decreases with slower
time scales. The largest possible dynamic range is necessary for
accurate kinetic measurements.

4. Results

Examples involving temperature changes can be found in the work of HO and
co-workers [2,9,18]. The system CO/Cu(100) will be used here to illustrate some pressure modulation experiments. This is an ideal test system
since previous EELS results have shown that the carbon-oxygen stretching
frequency is constant and the intensity linear over a wide coverage
range. In Fig.2 the adsorption of CO at constant temperature and
pressure is shown. A molecular beam was not used, instead the entire
chamber pressure was raised. The only advantage of TREELS here is the
ease at which these data can be taken. In the past, such data were
obtained by making many measurements after different doses. It should be
noted that at least in the present study the TREELS spectrometer was
remarkably insensitive to work function changes during dosing.

The linear coverage dependence is indicative of adsorption via a
mobile precursor. To confirm that this is not an artifact of the EELS
intensity, DUBOIS and ZEGARSKI [29] measured the integrated thermally desorbed flux after many calibrated doses, and observed the same behavior.
Above 1 L the EELS intensity decreases, at the same time as the CO goes
into a compressed stage. Recent second harmonic generation results indicate that the adsorption kinetics are Langmuir-like up to a dose of 6 L
[30]. This apparent discrepancy cannot be explained from the available
data.

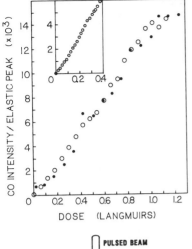

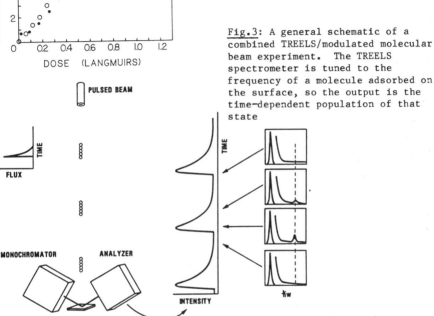

Fig.2: Adsorption kinetics of CO on
Cu(100) at 120 K. The intensity of
the CO stretching vibration (open
circles) measured in real-time is
compared to the integrated area
under the TPD peak (filled circles)
measured after discrete exposures.
The inset shows EELS data taken at
low exposure to demonstrate the
sensitivity of the technique

Fig.3: A general schematic of a
combined TREELS/modulated molecular
beam experiment. The TREELS
spectrometer is tuned to the
frequency of a molecule adsorbed on
the surface, so the output is the
time-dependent population of that
state

A pressure waveform experiment is shown schematically in Fig.3, and
results for CO/Cu(100) given in Fig.4. For these data, a differentially
pumped, collimated molecular beam was not available, thus the incident
waveform shape is determined by the pumping speed of the chamber. We
estimate that its decay is biexponential, with a large component at 120
msec. and a smaller one at 600 msec. Its uncertainty limits an accurate
analysis of the data. Measurements were easily performed over 3 orders
of magnitude in rates, although due to the aforementioned uncertainty we
cannot trust the beam deconvolution below ~100 milliseonds. Figure 4a
is for a single pressure pulse, and Fig.4f is signal averaged for only
20 pulses. Each pulse is on the order of 1 L in intensity. A complete
analysis of the data would require measurements as a function of dose and
pulse rate, but the beam here is not adequate for this. We did however
fit these data to simple first order desorption kinetics, using the model
for the adsorption kinetics discussed above. The results are the solid
lines in Fig.4. The fit to first order kinetics for these data was
confirmed by the Fourier transform method. Figure 5 shows the transfer
function [31] for three temperatures.

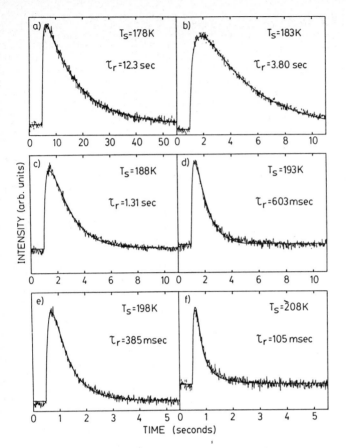

Fig.4 : Measurement of the surface CO population after pulsed beam dosing. Note the change in time scale for the different temperatures. The smooth line is a fit to the data as described in the text. The intensity of the pulse was not calibrated, so the intensity was a free parameter in the fit

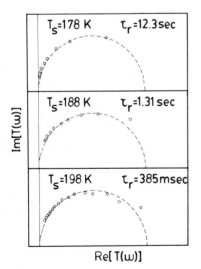

Fig.5: For three of the temperatures in Fig.4, the data were Fourier transformed and the transfer function plotted. A semi-circle indicates first order desorption. The dashed line is the theoretical transfer function using the mean residence time (the reciprocal of the desorption rate constant) obtained from the fits to the real-time waveforms

From these data we can construct an Arrhenius plot and abstract an activation energy of 14+/-1 kcal/mole. This is consistent with the heat of adsorption as determined by TRACY [32], 13.3 kcal/mole, obtained from equilibrium measurements. Our measured pre-exponential factor of $10^{15+/-1}$ is similar to the values found for CO desorption from several other surfaces [27]. Finally, we obtained some data which could not be fit with simple first order kinetics. One explanation could be that the data in Fig.4 are for a surface where the steps and defects have been saturated. This will be studied in more detail when the TREELS spectrometer is coupled with a true molecular beam source.

5. Summary

Time-resolved surface spectroscopies show great promise for the study of surface kinetics, and there is now considerable effort being directed towards this end. There are many possible experimental configurations for these studies, several of which have been tested already. The full potential of the technique will only be reached when these spectrometers are coupled with molecular beam techniques of the type and quality which are currently supplying the field with the most accurate kinetic measurements [33].

6. References

1. S.D. Kevan, L.H. Dubois: Rev. Sci. Instrum. 55, 1604 (1985)
2. W. Ho: J. Vac. Sci. Technol. A 3, 1432 (1985)
3. K. Siegbahn, D. Hammond, H. Fellner-Feldegg, E.F. Barnett: Science 176, 245 (1972)
4. W. Bertozzi, M.V. Hynes, C.P. Sargent, W. Turchinetz, C. Williamson: Nucl. Instrum. Methods, 162, 211 (1979)
5. U. Gelius, E. Basilier, S. Svansson, T. Bergmark, K. Siegbahn: J. Electron Spectrosc. Relat. Phenom., 2, 405 (1974)
6. K. Karlsson, L. Mattsson, R. Jardny, R.G. Albridge, S. Pinchas, T. Bergmark, K. Siegbahn: J. Chem. Phys. 62, 4745 (1975)
7. R. Franchy and H. Ibach: Surface Sci. 155, 15 (1985)
8. T.H. Ellis, L.H. Dubois, S.D. Kevan, M.J. Cardillo: Science, 223, 256 (1985)
9. J.S. Villarrubia, L.J. Richter, B.A. Gurney, W. Ho: J. Vac. Sci. Technol. A4, 1487 (1986)
10. M.A. Chesters: J. Electron Spectrosc. Relat. Phenom. 38, 123 (1986)
11. J.B. Benziger, G.R. Schoofs: Surface Sci. 171, L141 (1986)
12. M. Balooch, D.R. Orlander, J. Abrefah, W.J. Siekhaus: Surface Sci. 149, 285 (1985)
13. B. Polsema, L.K. Verheij, G. Comsa: Phys. Rev. Letters 51, 2410 (1983)
14. B.A. Sexton: Surface Sci. 88, 319 (1979)
15. L.H. Dubois, T.H. Ellis, B.R. Zegarski, S.D. Kevan: Surface Sci. 172, 385 (1986)
16. P.A. Redhead: Vacuum 12, 203 (1962)
17. J.T. Yates, Jr.: In Solid State Physics: Surfaces, ed. by R.L. Park and M.G. Legally, Methods of Experimental Physics, Vol. 22 (Academic Press, Orlando 1985)
18. L.J. Richter, B.A. Gurney, J.S. Villarrubia, W. Ho: Chem. Phys. Lett. 111, 185 (1984)
19. G. Ehrlich: J. Chem. Phys. 34, 29 (1961)
20. J.R. Engstrom, W.H. Weinberg: Phys. Rev. Letters 55, 2017 (1985)
21. C.H. Mak, J.L. Brand, A.A. Deckert, S.M. George: J. Chem. Phys. 85, 1676 (1986)

22. D.A. King, M.G. Wells: Surface Sci. $\underline{29}$, 454 (1972)
23. J.A. Barker, D.J. Auerbach: Surface Sci. Reports $\underline{4}$, 1 (1985)
24. L.H. Dubois, T.H. Ellis, S.D. Kevan: J. Electron Spectrosc. Relat. Phenom. $\underline{39}$, 27 (1986)
25. S.L. Tang, M.B. Lee, J.D. Beckerle, M.A. Hines, S.T. Ceyer: J. Chem. Phys. $\underline{82}$, 2826 (1985)
26. R.H. Jones, D.R. Orlander, W.J. Siekhaus, J.A. Schwarz: J. Vac. Sci. Technol. $\underline{9}$, 1429 (1972)
27. M.P. D'Evelyn, R.J. Madix: Surface Sci. Reports $\underline{3}$, 413 (1984)
28. H. Froitzheim, U. Kohler, H. Lammering: Phys. Rev. B $\underline{34}$, 2125 (1986)
29. L.H. Dubois, B.R. Zegarski: (unpublished results)
30. X.D. Zhu, Y.R. Shen, R. Carr: Surface Sci. $\underline{163}$, 114 (1985)
31. H.H. Sawin, R.P. Merrill: J. Vac. Sci. Technol. $\underline{19}$, 40 (1981)
32. J.C. Tracy: J. Chem. Phys. $\underline{56}$, 2748 (1972)
33. M.J. Cardillo: Ann. Rev. Phys. Chem. $\underline{32}$, 331 (1981)

Isotopic Effects in the Adsorption and Desorption of Hydrogen by Ni(111)

J.T. Yates, Jr., J.N. Russell, Jr., I. Chorkendorff, and S.M. Gates*

Surface Science Center, Department of Chemistry,
University of Pittsburgh, Pittsburgh, PA 15260, USA

The dynamics of the interaction of hydrogen with the Ni(111) surface have been investigated using several methods involving comparisons between H_2 and D_2 adsorbates. In addition, the desorption of hydrogen has been studied by methods which yield information about the angular distribution of the desorbing species. It has been found that on Ni(111), hydrogen molecules adsorb via an interaction with a single site on the Ni(111) surface. The adsorption process produces an activated complex which is located near the entrance channel (for adsorption at low coverages) on the potential energy surface describing the interaction. The adsorption process occurs through the formation of an intrinsic molecular precursor state. This state may be observed by TPD above one-half monolayer coverages of atomic deuterium in the case of molecular D_2. Adsorbed molecular H_2 is not observed. The D_2 molecular state desorbs via first order kinetics with an activation energy of 11.1 kJ/mole and a preexponential factor of 1.2×10^5 sec^{-1}. These conclusions are confirmed by angular distribution measurements of H_2 desorption which indicate that adsorbed H recombines over an exit channel (for desorption) barrier when the coverage is below one-half monolayer. For atomic hydrogen coverages above one-half monolayer, recombination occurs over an entrance channel (for desorption) barrier. These results suggest that the location of the activation barrier for H + H recombination is dependent on the coverage of adsorbed atomic hydrogen.

1. Introduction

The study of the adsorption and desorption of hydrogen by atomically clean metal surfaces offers an opportunity for the observation of isotope effects in these processes, permitting one to obtain a deeper level of understanding of the molecular-dynamic aspects of chemisorption. Through the study of differences in the behavior of hydrogen and deuterium, one may gain insights into the character of hydrogen-metal interactions which also apply in a more general way to surface processes as a whole. The work to be discussed here utilizes differences between hydrogen and deuterium as a probe of the general aspects of the potential surface which govern the adsorption-desorption process for hydrogen. In addition, a study of the angular distribution of desorbing H_2 has been used to further understand the details of the potential surface.

It is well established that a barrier to hydrogen adsorption exists on the Ni(111) surface [1-3]. ROBOTA, et.al [1] employed translationally excited molecular beams of hydrogen and found an enhancement of the sticking coefficient above a threshold kinetic energy level. They estimated a barrier height of 10 kJ/mole. HAYWARD and TAYLOR [3] estimated the threshold energy for adsorption to be only 1.5 kJ/mole. Recent calcu-

lations indicate that a molecular precursor state may exist for dissociative chemisorption of hydrogen on Ni(111) [4-6]. The barrier between
the molecular precursor and the chemisorbed atomic state leads to a rather
low sticking coefficient (S_0 = 0.02-0.13) for adsorption of H_2 on Ni(111).
In contrast, the sticking coefficient for H_2 on Ni(110) is 0.40-0.96
[2b,7]. This difference between the two Ni planes indicates that structural factors at the surface are important in determining the dynamics of
hydrogen adsorption. The presence of a barrier to chemisorption of hydrogen has been seen previously by BALOOCH, et al. [8] who showed that there
is a ~ 20 kJ/mole barrier for H_2 adsorption on Cu(111). In their work,
differences in barrier heights between various Cu planes were observed.

According to VAN WILLIGEN [9], a barrier for adsorption in the entrance
channel will cause desorption trajectories to be preferentially directed
normal to the surface. A test of this concept is provided by the work of
STEINRÜCK, WINKLER, and RENDULIC [2a] who have compared the angular
distributions for desorbing H_2 from Ni(111) and Ni(110). They found that
for Ni(111), H_2 thermal desorption occurred in a $\cos^{4.5} \Theta$ angular distribution for hydrogen coverages of 0.5 H/Ni. In contrast, H_2 desorption
from Ni(110) exhibits a $\cos \Theta$ angular distribution. Related studies of CO
desorption indicated the presence of a $\cos \Theta$ angular distribution on both
Ni planes.

There is general agreement that H adsorbs dissociatively in three-
fold hole sites on the Ni(111) surface for Θ < 0.5 H/Ni. Thus, dynamic
LEED studies [10] indicate that at Θ_H = 0.5 H/Ni, both types of threefold
hole sites are occupied by H. HO, et al. [11] have studied a saturation
coverage H layer on Ni(111) using HREELS at 170 K, finding two vibrational
modes which could be assigned to H atoms located in both types of three-
fold hole sites. This assignment is consistent with cluster calculations [4].

2. Experimental

The ultrahigh vacuum chamber and our experimental methods have been
described elsewhere [12,13]. A multiplexed quadrupole mass spectrometer
(QMS) is located along the surface normal of the Ni(111) crystal, inside a
differentially-pumped cylindrical random flux shield, as shown in Fig. 1.
In the various experiments shown here, two different apertures, 1.6mm or
3.0mm in diameter, were used at the entrance to the differentially-
pumped mass spectrometer. During temperature programmed desorption
experiments [14], the crystal was located within 1mm of the random flux
shield aperture, assuring that gas desorption from the front face of the
crystal was being measured, and that efficient discrimination against gas
from the support leads, the crystal edges, and the crystal back face was
achieved. Adsorption of gases onto the crystal front face was carried out
using a close-coupled molecular beam doser oriented normal to the crystal
face. This doser permits the adsorption of gas from a well-collimated
beam of known absolute flux. The chamber contains an argon ion sputter
gun for surface cleaning, and a cylindrical mirror Auger analyzer for analysis of the surface for initial surface cleanliness. For temperature
programmed desorption experiments, the crystal temperature was linearly
increased at 2K/s using an electronic temperature programmer of our own
design [15].

Two Ni(111) single crystals were used in these experiments. Each was
oriented, cut and polished to within ± 0.5° of the (111) surface.

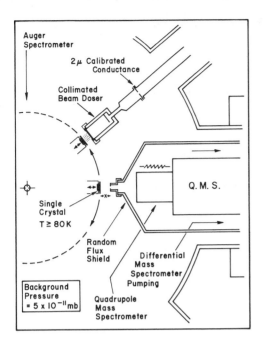

Fig. 1. Ultrahigh vacuum apparatus for quantitative studies of adsorption and desorption.

Tungsten heating leads and chromel-alumel thermocouple wires were spot welded to the back of the crystals, effectively shielding them from the QMS aperture.

Bulk carbon was removed by oxygen treatment of the crystal followed by heating to 1000 K in a molecular beam of H_2. This procedure was repeated several times. The crystal was cleaned daily by argon ion sputtering, followed by annealing at 1000 K for 10 minutes. No impurities were detected within the sensitivity limits of AES. (The sensitivity limits for surface C, O, and S are 0.14, 0.07, and 0.04 atomic percent respectively).

ROBOTA, et. al [16] determined that trace amounts of C, undetectable by AES, can affect the kinetics of H_2 adsorption. They devised a molecular beam procedure involving isotopic mixing between H_2 and D_2 as a means of showing that the crystal was carbon free. We have employed this technique to verify that the crystals used here were also free of carbon.

The backing pressures of H_2 and D_2 in our stainless steel molecular beam dosing system were adjusted $[P_{D_2}/P_{H_2} = \sqrt{2}]$ to ensure that in coadsorption experiments involving these isotopic species, a 50%-50% mixture of the isotopic species reached the Ni(111) crystal, compensating for the differential rate of effusion of the two isotopic gases through the 2 micron diameter orifice which controls the flow rate from the doser as shown in Fig. 1.

QMS sensitivity factors for H_2, HD, and D_2 were determined by comparison of integrated temperature programmed desorption curves for the H_2 and D_2 saturated surfaces, making small corrections for the production of HD when mixtures of the isotopic molecules were present. The relative QMS sensitivities were: H_2 = 1.00; HD = 0.91; and D_2 = 0.85.

Temperature Programmed Desorption of H_2 and D_2: Ni(III)

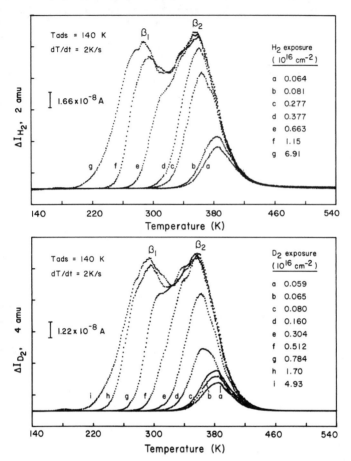

Fig. 2. Series of temperature programmed desorption spectra after exposure of the Ni(111) crystal held at 140 K to pure H_2 (top panel) or pure D_2 (bottom panel). The crystal temperature was increased at 2 K s^{-1}. The thermal desorption signals have not been corrected for the sensitivity of the QMS

3. Results

3.1. Adsorption of Hydrogen at 140 K

A sequence of temperature programmed desorption experiments for pure H_2 and D_2 adsorption are shown in Fig. 2. In agreement with the work of others [7,10,17], two desorption states are observed. Below $\Theta_H = 0.5$ H/Ni, the β_2-desorption state is observed with its peak maximum near 370 K. Above $\Theta_H = 0.5$ H/Ni, a lower temperature shoulder develops which is designated β_1, with peak maximum near 290 K. From a comparison of the two sets of desorption spectra in Fig. 2, little difference can be seen in

74

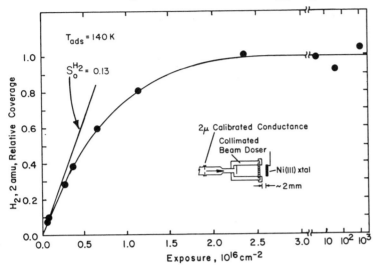

Relative Coverage of H(ads) on Ni(III) vs. H₂ Exposure

Fig. 3. Plot of the relative coverage of H(a) on Ni(111) as a function of
H₂ exposure to the crystal held at 140 K. The insert shows the position
of the crystal with respect to the doser when exposing the crystal to H₂.
The collimated beam results are not quantitatively comparable with experiments involving a random incident flux.

the behavior of the two hydrogen isotopes for adsorption at 140 K. The
desorption spectra develop to the same relative coverage levels for the
two states, and only slight differences exist in the temperature of the
desorption peak maxima. Figure 3 shows a plot of the hydrogen coverage
versus hydrogen exposure for the experiments shown in Fig. 2. The initial sticking coefficient for the adsorption from the beam is $S_0^{H_2} = 0.13$.

It was of interest to determine the kinetic rate law for H₂ adsorption
on Ni(111), as has been done before for other surfaces [18]. The rate of
adsorption can be expressed as involving one-site Langmurian kinetics
(equation 1) or two-site Langmurian kinetics (equation 2):

$$\frac{d\theta}{d\varepsilon} = \frac{2S_0^{H_2}(1-\theta)}{C_S} , \qquad (1)$$

$$\frac{d\theta}{d\varepsilon} = \frac{S_0^{H_2}(1-\theta)^2}{C_S} . \qquad (2)$$

Here $C_S = 1.88 \times 10^{15}$ sites cm⁻², $S_0^{H_2}$ = the sticking coefficient for an
H₂ molecule on an empty site, ε = exposure to H₂, and $(1-\theta)$ = the fractional coverage of empty sites.

Figure 4 shows a plot of the adsorption kinetic data using the kinetic
expressions in equations (1) and (2). From the linearity of the left
plot, it is clear that one-site Langmurian kinetics are controlling the

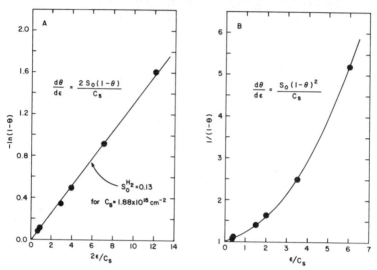

Fig. 4. Langmuir kinetic plots for hydrogen adsorption on Ni(111). the data in Fig. 2 has been replotted in terms of the kinetic rate expressions for one-site Langmuirian adsorption kinetics (A) and two-site Langmuirian adsorption kinetics (B). The expressions are described within the text

adsorption of H_2 on Ni(111). This is compelling evidence for the involvement of an H_2 molecule in the rate determining step for adsorption, i.e., for an early transition state for adsorption for the full range of hydrogen coverage at 140 K adsorption temperature. Just the opposite behavior has been observed for H_2 adsorption on Rh(111) [18].

3.2. Isotope Effects in Adsorption at 87 K

We have made adsorption kinetic measurements at 87 K in order to allow deuterium kinetic isotope effects to become more easily observable at the lower temperature. A comparison between H_2 and D_2 behavior at 140 K and 87 K is shown in Fig. 5. In the lower two panels it can be seen that the desorption state development at 140 K for H_2 and D_2 is essentially identical, in agreement with the results of Fig. 2. However, at the lower temperature, 87 K, it is evident that a retardation of D_2 adsorption occurs as indicated by the incomplete filling of the β_1 desorption state. By monitoring the competitive buildup in D and H coverage at 87 K, as shown in Fig. 6, it is clear that there is a retardation of D_2 adsorption relative to H_2 adsorption from an equal mixture of H_2 and D_2 over the entire coverage range. The data have been corrected for the small (<0.08 x 10^{15} H/cm^2) contribution due to ambient H_2 adsorption. From these data, we estimate $S_0^{D_2} \simeq 0.7\ S_0^{H_2}$.

A series of adsorption measurements was performed at 87 K using a mixture of H_2 and D_2 adsorbates in our molecular beam doser. The hydrogen isotope desorption spectra from this experiment are shown in Fig. 7 for increasing exposure to the $H_2 + D_2$ mixture. The ratio of H_2 and D_2 was

Effect of Surface Temperature on the Dissociative Adsorption Kinetics of H_2 and D_2 on Ni(III)

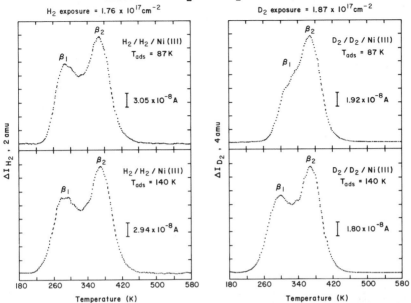

Fig. 5. Effect of surface temperature on the dissociative adsorption kinetics of H_2 and D_2 on Ni(111). The H_2 and D_2 saturation coverage TPD
spectra for H_2 or D_2 adsorbed at 87 K and 140 K are shown in the left and
right panels, respectively. A 1.6 mm diameter sampling aperture was used
on the QMS. Because of the smaller sampling angle, the intensity of the
β_1 feature was reduced relative to the intensity of the β_2 feature
(compared to data in Fig. 2). H_2 exposure = 1.76 x 10^{17} cm^{-2}; D_2 exposure
= 1.87 x 10^{17} cm^{-2}

Relative Coverage of H(ads) and D(ads) on Ni(III)

vs Exposure to the H_2:D_2 Mixture

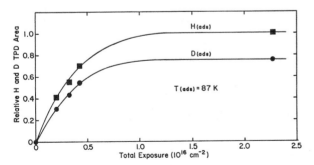

Fig. 6. Relative H and D coverages plotted as a function of H_2 + D_2 total
exposure. Ni(111) temperature = 87 K

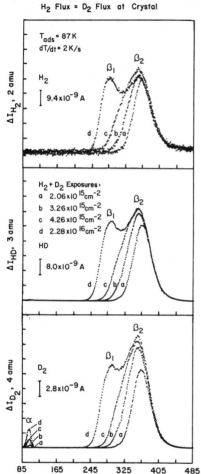

TPD Series for H_2/D_2 Mixture on Ni(III)

H_2 Flux = D_2 Flux at Crystal

T_{ads} = 87 K
dT/dt = 2 K/s

H_2

9.4×10^{-9} A

β_1

β_2

d c b a

ΔI_{H_2}, 2 amu

$H_2 + D_2$ Exposures:
a $2.06 \times 10^{15} cm^{-2}$
b $3.26 \times 10^{15} cm^{-2}$
c $4.26 \times 10^{15} cm^{-2}$
d $2.28 \times 10^{16} cm^{-2}$

HD

8.0×10^{-9} A

β_1

β_2

d c b a

ΔI_{HD}, 3 amu

D_2

2.8×10^{-9} A

β_1

β_2

α

d c b a

d c b a

ΔI_{D_2}, 4 amu

85   165   245   325   405   485
Temperature (K)

Fig. 7. Multiplexed H_2, HD, and D_2 temperature programmed desorption spectra for a series of exposures of a mixed beam of H_2 and D_2 to the Ni(111) surface held at 87 K. The H_2 flux equaled the D_2 flux at the crystal. The spectra have not been corrected for the sensitivity of the QMS

adjusted such that an equimolar flux of the two isotopic molecules arrived at the crystal surface [19]. Following various exposures to the mixture of the hydrogen isotopes, temperature programmed desorption experiments were performed using multiplexed mass spectrometer techniques. There are three basic observations:

1. There is kinetic retardation of D_2 adsorption relative to H_2 adsorption throughout the entire range of coverage. Initial values of $S_0^{D2}/S_0^{H2} \cong 0.7$ at 87 K. This fact may be seen in the rapid saturation of the β_2 state by H_2 compared to D_2. This trend is confirmed by the intermediate behavior of β_2-HD, compared to H_2 and D_2.

2. The development of the β_1 feature in D_2-desorption is incomplete, relative to the H_2 desorption in the β_1 state. This is best seen by com-

parison of the relative intensities for β_1 and β_2 for H_2 and D_2 desorption.

3. A new desorption feature is observed at T = 100 K in the D_2 desorption spectrum and is absent in the H_2 and HD desorption spectra. It will be referred to as $\alpha-D_2$.

These experiments confirm that retardation of D_2 adsorption occurs at 87 K throughout the entire coverage range. They also show that a low surface coverage of molecular D_2 is preferentially stabilized as the coverage of dissociated hydrogenic species increases.

3.3. Characterization of $\alpha-D_2$

The $\alpha-D_2$ desorption feature reproducibly appeared in the D_2-TPD spectra when β_2-D_2 was nearly saturated. $\alpha-D_2$ was observed in experiments using $H_2:D_2$ mixtures as well as for pure D_2. The low temperature molecular desorption state is characteristic only of the deuterium molecule and not of H_2.

In an attempt to investigate the character of the sites retaining $\alpha-D_2$, we carried out experiments on a disordered Ni(111) crystal which had been subjected to sputter roughening by Ar^+ ion bombardment (600 eV x 2.9 µA x 1200 sec.). From this initial disordered condition, three increasing degrees of surface order were achieved by annealing the crystal as shown in Fig. 8, where sequential D_2 adsorption experiments were carried out

Effect of Surface Smoothness on $\alpha-D_2$

Ar^+ sputter crystal: 600V, 1200 sec x 2.9 µA at T_{xtal}= 480 K

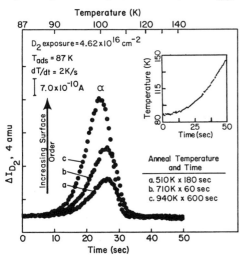

Fig. 8. Effect of induced surface roughness on the $\alpha-D_2$ state. The crystal surface has been roughened by Ar ion bombardment (600 eV x 1200 sec x 2.9 µA at $T_{surface}$ = 480 K). The $\alpha-D_2$ TPD spectra are generated after exposing the surface to D_2 after annealing the clean crystal at different temperatures. The insert is a digitized plot of the crystal temperature versus time during programmed desorption

following the annealing program. The quantity of α-D_2 adsorbed at 87 K for an exposure of 6.62×10^{16} D_2 molecules cm^{-2} increases as the crystal becomes smoother. This result suggests that the population of the α-D_2 state is inversely dependent upon the number of defect sites on the surface, and therefore indicates that α-D_2 originates from the ordered crystal surface and is not an artifact due to adsorption on defects, the heating leads, or the crystal edge. The α-D_2 state desorbs with first order kinetics, with $\nu = 1.2 \times 10^5$ sec^{-1}, and $E_A = 11.1$ kJ $mole^{-1}$ [19].

3.4. Angular Distributions of H_2 Desorbing Species

The apparatus was employed to make an estimate of the angular distribution of desorbing H_2 molecules using the principles illustrated in Fig. 9. The single crystal containing a saturated overlayer of chemisorbed hydrogen could be investigated using the shielded quadrupole mass spectrometer with the crystal being located at various distances, x, from the aperture. For the two examples shown, involving a $\cos\theta$ and a $\cos^{4.5}\theta$ angular dependence, it can be seen that as x increases and the solid angle sampled by the QMS decreases, the more directed distribution of desorbing species will be less attenuated compared to the $\cos\theta$ distribution.

A series of TPD experiments from the H-saturated overlayer is shown in Fig. 10 for a sequence of increasing values of x, the distance to the aperture. It is very clear that the relative peak intensities for the two hydrogen desorption states, β_1 and β_2, behave differently, with β_1 hydrogen exhibiting a broader angular distribution compared to β_2. The smooth dependence of the peak height ratios for the two thermal desorption states is shown in the inset to Fig. 10.

SCHEMATIC DIAGRAM OF APPARATUS USED
TO MEASURE DIFFERENCES IN
THERMAL DESORPTION ANGULAR DISTRIBUTIONS

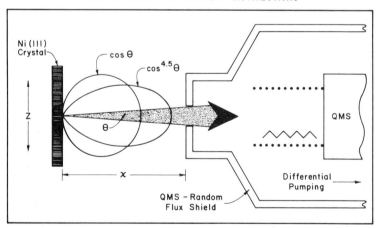

Fig. 9. Schematic diagram of the apparatus used to discriminate different angular distributions of thermally desorbing molecules. By varying the crystal position along either the x or z axes, the solid sampling angle of desorbing molecules could be varied. Angular distributions ($\cos\theta$ and $\cos^{4.5}\theta$) are illustrated. The shaded region illustrates the solid angle sampled by the QMS

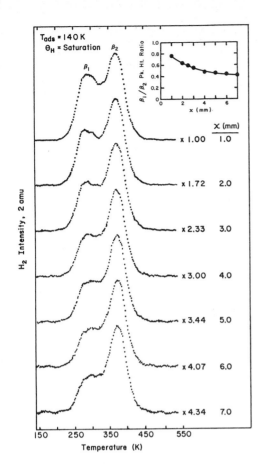

VARIATION OF FULL COVERAGE H_2/Ni(III)
THERMAL DESORPTION FEATURES AT
VARIOUS CRYSTAL TO SAMPLING
APERTURE DISTANCES

Fig. 10. Variation of the saturation coverage H_2/Ni(111) thermal
desorption features as the crystal to sampling aperture distance,
x, is systematically changed. The
inset shows the ratio of the β_1
and β_2 desorption peak intensities as the crystal position is
changed

An estimate of the angular distributions for β_1 and β_2 H_2 desorption
states can be made by comparison with other work. It is known from studies
by STEINRÜCK, et al. [2a] that CO desorbs from Ni(111) with a cos Θ distribution. We have therefore employed the desorption of CO as a calibrating
method for estimating the angular distribution of the desorbing hydrogen
from Ni(111). Fig. 11 shows that the behavior of β_1-H_2 is quite similar to
that observed for CO, indicating that the desorption of β_1-H_2 obeys a cos
Θ angular distribution law. In contrast to this behavior, the β_2-H_2
desorption state is highly peaked in the forward direction according to
our measurements, and in agreement with the work of STEINRÜCK, et al.[2],
who measured a $\cos^{4.5} \Theta$ angular distribution.

The qualitative difference between the angular distribution of β_1-H_2
and β_2-H_2 desorption was confirmed by translating the crystal in the z-
direction parallel to the plane of the sampling aperture, with x being
held constant at 2 mm. The results of this experiment are shown in Fig.
12 for z = +5, 0, and -5mm. The crystal position with respect to the
aperture is illustrated on the right side of the figure for each TPD
measurement. The desorption spectra were normalized to the β_1-feature.
The β_2-H_2 intensity decreases with respect to the β_1-H_2 intensity for the
crystal position below and above the line of sight of the QMS. By con-

COMPARISON OF β_1 & β_2-H$_2$ AND CO
DESORPTION PEAK INTENSITY
VERSUS CRYSTAL DISTANCE
FROM THE QMS SHIELD

VARIATION OF FULL COVERAGE H$_2$/Ni(III)
THERMAL DESORPTION ANGULAR DISTRIBUTION
AS THE CRYSTAL IS TRANSLATED PARALLEL
TO THE PLANE OF THE SAMPLING APERTURE

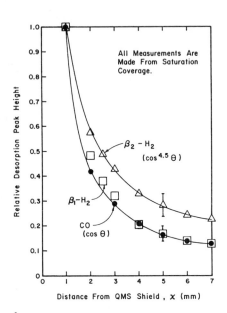

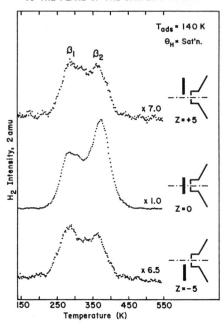

Fig. 11. Comparison of the relative $\overline{\beta_1}$-H$_2$, β_2-H$_2$, and CO desorption peak intensities for saturation coverage as a function of the crystal distance, x, from the QMS shield. Representative error bars are shown

Fig. 12. Variation of the saturation coverage H$_2$/Ni(111) thermal desorption angular distribution as the crystal is translated parallel to the plane of the sampling aperture. The positions of the crystal with respect to the aperture for each measurement are shown on the right

sidering the two schematic angular distributions shown in Fig. 9, one can easily determine that for larger angles with respect to the surface normal, more molecules will be sampled by the QMS for the cos Θ distribution, β_1, than for the cos$^{4.5}$ Θ distribution, β_2.

We believe that these results indicate that neither the solid angle subtended by the crystal nor desorption from the crystal edges is seriously influencing the experimental results reported here. Our estimate of cos Θ involves error bars which would allow cos Θ and cos^3 Θ to be clearly distinguished.

These results, obtained in an apparatus not designed to measure the angular distribution of desorbing molecules, indicate that the angular distribution of desorbing H$_2$ from Ni(111) is peaked significantly in the forward direction when the H-coverage is below Θ_H = 0.5H/Ni. At higher H coverages, the angular distribution of desorbing H$_2$ becomes unpeaked and exhibits a cos Θ angular distribution.

4. Discussion

4.1. Kinetics of H_2 Adsorption on Ni(111) - Evidence for the Involvement of Single Adsorption Sites

As shown in Fig. 4, the kinetics of H_2 adsorption follow single site Langmurian behavior accurately over the entire hydrogen coverage range. This observation is consistent with a rate determining step for adsorption which involves the interaction of H_2 with a single site on the surface, and a kinetic model involving the interaction of an H_2 molecule interacting with a single surface site in an early transition state for adsorption is envisioned. This behavior is not found for H_2 adsorption on all (111) surfaces of face-centered transition metals; Rh(111) exhibits just the opposite kinetic behavior with two-site Langmuirian kinetics being rate controlling [18].

4.2. Activated Dissociative Chemisorption- Evidence for an Activation Energy Barrier Based on Isotopic Studies

It is now well accepted that hydrogen dissociatively chemisorbs on Ni(111) via an activated process which is responsible for the relatively low sticking coefficient on the (111) plane compared to other Ni planes. To test whether the chemisorption of H_2 is activated, and whether the activated process proceeds from a molecular precursor state of H_2, we have performed a kinetic comparison for the two isotopic molecules, H_2 and D_2. These two molecules have rather large differences in their zero point vibrational energies (ZPE) which in the unperturbed gas phase molecules amounts to 7.67 kJ/mole.

In Fig. 13 is shown a potential energy plot for the interaction of the isotopic hydrogen species with the Ni(111) surface. Here the molecular

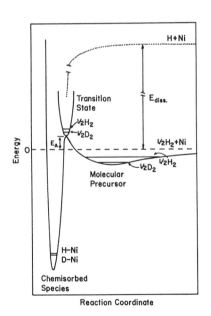

Schematic One Dimensional Potential Energy Diagram for Activated Dissociative Chemisorption : $H_2 + Ni(III)$

Fig. 13. Schematic one-dimensional potential energy diagram for activated dissociative chemisorption. Differences in the zero point energies for the hydrogen and deuterium surface species are indicated. From this diagram, $E_a(D_2) > E_a(H_2)$

precursor exhibits a ZPE splitting for H_2 and D_2, and a barrier for chemisorption from the molecular precursor state is indicated. Since the transition state will involve electronic perturbation of the hydrogen molecule, a splitting in the total ZPE for the isotopic transition complex species is shown which is smaller than that for the molecular precursor. This corresponds to the conversion of a vibrational normal mode in the molecular precursor to a translational mode leading to dissociation [20]. Since the H-H vibrational energy is so high, and since it is the H-H bond which is ultimately broken in dissociation, rather large deuterium kinetic isotope effects might be expected at low temperatures of adsorption.

The magnitude of the deuterium kinetic isotope effect will depend upon the location of the barrier. For a barrier located in the entrance channel for chemisorption (exit channel for desorption), translational energy effects will be dominant and Δ(ZPE) effects will be less important. For the barrier location in the exit channel for chemisorption, Δ(ZPE) effects will assume a larger role in determining the kinetics. The profound role of translational energy has already been demonstrated experimentally in the low coverage (β_2-H_2) region [2]. The work presented here shows that a significant deuterium kinetic isotope effect also exists in this low coverage region and suggests that vibrational energy is also important in overcoming the barrier to chemisorption.

An alternate view in which preferential H-tunneling through the barrier is invoked is also possible, since Δ(ZPE) effects are expected for thermal activation into tunneling states near the barrier top. HAMZA and MADIX [21] have recently studied the dynamics of H_2 adsorption on Ni(100), and have concluded that a tunneling mechanism is operative on the Ni(100) plane. On the basis of the functional dependence of the sticking coefficient on $E_\perp$ for both the Ni(111) [1] and the Ni(100) plane, it may be that tunneling or a combination of activated adsorption and tunneling is involved in the rate determining step for hydrogen dissociative adsorption on Ni(111).

The kinetic retardation of D_2 adsorption relative to H_2 adsorption as observed at 87 K in Figs. 5 and 6 can be explained by the potential energy diagram in Fig 13. From the zero coverage value of $S_0^{D_2}/S_0^{H_2} = 0.7$ at 87 K, we can estimate the difference in barrier height for the two isotopic species, $(E_A^{D_2} - E_A^{H_2}) = 0.2$ kJ/mole. This small difference in the estimated Δ(ZPE) compared to the maximum Δ(ZPE) expected for complete destruction of the H-H bond in the transition complex (7.67 kJ/mole) suggests that we are dealing with an early transition complex for dissociative adsorption, in agreement with the single adsorption site kinetics involving a molecular H_2 species discussed earlier.

4.3. α-D_2 - Molecular Precursor State

It has been postulated that prior to dissociatively chemisorbing on nickel, H_2 first accommodates on the surface in a weakly-bound, mobile molecular precursor state [7]. Recent calculations [6,22] of the H_2 + Ni(111) interaction have indicated that there may be an H_2 molecular adsorption well on this surface. HARRIS and ANDERSSON [22] have suggested that it may be possible to freeze the molecular precursor state on the surface by cooling the crystal to a low temperature.

Recently, molecular hydrogen has been successfully trapped and subsequently observed by electron energy loss spectroscopy (EELS) after

cooling a Cu(100) surface to about 15 K [23-25]. There are also examples
of molecular hydrogen desorption states in the literature for certain
transition metals. MADEY and YATES [26] noted the preferential formation
of a γ-H_2 state on W(111) compared to W(100). BENNINGHOVEN, et al. [27]
identified (with SIMS) the low temperature H_2 desorption feature observed
by WEDLER, et al. [28] to be a molecular species on amorphous Ni. Our
experiments comparing D_2 and H_2 adsorption were done at 87 K which is considerably lower than the experiments of others, and the use of this low
adsorption temperature has permitted the direct observation of molecular
D_2 by TPD methods, as shown in Figs. 7 and 8. In addition, the observation
of a molecular H_2 state on a stepped Ni crystal has recently been reported
using HREELS [29].

The presence of α-D_2 and the absence of α-H_2 implies that molecular D_2
is more stable on the surface than molecular H_2 at 87 K. The difference in
stability of H_2 and D_2 in the molecular precursor state currently is not
well understood. The abnormally low preexponential factor for first order
desorption suggests that a highly improbable trajectory is involved in
molecular D_2 desorption from the α-D_2 state.

It is of interest to note that the α-D_2 species is first observed in
Fig. 7 when the surface hydrogen coverage nears Θ_H = 0.5 H/Ni, where the
β_2-H_2 desorption feature saturates [19]. This is probably related to
changes in the potential energy surface for hydrogen adsorption as hydrogen coverage increases, and will be discussed in the following section.

4.4. Angular Distribution Effects in H_2 Desorption From Ni(111)

Figure 14 shows schematic one-and two-dimensional representations of the
potential energy surface for the interaction of H_2 with a Ni(111) surface.
In the one-dimensional representation, the heat of desorption for β_1-H is
represented as being smaller than that for β_2-H, in accordance with the
commonly invoked repulsive forces between adsorbate atoms in this system
[10]. It can be seen that in Fig. 14 A, as the heat of adsorption
decreases, the barrier for activated adsorption increases and moves
toward the entrance channel for desorption. Figure 14 B and C illustrate
this motion of the barrier from the desorption exit channel to the desorption entrance channel as hydrogen coverage increases from the β_2-coverage
region to the β_1-coverage region. The ideas inherent in the arguments to
be presented below are consistent with the model of VAN WILLIGEN [9] as
well as the more general ideas of POLANYI and WONG [30].

The sharply-forward focused H_2 desorption associated with β_2 hydrogen
is consistent with the barrier for the desorption process being in the
exit channel for desorption. An adsorbate atom surmounting this barrier
converts its extra potential energy into forward translational momentum,
and escapes with an excess kinetic energy normal component, resulting in
the $\cos^{4.5}\theta$ angular distribution for desorption. As the initial H
coverage increases above Θ_H = 0.5 H/Ni, the barrier moves toward the
desorption entrance channel, and the excess translational energy is converted into other degrees of freedom (vibrational or rotational) as the
desorbing molecule escapes around the corner of the potential energy surface. This will produce desorbing H_2 molecules which are not preferentially forward scattered ($\cos\theta$ angular distribution).

SCHEMATIC ONE AND TWO DIMENSIONAL POTENTIAL ENERGY
DIAGRAMS DESCRIBING THE H_2 + Ni(III) INTERACTION
AT ZERO AND SATURATION H COVERAGE

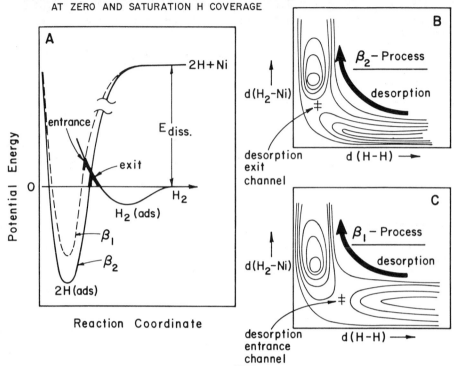

Fig. 14. Schematic potential energy diagrams proposed to describe the H_2 + Ni(111) β_2 (A,B) and β_1 (A,C) desorption processes. The symbol ‡ illustrates the location of the transition complex for adsorption/desorption

5. Summary and Conclusions

Through the combination of adsorption kinetic measurements and the study of the angular distribution of desorbing H_2 from Ni(111) it has been possible to determine the following details regarding the interaction between the Ni(111) surface and hydrogen.

1. Hydrogen molecules adsorb by means of an activated process via interaction with a single site on the Ni(111) surface.

2. For atomic hydrogen coverages below one-half monolayer, adsorption involves the production of an activated complex which is near the entrance channel for adsorption on the potential energy surface describing the interaction.

3. Measurements of the kinetics of D_2 versus H_2 adsorption show that adsorption occurs via an intrinsic molecular precursor state.

4. The molecular precursor state of D_2 is sufficiently stable to be observable above one-half monolayer of atomic hydrogen adsorption. It desorbs with first order kinetics with an activation energy of 11.1 kJ/mole and a preexponential factor of 1.2×10^5 sec^{-1}. A comparable H_2 state is not observed.

5. Angular distribution measurements indicate that adsorbed atomic hydrogen recombines over an exit channel barrier for desorption when the coverage is below one-half monolayer.

6. For atomic hydrogen coverages above one-half monolayer, recombination occurs over an entrance channel barrier for desorption, suggesting that the location of the activation barrier for H + H recombination is dependent on the coverage of adsorbed atomic hydrogen.

7. The mechanistic arguments presented here form one rationalization for the experimental results. As with all kinetic mechanisms, alternate explanations are probably possible.

6. Acknowledgements

We gratefully acknowledge support of this work by the NSF Division of Materials Science under Contract No. DMR-8414362. The authors thank Dr. Humayun Siddiqui for many valuable discussions.

7. References

‡ Present Address: IBM, Thomas J. Watson Research Center, P. O. Box 218, Yorktown Heights, NY 10598 USA.

1. H. J. Robota, W. Vielhaber, M. C. Lin, J. Segener, and G. Ertl: Surf. Sci., 155, 101 (1985).
2. a. H. P. Steinrück, A. Winkler, and K. D. Rendulic: J. Phys. C. 17, L311 (1984).
   b. H. P. Steinrück, M. Luger, A. Winkler, and K. D. Rendulic: Phys. Rev. B 32, 5032 (1985).
   c. H. P. Steinrück, K. D. Rendulic, and A. Winkler: Surf. Sci. 154, 99 (1985).
   d. H. P. Steinrück, K. D. Rendulic, and A. Winkler: Surf. Sci. 152/153 (1985) 323.
3. D. O. Hayward and A. O. Taylor: Chem. Phys. Lett, 124, 264 (1986).
4. T. H. Upton, W. A. Goddard III, and C. E. Melius: J. Vac. Sci. Technol., 16, 531 (1979).
5. V. I. Avdeev, T. H. Upton, W. H. Weinberg, and W. A. Goddard III: Surf. Sci. 95, 391 (1980).
6. a. C. Y. Lee and A. E. DePristo: J. Chem. Phys., 84, 485 (1986).
   b. C. Y. Lee and A. E. DePristo: J. Chem. Phys., 85, (1986) in press.
7. A. Winkler and K. D. Rendulic: Surf. Sci., 118, 19 (1982).
8. a. M. Balooch and R. E. Stickney: Surf. Sci., 44, 310 (1974).
   b. M. Balooch, M. J. Cardillo, D. R. Miller, and R. H. Stickney: Surf. Sci., 46, 358 (1974).
9. W. Van Willigen: Phys. Lett., 28A, 80 (1968).
10. K. Christmann, R. J. Behm, G. Ertl, M. A. Van Hove, and W. H. Weinberg: J. Chem. Phys., 70, 4168 (1979).
11. W. Ho, N. J. DiNardo, and E. W. Plummer: J. Vac. Sci. Technol, 17, 134 (1979).

12. S. M. Gates, J. N. Russell, Jr., and J. T. Yates, Jr.: Surf. Sci., 159, 233 (1985).
13. S. M. Gates, J. N. Russell, Jr., and J. T. Yates, Jr.: Surf. Sci., 146, 199 (1984).
14. J. T. Yates, Jr: In Methods of Experimental Physics, Vol. 22, ed. by R. L. Park and M. Lagally, (Academic Press, Inc., New York, 1985) p.425.
15. R. J. Muha, S. M. Gates, J. T. Yates, Jr., and P. Basu: Rev. Sci. Instrum., 56, 613 (1985).
16. H. Robota, W. Vielhaber, and G. Ertl: Surf. Sci., 136, 111 (1984).
17. K. Christmann, O. Schober, G. Ertl, and M. Neuman: J. Chem. Phys., 60, 4528 (1974).
18. J. T. Yates, Jr., P. A. Thiel, and W. H. Weinberg: Surf. Sci., 84, 427 (1979). See also references therein.
19. J. N. Russell, Jr., S. M. Gates, and J. T. Yates, Jr.: J. Chem. Phys., 85, (1986), in press.
20. L. Melander and W. H. Saunders: Reaction Rates of Isotopic Molecules, (Wiley, New York, 1980).
21. A. V. Hamza and R. J. Madix: J. Phys. Chem., 89, 5381 (1985).
22. J. Harris and S. Andersson: Phys. Rev. Lett., 55, 1583 (1985).
23. S. Andersson and J. Harris: Phys. Rev. Lett., 48, 545 (1982).
24. S. Andersson and J. Harris: Phys. Rev. B, 27, 9 (1983).
25. S. Andersson, L. Wilzen, and J. Harris: Phys. Rev. Lett., 55, 2591 (1985).
26. T. E. Madey and J. T. Yates, Jr.: Surf. Sci., 63, 203 (1977).
27. A. Benninghoven, P. Beckmann, D. Greifendorf, K.-H. Müller, and M. Schemmer: Surf. Sci., 107, 148 (1981).
28. G. Wedler, H. Pappa, and G. Schroll: J. Catal., 38, 153 (1975).
29. A.-S. Martensson, C. Nyberg, and S. Andersson, Phys. Rev. Lett. 57 (1986) 2045.
30. J. C. Polanyi and W. H. Wong: J. Chem. Phys., 51, 1439 (1969).

Adsorption-Desorption Kinetics: Some Comments

H.J. Kreuzer

Department of Physics, Dalhousie University,
Halifax, B3H 3J5, Nova Scotia, Canada

The terms most often used in this session were, no doubt, equilibrium, nonequilibrium, and detailed balance. The equilibrium of a system is operation-
ally defined as that state in which a small number of macroscopic variables,
such as internal energy, entropy, volume and mole numbers, are independent
of time and of the previous history of the system. At the microscopic level
myriads of molecular collisions take place whose dynamics follow quantum
(or approximately Newtonian) mechanics. The mechanical equations of motion
(Schrödinger or Newton) are invariant under time reversal. It has been
proved rigorously by Stückelberg [1] that the time reversal invariance of
the microscopic equations of motion implies detailed balance of the equilibrium rates. An example might clarify the point: let us consider a sytem of
r chemically reacting components which, at temperature T and pressure P,
have mole numbers N_a^0. Detailed balance then says that in equilibrium the
rates between the various components balance, i.e.,

$$R_{ab}N_b^0 = R_{ba}N_a^0 ,$$

(1)

where R_{ab} is the equilibrium rate constant for the reaction that turns component b into component a. Detailed balance thus precludes the possibility
that the system is maintained in equilibrium by a cyclic reaction
$1 \rightarrow 2 \rightarrow 3 \rightarrow ... \rightarrow r \rightarrow 1$.

Let us next disturb the equilibrium by manipulating the external constraints such as changing temperature or pressure or adding chemicals. As long
as the system remains "close" to equilibrium its mole numbers will evolve
according to linear constitutive relations

$$\frac{dN_a}{dt} = \sum_b L_{ab} \frac{\partial S}{\partial N_b} .$$

(2)

Writing down these rate equations obviously presumes that an entropy S can
be defined. This is possible as long as the system remains in local equilibrium, the general criteria are discussed in detail in [2]. It suffices for
the present discussion to consider a gas as an example for which local equilibrium can be assumed as long as relative changes in the macroscopic
observables, such as mole numbers, are small over the time scale of molecular collisions. In addition possible spatial gradients must be small over
the length scale of the mean free path. Under such conditions one can prove
the symmetry of the Onsager coefficients, $L_{ab}=L_{ba}$, and express them in
terms of the equilibrium rate constants R_{ab}. This is essentially the regime
of quasi-equilibrium, as discussed by Menzel, or local equilibrium in the
parlance of nonequilibrium thermodynamics. As one consequence of this discussion we note that in order to make use of equilibrium properties of a 2-D
gas phase (in coexistence with a 2-D condensate or not) in the analysis of

kinetic data one must make sure that the desorption times are much longer than 2-D collision times; the time scales discussed by Brenig only suffice at zero coverage.

What complications arise in situations where local equilibrium is not maintained? Far from equilibrium one must deal with the full density matrix of the system; it satisfies a generalized master equation that incorporates the evolution of many-particle correlations and contains memory effects. Little has been done along these lines for desorption kinetics except some formal manipulations [3]. However, if the time scale over which we want to follow the time evolution of the system (i.e., desorption times) is much longer than the internal relaxation times, it is possible to treat some macroscopic quantities in the system (such as the energy of a particle in the zero coverage regime) as stochastic variables whose joint probability distribution function evolves according to Kolmogorov's equation. If the system does not exhibit intrinsic memory effects, we are dealing with Markov processes which, with some further simplifications, follow a master equation (for a detailed discussion see [4])

$$\frac{dn_i}{dt} = \sum_k [W_{ik}(t) \, n_k - W_{ki}(t) \, n_i], \tag{3}$$

where $n_i(t)$ is the probability that the system is in a state "i" at time t. For a single particle (in the zero coverage limit) the set of variables "i" might reduce to just the energy or momentum of the molecule in the adsorption potential or the gas phase. In this situation the stochastic process is a random walk of the molecule through these energy or momentum states. We note that the transition probabilities in (3) can still be time dependent; they are independent of time for a stationary Markov process for which (3) then reduces to

$$\frac{dn_i}{dt} = \sum_k [W_{ik} \, n_k - W_{ki} \, n_i]. \tag{4}$$

This is the equation used by several groups [4] to deal with adsorption-desorption kinetics. It has the great advantage that (for stationary Markov processes) detailed balance can again be invoked in a straightforward manner; it gives

$$W_{ik}n_k^{\,o} = W_{ki}n_i^{\,o} \tag{5}$$

and thus is a useful restriction on the structure of the transition probabilities W_{ik}. For a nonstationary Markov process detailed balance only puts a restriction on the shape of the transition probabilities in (3) for $t \to \infty$. If, however, the time evolution, proceeding according to (3), is so slow that local equilibrium is maintained, i.e. that the probability distribution n_i evolves through a sequence of equilibrium states, detailed balance (5) can be invoked at each instant of time.

To deal with finite coverage effects it is most unlikely that (4) is adequate. A microscopic theory to study adsorption and desorption of helium [4,5] has been based on (3) in which the transition probabilities become explicitly dependent on n_i; it, among other features, exhibits a compensation effect between the pre-exponential factor and the desorption energy. An extension to other adsorption systems is unfortunately not straightforward.

90

To study the coverage dependence of the sticking probability one could follow the Iche-Noziere approach [6] as outlined by Brenig, but have the incoming particles scatter off an adsorbate-covered surface. In this situation it is, however, no longer possible to reduce the problem to one dimension, as lateral energy loss within the adlayer, particularly if the latter is mobile, becomes very important.

The kinetic data presented by Menzel for rare gases on metals demand that explicit account is taken of the existence of a 2-D condensate. Short of resorting to a purely phenomenological approach (2), where the N_a might be the mole numbers in the 2-D condensate, the 2-D gas phase, and the 3-D gas phase [7,8], one can set up a theory in which one follows the time evolution of the number n_ℓ of clusters of size ℓ, with 2-D condensation and evaporation included, in addition to adsorption and desorption from the 2-D gas phase and from the top of the clusters. Such an approach, modeled after the droplet model of 3-D condensation, would certainly reproduce most of the features in Menzel's data; the challenge for theory would be to determine the rate constants from microscopic considerations.

Some comments about the 2-D gas phase might be in order. Let us study a system at a coverage of about half a monolayer mostly in islands of a 2-D condensate. Their average size ℓ might be a few hundred angstroms, which at a coverage of half a monolayer is then also the size of the intertwining patches of 2-D gas. To speak about a gas phase the particle density must be so low that their average separation is 2-3 adsorption sites apart, i.e. more than 10Å. A patch of 2-D gas of, say, about $(100Å)^2$, surrounded by 2-D condensate islands, will then contain only a few hundred adsorbed particles. This implies that the statistical fluctuations within this gas are of the order of 10%, making arguments about density gradients and particle depletion close to the island edges rather dubious. If this argument is not convincing one should look at some time scales involved. Let us assume that the gas particles are highly mobile along the surface. At a temperature T they will have an average velocity $v = \sqrt{kT/2m}$ and will traverse the gas region of size ℓ in a time $\tau = \ell/v$. For an argon atom at 50K we get $\tau \approx 10^{-10}$s. Only for shorter times is it meaningful to talk about density gradients in the gas phase; for immobile adsorbates similar considerations can be advanced.

I would like to add a further comment on a way maximum information can be extracted from desorption experiments. Let us prepare a system at temperature T_0 at which the adsorbate consists of two co-existing phases. In a first, temperature programmed desorption experiment we raise the temperature so slowly that the adsorbate remains in local equilibrium. This, in particular, means that the desorption rate must be slow as compared to the 2-D evaporation and condensation. This experiment can be described by (2) and yields information on the equilibrium rates. Next, in a second experiment again starting from T_0, we raise the temperature suddenly and study isothermal desorption at the elevated temperature T, far from equilibrium. The following scenario might take place: if T is only high enough, the 2-D gas phase might desorb rapidly, leaving 2-D condensate clusters behind. Their coupling to the substrate is typically effective enough to maintain them at T so that 2-D evaporation becomes the rate-determining step. A series of high T isothermal experiments will then allow the determination of the temperature dependence of the evaporation rate constant. Different initial temperatures T_0 will produce different cluster distributions within the coexistence region and thus allow a study of the cluster size dependence of these rate constants. Numerical modeling of such a scenario is under way.

References

1. E.C.G. Stückelberg: Helv. Phys. Acta 25, 577 (1952); for a discussion see also W. Heitler: The Quantum Theory of Radiation (Oxford University Press, Oxford 1970)
2. H.J. Kreuzer: Nonequlilibrium Thermodynamics and Its Statistical Foundations (Oxford University Press, Oxford 1981, 1983)
3. S. Efrima, C. Jedrzejek, K.F. Freed, E. Hood, H. Metiu: J. Chem. Phys. 79, 2436 (1983)
4. H.J. Kreuzer and Z.W. Gortel: Physisorption Kinetics, Springer Ser. Surface Sci., Vol.1 (Springer, Berlin, Heidelberg 1986)
5. E. Sommer and H.J. Kreuzer: Phys. Rev. Lett. 49, 61 (1982); Phys. Rev. B 26, 4094 (1982)
6. G. Iche, P.Noziere: J. Phys. (Paris) 37, 1313 (1976)
7. J.R. Arthur, A.Y. Cho: Surface Sci. 36, 641 (1973); 38, 394 (1973)
8. G. Le Lay, M. Manneville, R. Kern: Surface Sci. 65, 261 (1977);
   J. A. Venables, G. D. T. Spiller, M. Hanbücken, Rep. Prog. Phys. 57, 399 (1984).

Part II

Precursors: Myth or Reality?

Precursor Intermediates and Precursor-Mediated Surface Reactions: General Concepts, Direct Observations and Indirect Manifestations

W.H. Weinberg

Division of Chemistry and Chemical Engineering,
California Institute of Technology, Pasadena, CA 91125, USA

A brief review is presented of precursor-mediated chemisorption on and desorption from solid surfaces. A heuristic concept of the precursor is put forward followed by a delineation of the rationale for the original proposal of the existence of these species. A number of cases where the direct experimental observation of precursor states has been presented are discussed, and a number of rate expressions describing precursor-mediated adsorption and desorption are presented, all of which are based either on "kinetic" or "statistical" models. The approximations and assumptions that are implicit in each case are delineated. The limitations that become evident in this description of the surface phenomena lead naturally to a discussion of recently developed time-dependent Monte Carlo simulations of precursor-mediated chemisorption, desorption and ordering of chemisorbed overlayers. Comparison with the large base of experimental data that are often available for particular gas-surface interactions render this approach both feasible and extremely pedagogic insofar as a microscopic understanding of the elementary surface reactions is concerned. Next, two examples are considered that demonstrate how recent experimental observations might have been interpreted in terms of the presence of a precursor state on the surface. In some cases, the inclusion of a precursor can result in a qualitatively different concept of the nature of the gas-surface interaction. Finally, a critical synopsis of the importance of precursor-mediated surface reactions is presented.

I. Introduction

The question of whether or not "precursor intermediates" exist at surfaces and, if they do exist, how they influence the rates of chemisorption, desorption and other types of surface "reactions" is of fundamental importance in understanding kinetic phenomena at surfaces. The issue of the existence of a "short-lived" precursor state in chemisorption, for example, is not just semantic. If the precursor state does not exist and chemisorption is a "direct" reaction, then one would expect the probability of chemisorption to be a stronger function of the gas temperature than the surface temperature. On the other hand, if the chemisorption reaction is mediated by a precursor state that has accommodated to the temperature of the surface, then one would expect the surface temperature to be of primary importance in dictating the measured probability of chemisorption. Note that in the latter case the gas temperature may also play a role, since the probability of "trapping" of the incident molecule into the precursor state is, in general, a function of both the gas and surface temperature. Based on these arguments, one would expect clearly *observable* differences between "direct chemisorption" and "precursor-mediated chemisorption".

One can qualitatively picture the trapping of an incident molecule into the precursor state as a consequence of one or more inelastic collisions of the incident molecule with the surface (either phonon assisted or via electron-hole pair creation) resulting in a negative total energy of the molecule (with respect to a zero of energy of the molecule infinitely far from the surface and at rest). Once trapped in this sense, the precursor is usually assumed to accommodate fully to the temperature of the surface. This appears to require that the lifetime of the precursor be sufficiently long to make only a very small number of hops on the surface, possibly as few as two [1]. Consistent with this estimate, TULLY [2] has found, for example, that nearly all of the xenon atoms with an incident translational energy of 1960 K that are trapped at a Pt(111) surface, the temperature of which is 773 K, have accommodated to the temperature of the surface (for motion normal to the surface) after approximately 20 ps. One would expect this to correspond to no more than approximately ten hops on the surface. For motion parallel to the surface, essentially complete accommodation was found to occur in less than 100 ps. These results are quite useful, since they provide the time scale for the lifetime of a precursor at a surface prior to its accommodation to the surface temperature.

The point of view adopted here is that the precursor is a "real" intermediate (i.e. a bound state), and the temperature that dictates its motion is that of the surface. Although the lifetime (and the concentration) of the precursor *may* be very short (and very low), its existence must, nevertheless, be taken into account in rate expressions. Precursors are frequently imagined to be physically adsorbed species with both a short lifetime and a low concentration. These assumptions permit both the neglect of the precursor concentration in the "final state" of various elementary surface reactions and the applicability of the pseudo-steady-state approximation in the formulation of rate expressions, i.e. the time derivative of the surface coverage of the precursor is zero (cf. Sect. III). Although these assumptions concerning the precursor are reasonable in many cases, they are clearly not correct in general. For example, at sufficiently low surface temperatures, one would expect to be able to stabilize physically adsorbed precursors if there is a finite barrier separating the physically adsorbed ground state and the chemisorbed state. The more relevant "precursor" to dissociative chemisorption, however, might well be a molecularly chemisorbed molecule (when molecular chemisorption can occur) rather than a physically adsorbed molecule. In this case, the surface temperature necessary for experimental isolation of the "precursor" would generally not be so low as that required for the isolation of a physically adsorbed precursor. Examples illustrating this idea are presented in Sect. II.

There are a variety of experimental techniques that can be used to observe precursors on surfaces directly. For example, at surface temperatures sufficiently low to stabilize the precursor, its existence can be verified with photoelectron spectroscopy, vibrationally inelastic electron scattering, and contact potential difference measurements. At higher surface temperatures, where the fractional surface coverage of the precursor is negligible, its existence can be verified by angular (and to some extent by velocity) distributions of scattered molecular beams. Under experimental conditions where desorption from the chemisorbed state is negligible, those molecules that have been trapped in the potential well for physical adsorption and then "desorb" have a cosine angular distribution peaked in the direction of the surface normal, whereas those molecules that have undergone direct inelastic scattering have a quasi-specular angular distribution. In this connection, it should be noted that molecular beam experiments with high translational energies in the incident beam would appear to offer the best opportunity to observe direct chemisorption. There are two different reasons for this expectation. First, the probability of trapping into the physically adsorbed state on the surface decreases as the translational energy of the incident gas molecules increases, and this will obviously reduce the rate of precursor-mediated chemisorption. Moreover, as the translational energy of the incident molecules increases, their probability of surmounting any activation barrier to chemisorption increases. Since the potential energy surface leading to chemisorption is almost certainly not one-dimensional, however, the apparent (i.e. measured) barrier to direct chemisorption of the translationally "hot" molecules will not reflect, in general, the minimum activation energy of chemisorption of which the latter would be sampled by a precursor intermediate.

Indirect manifestations of the presense of precursor states would be expected in the measured rates of chemisorption and desorption, and in the observation of island growth in the chemisorbed overlayer (at surface temperatures at which the rate of hopping of chemisorbed species is negligible). Indeed, it was the observation that the probability of chemisorption of cesium on tungsten remained constant up to nearly saturation coverage that prompted Langmuir to visualize the original concept of a mobile precursor state to chemisorption on a surface [3,4]. This criterion remains one of the most straightforward indications of precursor-mediated adsorption, i.e. the presence of a precursor is implicated if the measured probability of first-order (molecular) chemisorption does not scale with the fraction of unoccupied surface sites, or if the measured probability of second-order (dissociative) adsorption does not scale with the square of the fraction of unoccupied surface sites. One would generally expect that in those cases in which chemisorption is precursor-mediated, desorption will also be mediated by the precursor state. An exception could be, for example, in connection with extremely high heating rates of the surface, e.g. by a pulsed laser, and when the preexponential factor of the desorption rate coefficient for direct desorption is greater than that for excitation of the chemisorbed molecule into the precursor state.

Nearly concurrent with the suggestion of Langmuir of the existence of a precursor to chemisorption to explain his experimentally measured rates of chemisorption of alkali metals on tungsten, LENNARD-JONES [5] presented plausibility arguments for the existence of a physically adsorbed intermediate to chemisorption on a surface. The idea of precursor-mediated chemisorption was revived by BECKER [6,7] in order to explain his measured rates of chemisorption of nitrogen on tungsten. Shortly thereafter, two extremely important mathematical descriptions of the rates of precursor-mediated chemisorption appeared. First, EHRLICH [8] refined the *kinetic model* of Becker making use of the pseudo steady-state approximation and assuming that a single type of precursor state exists on the surface. Then KISLIUK [9] presented a *statistical model* of the probability of chemisorption based on the premise (subsequently verified experimentally, cf. Sect. II) that two different kinds of precursor states can exist on the surface: the so-called *intrinsic* precursor state existing above an unoccupied surface site, and the *extrinsic* precursor state existing above a surface site occupied by a chemisorbed species. Kisliuk considered the hopping of precursor molecules from site to site on the surface, and accounted for the probabilities of chemisorption, desorption and migration at each site (with the assumption that the chemisorbed species were randomly distributed on the surface). An expression for the probability of chemisorption was obtained in closed form by summing the resulting geometric series to an infinite number of hops of the precursor on the surface. As discussed in Sect. III, these two formulations are exactly equivalent when the same model assumptions are employed.

Since this seminal work of Becker, Ehrlich and Kisliuk, there have been a number of other important refinements to the description of precursor-mediated chemisorption [e.g. (10)-(13)]. Subsequently, the influence of precursors on the rates of desorption has been formulated within the framework of both the kinetic and the statistical models (which are equivalent in this case also, cf. Sect. III) [e.g. (14)-(16)]. This work has culminated in the presentation of a unified description of both precursor-mediated adsorption and desorption for both the kinetic [16] and the statistical [17] models. Both of these models of precursor-mediated adsorption and desorption are *macroscopic* descriptions in the sense that the calculated rates are averages over a macroscopic number of particles. In an effort both to derive a better *microscopic* understanding of precursor-mediated phenomena at surfaces and to render the approximations in the model somewhat more physically realistic, HOOD et al. [18] have developed a time-dependent Monte Carlo simulation technique to describe precursor-mediated adsorption, desorption and ordering of chemisorbed overlayers.

This progress in theoretical formulations of the expected manifestations of precursor states on the measured rates of chemisorption and desorption has led to extensive recent efforts to stabilize (at low surface temperatures) and to observe directly both intrinsic and extrinsic precursor intermediates to chemisorption. In addition, important insight into the nature of precursor states at surface temperatures at which their steady-state concentration is exceedingly small has been gained from angular and velocity distributions of molecular beams scattered from surfaces.

In Sect. II, selected examples of the experimental isolation and observation of intrinsic and extrinsic precursor states on metallic surfaces are discussed. The nature of these states that is revealed by molecular beam scattering is reviewed elsewhere in this volume [19]. In Sect. III, expressions are presented for the rates of precursor-mediated adsorption and desorption, based on the kinetic and statistical models described above. Both first- and second-order adsorption and desorption are considered for a variety of different assumptions concerning the nature of the precursor state(s). The equivalence of these two models is discussed, as are various limiting cases of the rate expressions, the necessity of each model being consistent with a Langmuir isotherm at equilibrium for the same set of assumptions involving adsorption and desorption (neglecting lateral interactions and assuming a single type of site for chemisorption), and the constraints placed on certain ratios of rate coefficients by detailed balancing (equilibrium) arguments. This discussion is followed in Sect. IV by a comparison of these macroscopic models with the microscopic model embodied by the simulation of HOOD et al. [18]. In Sect. V, two specific (of many possible) examples from the recent literature are presented in which the experimental results are reinterpreted in terms of the presence of a precursor state. It is shown that this can lead to a qualitatively different picture of the relevant elementary surface reactions: a picture that is consistent with a variety of experimental results, some of which were considered heretofore to be mutually imcompatible. Finally, in Sect. VI, a critical synopsis of precursor-mediated surface reactions is put foward.

II. Experimental Isolation and Characterization of Precursors

Since precursors are considered to be real surface intermediates, at sufficiently low temperatures, one would expect to be able to isolate and observe the precursor state(s) directly. Here, a number of attempts designed to accomplish this will be discussed. Since a comprehensive review of this topic is well beyond the purpose of this chapter, the selected examples that are considered will be restricted to the dissociative chemisorption of oxygen and the molecular chemisorption of carbon monoxide and nitrogen on metallic surfaces.

NORTON et al. [20] have used X-ray and UV-photoelectron spectroscopy to study the adsorption of CO on a polycrystalline copper surface at surface temperatures of 20 K and above. At 20 K they observed only physically adsorbed CO on the surface in submonolayer concentrations of which, upon heating to 30 K, approximately 40% was desorbed and 60% was converted to a molecularly chemisorbed state (which was stable on the surface to temperatures above 100 K). This is a clear example of the physically adsorbed CO being an intrinsic precursor to chemisorption on the copper surface. Assuming that the preexponential factors of the rate coefficients of chemisorption and of desorption of carbon monoxide from the precursor are equal, these results imply that the activation energy of molecular chemisorption from the precursor is only approximately 25 cal/mol less than that of desorption from the precursor. If the prexponential factor of the desorption rate coefficient from the precursor is assumed to be 10^{13} s^{-1}, the activation energy of desorption is approximately 1800 cal/mol.

HOFMANN et al. [21] have shown convincingly with UV-photoelectron spectroscopy that oxygen chemisorbs dissociatively on the Al(111) surface at 30 K up to a coverage of approximately half a monolayer, at which point molecular oxygen is observed to be stabilized on the surface. This molecularly adsorbed oxygen was found to convert slowly (on the order of minutes) to dissociatively chemisorbed oxygen, implying that the molecularly adsorbed oxygen is a precursor to dissociative chemisorption. This implication is substantiated by the observation that the initial probability of dissociative chemisorption of oxygen on the single crystalline aluminium surface increases from approximately 0.01-0.05 at a surface temperature of 300 K to essentially unity at 30 K [21]. Certainly the most straightforward, and possibly the only reasonable explanation of these results is that there is a molecularly adsorbed precursor to the dissociative chemisorption of O_2 on Al(111), and that the preexponential factor of the desorption rate coefficient from this state is much greater (order of a factor of 100) than that of the rate coefficient for dissociative chemisorption from this state. Furthermore, the difference between the activation energies of desorption and chemisorption from the molecular precursor state (which is always positive) must decrease with increasing surface coverage of dissociatively chemisorbed oxygen, until the molecularly adsorbed oxygen is stabilized sufficiently to be observable spectroscopically at 30 K at a half-monolayer surface coverage of oxygen adatoms.

JACOBI et al. have used UV-photoelectron spectroscopy to investigate both the adsorption of oxygen on polycrystalline films of copper [22] and the oxygen-copper interaction for copper clusters isolated in oxygen matrices [23]. In both cases only physically "adsorbed" molecular oxygen is observed at temperatures between 7 and 30 K. After the desorption of physically adsorbed (and condensed) oxygen at 30 K, only molecularly chemisorbed oxygen remains, which then converts to dissociatively chemisorbed oxygen between approximately 70 and 300 K. These results show clearly that molecularly chemisorbed oxygen is a precursor to dissociative chemisorption. Furthermore, they imply that physically adsorbed oxygen is an intrinsic precursor intermediate to molecular chemisorption below 30 K. The latter is somewhat less unambiguous, since the presence of any molecularly chemisorbed oxygen below 30 K could be masked in the UV-photoemission spectra by the physically adsorbed and the condensed oxygen that are present in much higher concentrations.

JACOBI et al. [24] have also applied UV-photoelectron spectroscopy to study the adsorption of nitrogen, carbon monoxide and oxygen on Ni(110) at surface temperatures as low as 20 K. In all cases, only the chemisorbed states (molecular nitrogen and carbon monoxide, and atomic oxygen) were observed at low surface coverages at 20 K. In fact, the only evidence of any stable precursor state was the observation of a mixture of molecularly chemisorbed and physically adsorbed nitrogen at submonolayer surface coverages, which suggests that the physically adsorbed nitrogen is an extrinsic precursor to molecular chemisorption (i.e. it exists above the chemisorbed nitrogen rather than on bare patches of the metallic surface).

Similarly, GRUNZE et al. [25] have employed X-ray photoelectron spectroscopy in an attempt to isolate physically adsorbed precursors to the molecular chemisorption of nitrogen on the Ni(100), Re(001) and W(100) surfaces at temperatures as low as 20 K. No intrinsic precursor to molecular chemisorption was observed on any of the three surfaces, even at a temperature of 20 K. However, on each surface, the presence of a molecular (extrinsic) precursor state of nitrogen was observed prior to saturation of the molecularly chemisorbed overlayer. This physically adsorbed extrinsic precursor desorbed at surface temperatures of approximately 50-60 K, well above the desorption temperature of condensed nitrogen. Assuming that the preexponential factor of the rate coefficient for conversion of the intrinsic precursor to the molecularly chemisorbed state is 10^{13} s^{-1}, then the fact the former was not isolated at 20 K during the 300 s measurement period implies that the activation barrier separating any intrinsic precursor from molecularly chemisorbed nitrogen is less than 1420 cal/mol, cf. the activation energy of desorption of the extrinsic precursor of approximately 2970-3570 cal/mol (assuming the same value as above for the preexponential factor). Obviously, surface temperatures of 20 K are not sufficiently low to probe the existence of intrinsic precursor states of "reactive adsorbates" on surfaces of "reactive transition metals".

Accordingly, GLOVER et al. [26-28] have carried out a series of extremely elegant measurements of the adsorption of oxygen and carbon monoxide on the Ni(111) and Ni(100) surfaces at temperatures as low as 5.5 K. Contact potential difference measurements, which might be expected to be more sensitive than photoemission spectra to the transition between physical adsorption and chemisorption, were employed in this work. In the case of oxygen adsorption on Ni(111), a molecularly adsorbed intrinsic precursor state (with a work function change at saturation coverage $\Delta\phi_s$ of 70 meV) was identified clearly prior to dissociative chemisorption ($\Delta\phi_s \triangleq 320$ meV) at higher surface temperatures [26]. The measured rate of conversion from the molecular precursor state to the dissociative chemisorption state at 25 K corresponds to an activation barrier of approximately 1650 cal/mol if the preexponential factor of the reaction rate coefficient is 10^{11} s^{-1} [29]. In the case of oxygen adsorption on Ni(100), a molecularly adsorbed intrinsic precursor state was also identified as an intermediate to dissociative chemisorption, although the barrier separating the molecular precursor state from the dissociative chemisorption state is apparently lower on the Ni(100) surface compared to the Ni(111) surface [28,30].

Of even more significance perhaps, GLOVER et al. [27] have reported the isolation of an intrinsic precursor intermediate to the molecular chemisorption of carbon monoxide on the Ni(111) surface. The first indication of chemisorption of the carbon monoxide occurs at approximately 15 K which, assuming a measurement time of one minute and a preexponential factor of the rate coefficient of chemisorption of 10^{13} s^{-1}, corresponds to an activation barrier between the precursor and chemisorbed states of approximately 1000 cal/mol. Of equal significance, it was found that no intrinsic precursor to the molecular chemisorption could be stabilized on the Ni(100) surface at a surface temperature as low as 6 K [28]. For the same measurement time and preexponential factor assumed above for the chemisorption of carbon monoxide on the Ni(111) surface, the activation barrier between any intrinsic precursor state and the molecularly chemisorbed state of carbon monoxide on Ni(100) is inferred to be less than approximately 400 cal/mol.

To summarize, the experimental observations (or lack thereof) of precursors to chemisorption are entirely consistent with that which would have been expected a priori. In the case of molecular chemisorption, lower surface temperatures are necessary to stabilize physically adsorbed intrinsic precursor states on those surfaces with higher heats of molecular chemisorption, e.g. the initial heat of molecular chemisorption of carbon monoxide on Ni(110) ~ Ni(100) > Ni(111) > Cu. Likewise, the heat of dissociative chemisorption of oxygen on nickel and aluminum is greater than that on copper, resulting in a stabilization of the molecular precursor on the latter. Furthermore, lower surface temperatures are generally required to stabilize physically adsorbed precursors to molecular chemisorption compared with dissociative chemisorption. Finally, it might be expected that in the case of the dissociative chemisorption of those reactants capable of molecular chemisorption, a more useful picture of the precursor to dissociative adsorption is often the molecularly chemisorbed state. This expectation obviously depends on the rate coefficients for occupation and depletion of the two types of molecularly adsorbed species, and the possibility of this occurrence should be considered on a case-by-case basis in the analysis of relative experimental data. In fact, this point of view leads naturally to the correct conclusion that "stable" molecularly chemisorbed states can be considered to be "precursors" to dissociative chemisorption, e.g. in the case of the dissociative chemisorption of carbon monoxide on transition metal surfaces. In kinetic modeling, the rate of change of the molecularly chemisorbed intermediate is exactly zero at steady-state, and both its concentration and the time derivative of its concentration can be approximated as zero at sufficiently high surface temperatures (dictated by the stability of the precursor and the incident molecular flux at the surface). These ideas are used in Sect. III where rates of precursor-mediated adsorption and desorption are presented for most cases that one would expect to encounter experimentally.

III. Rates of Precursor-Mediated Chemisorption and Desorption Derived from Pseudo-Steady-State and Statistical Models

Rate expressions of both precursor-mediated adsorption and desorption based on either pseudo-steady-state or statistical models can be derived formally for a wide variety of approximations concerning the nature of the precursor state [31]. As would be expected and as is necessary, the rate expressions derived from each model are exactly equivalent when the same approximations are made. The rates of precursor-mediated chemisorption and desorption are presented here for all models that might be anticipated to be physically realistic. Unless explicitly stated otherwise, lateral interactions in the adsorbed overlayers are neglected and all adsites for chemisorption are assumed to be equivalent. These two assumptions imply that all derived adsorption isotherms must be of the Langmuir form, and this is verified in all cases. The ad hoc inclusion of lateral interactions between chemisorbed species is discussed very briefly for the case of dissociative chemisorption.

One-dimensional potential energy diagrams, which depict schematically precursor-mediated adsorption and desorption, are presented in Fig. 1. The case of molecular chemisorption is illustrated in Fig. 1(a), while that of dissociative chemisorption is shown in Fig. 1(b). The intrinsic precursor is denoted by an asterisk, whereas the extrinsic precursor is designated by a prime. The fact that neither chemisorption from the extrinsic precursor nor excitation from the chemisorbed state into the extrinsic precursor can occur (by construction) is indicated schematically by potential curves that cross rather than repel. The activation energies of migration (hopping) of the precursors on the surface cannot be shown on these one-dimensional diagrams, since surface diffusion occurs along orthogonal coordinates to that of Fig. 1. Here, the activation energy for migration *from* the intrinsic precursor will be defined as E_m^*, while that for migration *from* the extrinsic precursor will be defined as E_m'. *Direct* migration of all chemisorbed species is ignored. All rate coefficients of adsorption, desorption and migration are assumed to be of the form $k_i = k_i^{(0)}e^{-E_i/k_BT}$, where i = a, d and m.

Prior to a presentation of the various rate expressions appropriate for precursor-mediated chemisorption and desorption, rate expressions are delineated for the case of "direct" or "Langmuirian" chemisorption (R_a) and desorption (R_d) of the reactant A_2. For molecular (first-order) chemisorption and desorption,

$$R_a = k_a F_{A_2}(1 - \theta_{A_2}), \quad \text{and} \tag{1}$$

$$R_d = k_d \theta_{A_2}; \tag{2}$$

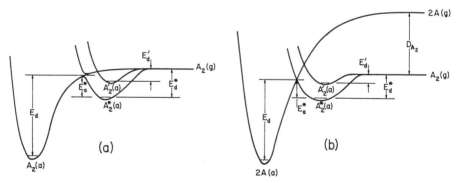

Figure 1. Schematic one-dimensional potential energy diagrams depicting precursor-mediated chemisorption and desorption. (a) Molecular chemisorption, and (b) dissociative chemisorption. Note that $E_d^* - E_d'$ is equal to $E_m^* - E_m'$, and the activation energies are defined in the text.

whereas for dissociative (second-order) chemisorption and recombinative desorption,

$$R_a = k_a F_{A_2}(1 - \theta_A)^2 , \quad \text{and} \tag{3}$$

$$R_d = k_d \theta_A^2 . \tag{4}$$

In (1) and (3), k_a is the rate coefficient of adsorption (which is unfortunately often denoted in scientifically unsatisfactory language as the "sticking coefficient"), and F_{A_2} is the flux of A_2 molecules incident on the surface $[p_{A_2}/(2\pi m_{A_2} k_B T_g)^{1/2}]$. In (2) and (4), k_d is the rate coefficient for direct desorption from the chemisorbed state to the gas phase. Finally, in (1)-(4), θ_{A_2} and θ_A are the fractions of occupied surface sites, and $1 - \theta_{A_2}$ and $1 - \theta_A$ are the fractions of unoccupied surface sites for molecular and dissociative chemisorption, respectively. The scaling of the rates of chemisorption and desorption with the concentration of occupied and unoccupied surface sites should be noted in particular, since it will be seen that this scaling can be modified qualitatively by precursor-mediated reactions.

A. Molecular Chemisorption and Desorption: "Single-Precursor" Models

In the case of chemisorption, this model is defined by the following set of reactions:

$$A_2(g) \underset{k_d^*}{\overset{\xi^* F_{A_2}}{\rightleftarrows}} A_2^*(a) \overset{k_a^*}{\rightarrow} A_2(a) ,$$

where ξ^* is the probability of *trapping* of the incident molecule A_2 into the precursor state A_2^*, F_{A_2} is the incident flux of A_2 molecules [proportional to the pressure of $A_2(g)$], k_d^* is the rate coefficient of desorption of the molecular precursor, and k_a^* is the rate coefficient of chemisorption of the molecular precursor. Similarly, in the case of desorption, this model may be described by

$$A_2(a) \underset{k_a^*}{\overset{k_d}{\rightleftarrows}} A_2^*(a) \overset{k_d^*}{\rightarrow} A_2(g) ,$$

where k_d is the rate coefficient of "desorption" of the molecularly chemisorbed species *into the precursor state*, and the other rate coefficients were defined above [cf. Fig. 1(a) also].

   1. *If only intrinsic precursor states can exist on the surface*, this model implies that the rates of chemisorption and desorption are given, respectively, by

$$R_a = \frac{\xi^* F_{A_2} k_a^*(1 - \theta_{A_2})}{k_a^* + k_d^*} , \tag{5}$$

and

$$R_d = \frac{k_d k_d^* \theta_{A_2}}{k_a^* + k_d^*} , \tag{6}$$

where θ_{A_2} is the fractional surface coverage of molecularly chemisorbed A_2.

In the limit that $k_a^* \gg k_d^*$, then

$$R_a \to \xi^* F_{A_2}(1 - \theta_{A_2}) , \quad \text{and} \tag{5a}$$

$$R_d \to \frac{k_d k_d^*}{k_a^*} \, \theta_{A_2} . \tag{6a}$$

This is the "pseudo-Langmuirian limit" of unactivated adsorption, cf. (1) and (2). Notice, however, that although the apparent activation energy of desorption corresponds to the difference in energy between the chemisorbed state and the energy zero that has been adopted, the apparent preexponential factor of desorption is a ratio of preexponential factors involving three different elementary reactions, rather than that of a single elementary surface reaction.

In the limit that $k_a^* \ll k_d^*$, then

$$R_a \to \xi^* \left[\frac{k_a^*}{k_d^*} \right] F_{A_2}(1 - \theta_{A_2}) , \quad \text{and} \tag{5b}$$

$$R_d \to k_d \theta_{A_2} . \tag{6b}$$

This is the "pseudo-Langmuirian limit" to be expected for activated (with respect to the energy zero) adsorption, cf. (1) and (2). Notice, however, that although the apparent activation energy of adsorption is the same as that embodied in k_a of (1), the preexponential factor here is a ratio of those of two different elementary reactions, rather than that of a single elementary reaction as implied by (1).

2. If the intrinsic and extrinsic precursors are equivalent energetically on the surface (and both are denoted by an asterisk in this special case), then

$$R_a = \frac{\xi^* F_{A_2} k_a^*(1 - \theta_{A_2})}{k_d^* + k_a^*(1 - \theta_{A_2})} , \quad \text{and} \tag{7}$$

$$R_d = \frac{k_d k_d^* \theta_{A_2}}{k_d^* + k_a^*(1 - \theta_{A_2})} . \tag{8}$$

In the limit that $k_a^*(1 - \theta_{A_2}) \gg k_d^*$, then

$$R_a \to \xi^* F_{A_2} , \tag{7a}$$

and

104

$$R_d \rightarrow \left[\frac{k_d k_d^*}{k_a^*}\right]\left[\frac{\theta_{A_2}}{1-\theta_{A_2}}\right] ; \qquad (8a)$$

whereas in the limit that $k_a^*(1-\theta_{A_2}) \ll k_d^*$, then

$$R_a \rightarrow \xi^* \left[\frac{k_a^*}{k_d^*}\right] F_{A_2}(1-\theta_{A_2}) , \quad \text{and} \qquad (7b)$$

$$R_d \rightarrow k_d \theta_{A_2} . \qquad (8b)$$

Equations (7a) and (8a) represent the limit in which effects of the precursor are manifest to the greatest extent [note especially the $1-\theta_{A_2}$ denominator of (8a) and the independence of R_a on θ_{A_2} in (7a)], whereas (7b) and (8b) represent the "pseudo-Langmuirian limit" which would be expected in the case of activated chemisorption.

3. It should be noted that, as is required thermodynamically, the *adsorption isotherms* implied by both the above interpretations of the precursor (derived either by considering adsorption-desorption equilibrium or by equating the rates of adsorption and desorption far from equilibrium) are of the Langmuir form. In particular, in all cases

$$\frac{\theta_{A_2}}{1-\theta_{A_2}} = \left[\frac{\xi^* k_a^*}{k_d k_d^*}\right] F_{A_2} . \qquad (9)$$

It is clearly necessary, however, that rates of adsorption and desorption be compared only when employing the same model of the precursor state(s). This obvious restriction applies also to all other models of precursor-mediated chemisorption and desorption that are presented in this section.

B. Dissociative Chemisorption and Recombinative Desorption: "Single-Precursor" Models

In the case of chemisorption, this model may be described by

$$A_2(g) \underset{k_d^*}{\overset{\xi^* F_{A_2}}{\underset{\longleftarrow}{\longrightarrow}}} A_2^*(a) \overset{k_a^*}{\rightarrow} 2A(a) ;$$

whereas in the case of desorption,

$$2A(a) \underset{k_a^*}{\overset{k_d}{\underset{\longleftarrow}{\longrightarrow}}} A_2^*(a) \overset{k_d^*}{\rightarrow} A_2(g) ,$$

where the rate coefficients are the analogues of those that have been defined earlier. In this case, there is an additional complication compared with molecular chemisorption, namely the occupation of the precursor by either a single surface site (for chemisorption) or by a nearest-neighbor (nn) pair of surface sites must be considered.

1. *If only intrinsic precursor states can exist on the surface and they occupy a single adsite*, then the rates of adsorption and desorption are given by

$$R_a = \frac{\xi^* F_{A_2} k_a^* (1-\theta_A)^2}{k_d^* + k_a^*(1-\theta_A)} \,, \tag{10}$$

and

$$R_d = \frac{k_d k_d^* \theta_A^2}{k_d^* + k_a^*(1-\theta_A)} \,. \tag{11}$$

On the other hand *if only intrinsic precursor states can exist on the surface and they occupy an nn pair of surface adsites*, then the rates of adsorption and desorption are given by

$$R_a = \frac{\xi^* F_{A_2} k_a^* (1-\theta_A)^2}{k_a^* + k_d^*} \,, \tag{12}$$

and

$$R_d = \frac{k_d k_d^* \theta_A^2}{k_a^* + k_d^*} \,. \tag{13}$$

In the limit that $k_a^*(1-\theta_A) \gg k_d^*$, (10) and (11) reduce to

$$R_a \rightarrow \xi^* F_{A_2}(1-\theta_A) \,, \tag{10a}$$

and

$$R_d \rightarrow \frac{k_d k_d^*}{k_a^*} \left[\frac{\theta_A^2}{1-\theta_A} \right] \,; \tag{11a}$$

whereas in the limit that $k_a^*(1-\theta_A) \ll k_d^*$, (10) and (11) reduce to

$$R_a \rightarrow \xi^* F_{A_2} \left[\frac{k_a^*}{k_d^*} \right] (1-\theta_A)^2 \,; \tag{10b}$$

and

$$R_d \rightarrow k_d \theta_A^2 \,. \tag{11b}$$

Again, (10a) and (11a) correspond to the case where the influence of the precursor is the most pronounced; whereas (10b) and (11b) correspond to the "pseudo-Langmuirian limit", which might be expected for activated adsorption. Notice that (10a) implies that the rate of dissociative chemisorption is *first-order* in the fraction of unoccupied surface sites, and that there is a $1-\theta_A$ denominator in the rate of desorption in this limiting case of this model. Equations (12) and (13), on the other hand, reduce to "pseudo-Langmuirian" rate expressions in the two limiting cases, just as (5) and (6) did for molecular adsorption and desorption. This is to be expected, since in both cases there is a one-to-one correspondence between the sites occupied by the chemisorbed species and the sites occupied by the intrinsic precursor state.

2. *If the intrinsic and extrinsic precursors are energetically equivalent on the surface and they occupy a single adsite*, then the rates of adsorption and desorption are given by

$$R_a = \frac{\xi^* F_{A_2} k_a^* (1-\theta_A)^2}{k_d^* + k_a^* (1-\theta_A)^2} , \tag{14}$$

and

$$R_d = \frac{k_d k_d^* \theta_A^2}{k_d^* + k_a^* (1-\theta_A)^2} . \tag{15}$$

If the intrinsic and extrinsic precursors are energetically equivalent and they occupy an nn pair of surface adsites, then the rates of adsorption and desorption are also given by (14) and (15), as one would have anticipated intuitively.

In the limit that $k_a^* (1-\theta_A)^2 \gg k_d^*$, (14) and (15) reduce to

$$R_a \to \xi^* F_{A_2} , \tag{14a}$$

and

$$R_d \to \left[\frac{k_d k_d^*}{k_a^*} \right] \left[\frac{\theta_A}{1-\theta_A} \right]^2 ; \tag{15a}$$

whereas in the limit $k_a^* (1-\theta_A)^2 \ll k_d$, they reduce to

$$R_a \to \xi^* F_{A_2} \left[\frac{k_a^*}{k_d^*} \right] (1-\theta_A)^2 , \quad \text{and} \tag{14b}$$

$$R_d \to k_d \theta_A^2 . \tag{15b}$$

As expected, (14a) and (15a) correspond to the limit in which the effects of the precursor are manifest the most, whereas (14b) and (15b) represent the "pseudo-Langmuirian limit". Notice that if the rate of adsorption is independent of surface coverage, (14a), then a $(1-\theta_A)^2$ term appears in the denominator of the rate of desorption, (15a). This is the second-order analogue of the $1-\theta_{A_2}$ denominator of (8a) for first-order desorption in this same "extreme precursor limit".

3. As required, the *adsorption isotherms* corresponding to each set of assumptions concerning the nature of the precursor state are identical and of the Langmuir form. In particular, each isotherm may be written as

$$\left[\frac{\theta_A}{1-\theta_A}\right]^2 = \left[\frac{k_a^*}{k_d k_d^*}\right]\xi^* F_{A_2} . \tag{16}$$

C. Molecular Chemisorption and Desorption: "Two-Precursor" Model

In this case where both intrinsic and extrinsic precursors are considered explicitly, the mechanistic model of molecular chemisorption may be written as

where ξ' is the probability of trapping of an incident molecule into the extrinsic precursor state, and k_d' is the rate coefficient of desorption from the extrinsic precursor. The other rate coefficients have been defined previously. Likewise, the mechanistic model of molecular desorption may be written as

where all rate coefficients have been defined previously.

For this model, the rates of adsorption and desorption are given by

$$R_a = \frac{k_a^*(1-\theta_{A_2})F_{A_2}\{\xi^* + \xi'k_m'\theta_{A_2}/[k_d' + k_m'(1-\theta_{A_2})]\}}{k_a^* + k_d^* + k_m^* k_d^* \theta_{A_2}/[k_d' + k_m'(1-\theta_{A_2})]} , \tag{17}$$

and

$$R_d = \left[\frac{k_d\theta_{A_2}}{k_a^* + k_d^* + k_m^* k_d^* \theta_{A_2}/[k_d' + k_m'(1-\theta_{A_2})]}\right]\left[k_d^* + \frac{k_d' k_m^* \theta_{A_2}}{k_d' + k_m'(1-\theta_{A_2})}\right] . \tag{18}$$

These expressions are consistent with the Langmuir isotherm of (9), and restrictions imposed by considering the rates at equilibrium (frequently termed *detailed balancing arguments*) require that

$$\frac{\xi^*}{\xi'} = \frac{k_d^* k_m'}{k_d' k_m^*} . \tag{19}$$

108

Mathematical limits of (17) and (18) can be taken which will reduce these expressions either to the "extreme precursor limit" or to the "pseudo-Langmuirian limit".

D. Dissociative Chemisorption and Recombinative Desorption: "Two-Precursor" Models

Considering both intrinsic and extrinsic precursor states explicitly, this model of dissociative chemisorption may be written mechanistically as

$$
A_2(g) \begin{array}{c} \xi^* F_{A_2} \\ \rightleftharpoons \\ k_d^* \end{array} A_2^*(a) \xrightarrow{k_a^*} 2A(a) .
$$

$$
\begin{array}{c} k_d^* \quad k_m^{*} \Big\updownarrow k_m^{-} \\ \xi^{'} F_A \\ k_d^{-} \qquad A_2^{-}(a) \end{array}
$$

Likewise, the mechanistic model for recombinative desorption may be written as

$$
2A(a) \begin{array}{c} k_d \\ \rightleftharpoons \\ k_a^* \end{array} A_2^*(a) \xrightarrow{k_d^*} A_2(g) ,
$$

$$
k_m^{*} \Big\updownarrow k_m^{-} \Big/ k_d^{-}
$$

$$
A_2^{-}(a)
$$

where all rate coefficients have been defined previously. As indicated earlier, two different cases must be considered here, depending on whether the precursors occupy a single adsite or a pair of nn adsites for chemisorption.

If the precursors occupy a single adsite on the surface, the rates of adsorption and desorption are given by

$$
R_a = \frac{k_a^*(1-\theta_A)^2 F_{A_2}\{\xi^* + \xi' k_m' \theta_A/[k_d' + k_m'(1-\theta_A)]\}}{k_d^* + k_a^*(1-\theta_A) + k_m^* k_d' \theta_A/[k_d' + k_m'(1-\theta_A)]} , \tag{20}
$$

and

$$
R_d = \left[\frac{k_d \theta_A^2}{k_d^* + k_a^*(1-\theta_A) + \dfrac{k_m^* \theta_A}{1 + \dfrac{k_m'}{k_d'}(1-\theta_A)}} \right] \left[k_d^* + \dfrac{k_m^* \theta_A}{1 + \dfrac{k_m'}{k_d'}(1-\theta_A)} \right] . \tag{21}
$$

If the precursors occupy a pair of nn adsites on the surface, the rates of adsorption and desorption are given by

$$
R_a = \frac{k_a^*(1-\theta_A)^2 F_{A_2}\{\xi^* + \xi' k_m'(2\theta_A - \theta_A^2)/[k_d' + k_m'(1-\theta_A)^2]\}}{k_d^* + k_a^* + k_m^* k_d'(2\theta_A - \theta_A^2)/[k_d' + k_m'(1-\theta_A)^2]} , \tag{22}
$$

and

$$R_d = \left[\frac{k_d \theta_A^2}{k_d^* + k_a^* + \dfrac{k_m^*(2\theta_A - \theta_A^2)}{1 + \dfrac{k_m'}{k_d'}(1-\theta_A)^2}} \right] \left[k_d^* + \frac{k_m^*(2\theta_A - \theta_A^2)}{1 + \dfrac{k_m'}{k_d'}(1-\theta_A)^2} \right] . \quad (23)$$

As in the case of molecular chemisorption, (17) and (18), mathematical limits of (20) and (21), and (22) and (23) can be taken which will either emphasize or deemphasize the influence of the precursor. Both (20) and (21), and (22) and (23) are consistent with the second-order Langmuir isotherm given by (16); and both of these sets of equations also imply the equilibrium restriction of (19).

E. Summary

At this point, it is useful to summarize the results that have been presented in this section. Rates of adsorption and desorption have been given for both first- and second-order adsorption and desorption corresponding to a variety of (explicitly stated) assumptions concerning the nature of the precursor state(s). In all cases, it has been assumed that the concentration of precursor molecules is negligible (compared both with the concentration of chemisorbed species and that of unoccupied surface sites), that all adsites for chemisorption are equivalent, and that there are no lateral interactions within the adsorbed overlayers. For self-consistent model approximations, all of the results that have been presented can be derived either from the point of view of pseudo-steady-state (kinetic) modeling or from the point of view of statistical (successive site) modeling. For completeness, it should be noted that KING et al. [13,17] have included the occurrence of nn lateral interactions between chemisorbed species in the case of dissociative chemisorption and recombinative desorption. In this case, the probabilities of occurrence of two nn occupied sites, two nn unoccupied sites and two nn sites one of which is occupied and one of which is not are not simply proportional to θ^2, $(1 - \theta)^2$ and $2\theta(1 - \theta)$, respectively. However, these probabilities (or local surface concentrations) can be calculated easily in terms of the (assumed) nn interaction energy and the surface temperature. When this is done, the expressions for the rates of adsorption and desorption are once again found to be identical when evaluated via either the pseudo-steady-state or the statistical approach for the same assumptions concerning the nature of the precursor states. These expressions simply involve one additional parameter, the nn pairwise additive interaction energy between chemisorbed species.

Hence, these two (superficially) different approaches are in fact completely equivalent. Neither one offers any advantages over the other or yields any additional information compared with the other. Furthermore, neither approach gives rise to any detailed microscopic information in the sense that the rate coefficients or probabilities (the ratios of a rate coefficient to a sum of rate coefficients) are first averaged, and then the macroscopic averages are used to derive expressions for rates of adsorption and desorption. Obviously, one could engage in curve-fitting exercises aimed at achieving agreement between measured rates of adsorption and desorption (as a function of surface coverage) and the various self-consistent pairs of calculated rate expressions presented above. The ubiquitous presence of adsorbate-adsorbate interactions, however, would appear to render this sort of fitting of experimental measurements to macroscopic rate parameters of limited value (although, obviously, there may be exceptions to this generalization). In particular, experimental data that apply for low surface coverages can be very usefully analyzed within the context of the rate expressions presented above. Trivial examples of this are presented in Sect. V.

A more useful and more general approach would appear to be a comparison of experimental data to microscopic simulations in which rate parameters are allowed to be a function of *local* surface coverage via lateral interactions in the overlayer. This approach is fundamentally different from those discussed above, since microscopic rate coefficients are used to evaluate microscopic rates which are *then* averaged to compare with experimental data, i.e. the averaging does not destroy the microscopic insight. Although this model contains no fewer parameters (and frequently more) than those discussed above, it offers the advantage that each parameter is well defined at the molecular level. Consequently, one can in favorable cases make use of either independently measured or calculated values of some of the rate parameters that appear in the model. Furthermore, one can always place a priori limits on the values of the rate parameters, i.e. they are not "freely" variable. Preliminary efforts to codify and apply these ideas are summarized in Sect. IV.

IV. Monte Carlo Simulations of Precursor-Mediated Molecular Chemisorption, Desorption, and Ordering in Chemisorbed Overlayers

A. Formalism

A combination of Monte Carlo simulations and deterministic rate expressions has been applied by HOOD et al. [18,32] to describe precursor-mediated molecular chemisorption, desorption and ordering in the chemisorbed overlayer. All adsites for chemisorption are assumed to be geometrically identical, but they are rendered inequivalent by the presence of lateral interactions in the overlayer.

In the description of precursor-mediated molecular chemisorption, molecules are placed sequentially onto randomly chosen surface sites to simulate trapping from the gas phase into either a physically adsorbed intrinsic or extrinsic precursor state. The lattice is typically (96x96) in units of lattice spacings, and periodic boundary conditions are applied. Each precursor can either chemisorb, desorb or migrate across the surface in single hops to nn sites; the precursor "lives" until it either chemisorbs or desorbs. The physically adsorbed intrinsic precursors experience lateral nn and next-nearest-neighbor (nnn) interactions with *chemisorbed* molecules such that the binding energy of the intrinsic precursor is given by

$$\varepsilon_p = \varepsilon_p^o - \sum_j n_j \varepsilon_r' + \sum_k n_k \varepsilon_a , \tag{24}$$

where ε_p^o is the binding energy of an isolated intrinsic precursor, n_j is the occupation number of nn sites by chemisorbed molecules, n_k is the occupation number of nnn sites, ε_r is the pairwise additive (repulsive) nn interaction energy, and ε_a is the pairwise additive (attractive) nnn interaction energy. The sign convention implied by (24) should be noted, as should the fact that precursor-precursor interactions can be neglected since their concentration on the surface is small (except perhaps when adsorption on a very low-temperature surface with a high gas pressure is being simulated, a complication that is ignored here). The extrinsic precursors are assumed to have a constant binding energy, 85% of ε_p^o, for which there is some experimental justification [33]. Furthermore, it is reasonably assumed that the precursors are bound parallel to the surface by a periodic potential that can be represented by adjacent, intersecting harmonic wells of which the minima are dictated by (24). It is easy to show that the resulting barrier to precursor migration is given by

$$\varepsilon_m = \varepsilon_m^o + \Delta\varepsilon_p(i,j)/2 + [\Delta\varepsilon_p(i,j)]^2/16\varepsilon_m^o , \tag{25}$$

where ε_m^o is the barrier to migration of an isolated precursor, and $\Delta\varepsilon_p(i,j) \equiv \varepsilon_p(j) - \varepsilon_p(i)$ is the difference in binding energies of the precursor at the jth and ith sites.

Suppose the probabilities of migration, desorption and chemisorption of the precursor are denoted by P_m, P_d and P_c; and the rate coefficients corresponding to these three elementary reactions are denoted by k_m, k_d and k_c. It is easy to show that

$$P_m = \left[\frac{k_m^{(0)}}{k_m + k_d + k_c} \right] e^{-\varepsilon_m/k_B T} \equiv N e^{-\varepsilon_m/k_B T} , \tag{26a}$$

$$P_d = \left[\frac{k_m^{(0)}}{k_m + k_d + k_c} \right] \left[\frac{k_d^{(0)}}{k_m^{(0)}} \right] e^{-\varepsilon_p/k_B T} \equiv N\chi e^{-\varepsilon_p/k_B T} , \tag{26b}$$

and

$$P_c = \left[\frac{k_m^{(0)}}{k_m + k_d + k_c} \right] \left[\frac{k_d^{(0)}}{k_m^{(0)}} \right] \left[\frac{S_o}{1 - S_o} \right] e^{-\varepsilon_p^o/k_B T} \equiv \left[\frac{N\chi S_o}{1 - S_o} \right] e^{-\varepsilon_p^o/k_B T} , \tag{26c}$$

where S_o is the probability of chemisorption in the limit of low coverage at surface temperature T. Note that P_m, P_d and P_c are all functions of the local environment of the precursor on the surface. As described below, one can now evaluate probabilities (rates) of chemisorption as a function of surface coverage if one knows the values of the various elementary reaction rate coefficients. First, however, the methodology used to calculate desorption rates will be outlined.

Rates of desorption (e.g. thermal desorption spectra) are computed by combining the Monte Carlo formalism with deterministic rate equations. This calculation requires knowledge of the binding energy of a chemisorbed molecule at all sites on the surface, which, for the ith site, is given by

$$\varepsilon_b(i) = \varepsilon_b^0 - \sum_j n_j \hat{\varepsilon}_r + \sum_k n_k \hat{\varepsilon}_a , \qquad (27)$$

where ε_b^0 is the binding energy of an isolated, chemisorbed molecule, $\hat{\varepsilon}_r$ and $\hat{\varepsilon}_a$ are the nn (repulsive) and the nnn (attractive) pairwise additive interaction energies between chemisorbed molecules, and n_j and n_k have been defined in connection with (24). To evaluate desorption rates, one then calculates in a straightforward way rates of excitation from the chemisorbed state into the precursor state, rates of deexcitation from the precursor into the chemisorbed state, and rates of (ultimate) desorption from the precursor. However, for the example discussed in Sect. IV.B, it was found that desorption could be equally well described by "direct" desorption from the precursor (with a considerable saving in computation time). Moreover, it was shown that the preexponential factor of the rate coefficient for direct desorption could be constructed equally well either as a function of the macroscopic surface coverage or as a function of the local surface coverage. In order to make a connection with other experimental data, the former approach is presented here. Obviously, the latter approach is preferred in general.

The rate of desorption from the ith chemisorption site of which the occupation number is $\sigma(i)$ may be written as

$$-\frac{d\sigma(i)}{dt} = \hat{k}_d^{(0)}(\theta)\sigma(i)e^{-\varepsilon_b(i)/k_B T} , \qquad (28)$$

with $\hat{k}_d^{(0)}(\theta) = \hat{k}_d^{(0)}(0)e^{\alpha\theta}$. The system of coupled differential equations describing desorption from all chemisorption states (of which there are 49, for example, for a hexagonally close-packed surface with nn and nnn interactions) is solved numerically. During each computational cycle ($\Delta T \leq 0.5$ K with a heating rate of approximately 6 K/s), the appropriate number of molecules is selected from each chemisorption site according to (28) and removed from randomly chosen lattice sites (to avoid the introduction of artificial correlation effects). The calculation continues until all chemisorbed molecules have been desorbed.

B. Applications: The Interaction of Molecular Nitrogen with the Ru(001) Surface

The formalism described in Sect. IV.A has been applied to describe both the (weak) molecular chemisorption of nitrogen on the Ru(001) surface as a function of coverage at 77 K, and the rate of desorption of nitrogen from this surface as a function of temperature for a variety of initial surface coverages. The ordering that occurs in the chemisorbed overlayer was also evaluated as a function of surface coverage both during adsorption at 77 K and after desorption, which was effected by heating the surface to progressively higher temperatures. The extensive experimental data base that was available for the N_2/Ru(001) system [34-37] proved to be decisively advantageous in determining what tentatively appears to be a unique set of microscopic rate parameters. The excellent agreement between the

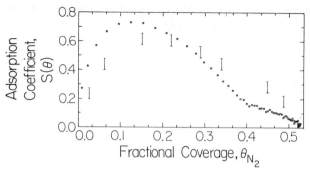

Figure 2. Probability of molecular chemisorption as a function of coverage for N_2 on Ru(001) at 77 K. Vertical bars represent experimental results [34,35], and squares represent averages of eight simulations [18].

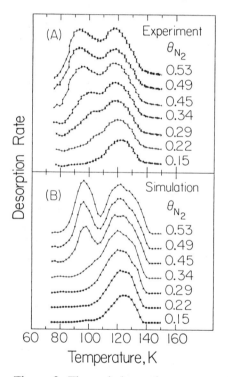

Figure 3. Thermal desorption spectra of N_2 on Ru(001) for various initial coverages. (a) Experimental results [34,35], and (b) simulation results [18].

theoretical and experimental probability of chemisorption and thermal desorption spectra as a function of surface coverage is obvious in Figs. 2 and 3, respectively. The set of parameters used in achieving this level of agreement is given in Table 1. Note that in this particular case of weak chemisorption ε_a and $\hat{\varepsilon}_a$ were found to

114

Table 1. Parameters characterizing the interaction of molecular nitrogen with the Ru(001) surface. All energies are in units of kcal/mol.

ε_p^o	Binding energy of an isolated intrinsic precursor	1.6
ε_m^o	Barrier to migration of an isolated precursor	0.3
ε_b^o	Binding energy of an isolated chemisorbed molecule	5.8
ε_r, $\hat{\varepsilon}_r$	Repulsive nearest-neighbor interaction energy	0.25
ε_a, $\hat{\varepsilon}_a$	Attractive next-nearest-neighbor interaction energy	0.45
χ	Ratio of preexponential factors of desorption to migration for a precursor	500-1000
$\hat{k}_d^{(0)}(0)$	Low coverage limit of preexponential factor of desorption of a chemisorbed molecule	10^{12} s^{-1}
α	Parameter describing coverage dependence of $\hat{k}_d^{(0)}(0)$	17

be equal, as were ε_r and $\hat{\varepsilon}_r$. It is unlikely that this "accidental degeneracy" will occur for the case of strong molecular chemisorption.

It is important to appreciate that the experimental data displayed in Figs. 2 and 3 that have been described accurately by these simulations could be characterized best as atypical. Even though the chemisorbed nitrogen occupies only "on-top" sites at all surface coverages [36,37], the probability of adsorption passes through a maximum as a function of surface coverage, and the thermal desorption spectra split into two peaks, which are of comparable intensity at saturation coverage. It is very unlikely that either the precursor models presented in Sect. III or models aimed at describing thermal desorption spectra by the ad hoc inclusion of lateral interactions could achieve the agreement with both the probability of adsorption and the rate of desorption that has been attained by the simulation discussed here. One should be cautiously optimistic that further work, together with what should be obvious refinements to this type of simulation, will lead to significant advances in our understanding of the elementary reactions associated with chemisorption, desorption, chemical reactions and overlayer ordering (or disordering) on surfaces.

V. Possible Examples of Precursor-Mediated Dissociative Chemisorption

Observations of precursor-mediated chemisorption (and, by implication, desorption) pervade the literature. Indeed, the observation of precursor-mediated rates of chemisorption are probably "the rule" rather than "the exception", when compared to Langmuirian kinetics (i.e. "direct" chemisorption). Moreover, even when Langmuirian kinetics *appear* to have been measured experimentally, the reaction mechanism could, nevertheless, involve a precursor intermediate [cf. the "pseudo-Langmuirian" rates of precursor-mediated chemisorption given by (5), (5a), (5b), (7b), (10b), (12), (14b), and the appropriate limits of (17), (20) and (22)]. Consequently, it is not possible (nor would it be very pedagogic) to review the vast literature concerning the manifestations of precursor intermediates on measured probabilities of chemisorption as a function of surface coverage. Rather, two examples from the recent literature have been chosen which concern rates (probabilities) of dissociative chemisorption in the limit of low surface coverages. These examples are particularly instructive because in neither case did the authors consider the participation of a precursor. The reader is therefore invited to judge the merits of the different interpretations, whereas additional experimental measurements are necessary to provide the final verdict.

A. Dissociative Chemisorption of Methanol on Ni(111)

GATES et al. [38] have carried out extensive experimental measurements concerning the decomposition of methanol to carbon monoxide and hydrogen on a Ni(111) surface. By employing CH_3OH, CH_3OD and CD_3OH, they have elegantly and convincingly shown that the rate–limiting step in the decomposition reaction is the cleavage of the OH or OD bond in the alcohol, i.e. dissociative chemisorption yielding an adsorbed methoxy and either a hydrogen or deuterium adatom. They found that the low coverage limit of the probability of reaction (dissociative chemisorption) at surface temperatures between 300 and 500 K was identical for CH_3OH and CD_3OH, and that it was greater than that for CH_3OD. Their data are shown in Fig. 4. They persuasively interpreted the lower reactivity of the OD compared with the OH bond as being due to differences in the zero point energy of the reactants. Furthermore, they attributed the decrease in the initial probability of reaction with increasing surface temperature to "less efficient accommodation of methanol with the higher temperature surface". On the other hand, this observed decrease in the initial probability of reaction with increasing surface temperature can be assessed quantitatively in order to test the possibility that there is a molecularly adsorbed precursor to dissociative chemisorption (and reaction).

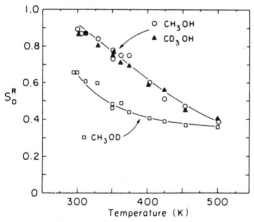

Figure 4. Plot of zero coverage reactive sticking coefficient, S_o^R, versus temperature for three methanol isotopes [38].

A tentative reason to question the proposal that the temperature dependence evident in Fig. 4 is due only to a decrease in the probability of trapping of the incident methanol is based on the expected relatively large heat of molecular adsorption of methanol, q; approximately 12 kcal/mol would be a reasonable guess, since that of water is on the order of 10 kcal/mol. In view of the fact that q/k_B is much greater than the surface temperature in all cases, one might expect the trapping probability not to vary significantly over this range of surface temperatures [39]. An alternate point of view is that molecularly adsorbed methanol is an intermediate in the reaction, and it can either react (resulting in dissociative chemisorption) with a rate coefficient k_r, or desorb with a rate coefficient k_d. In this picture, the initial probability of reaction may be written as

$$S_o^R = \frac{1}{1 + \dfrac{k_d}{k_r}} = \frac{1}{1 + \dfrac{k_d^{(0)}}{k_r^{(0)}} \, e^{-(E_d - E_r)/k_B T}} . \tag{29}$$

Consequently, plotting $\ln(1/S_o^R - 1)$ as a function of reciprocal temperature should yield a straight line of which the slope is $-(E_d - E_r)/k_B$ and the intercept is $\ln(k_d^{(0)}/k_r^{(0)})$. If the two sets of data in Fig. 4 are plotted in this fashion, one indeed obtains two straight lines, which imply that for CH_3OH and CD_3OH,

$$E_d - E_r \triangleq 3.7 \text{ kcal/mol} \quad \text{and} \quad \frac{k_d^{(0)}}{k_r^{(0)}} \triangleq 63 \, ;$$

whereas for CH_3OD,

$$E_d - E_r \triangleq 2.0 \text{ kcal/mol} \quad \text{and} \quad \frac{k_d^{(0)}}{k_r^{(0)}} \triangleq 16 \, .$$

The zero point energy difference of approximately 1.7 kcal/mol that is suggested is in very good agreement with that which would be expected for the OH(OD) stretching mode, and it would also be expected that $k_d^{(0)} > k_r^{(0)}$ in the case of a dissociative chemisorption reaction. Hence, these results provide strong evidence for the participation of a molecularly adsorbed precursor in the decomposition of methanol on Ni(111).

The temperature at which molecular methanol is observed to convert to a methoxy and a hydrogen adatom on Ni(111) implies an activation energy of reaction of CH_3OH of approximately 9 kcal/mol [40]. This, in turn, implies an activation energy of desorption (heat of molecular adsorption) of approximately 12.7 kcal/mol, making use of the results presented above that were derived from the data of GATES et al. [38]. Furthermore, one would predict the activation energy of reaction of CH_3OD to be approximately 10.7 kcal/mol, and the threshold temperature for dissociation of CH_3OD to be approximatly 20 K higher than that of CH_3OH. Finally, on the basis of the precursor-mediated picture of the reaction, one would expect that S_o^R of CH_3OH would be smaller than that of CH_3OD at surface temperatures higher than those investigated by GATES et al. [38], cf. Fig.

4. Experiments designed to test this prediction of the precursor-mediated reaction mechanism are planned [41].

B. Dissociative Chemisorption of Carbon Monoxide on Nickel

The dissociative chemisorption of carbon monoxide on various surfaces of nickel has been the subject of a number of experimental studies, most recently those of STEINRÜCK et al. [42]. STEINRÜCK et al. [42] measured the initial probability of dissociative adsorption of CO (S_o^R) on both a Ni(100) surface and a "sputter-damaged" Ni(100) surface. At a surface temperature of 500 K, S_o^R was found to be equal to approximately 0.02 on Ni(100) and approximately 0.40 on sputter-damaged Ni(100). By using molecular beam scattering techniques, they demonstrated convincingly that S_o^R is not a function of the translational energy of the CO, for translational energies up to at least 20 kcal/mol.

STEINRÜCK et al. [42] drew a number of significant and provocative conclusions on the basis of their experimental results. For example, they concluded that the dissociative chemisorption of CO occurs *only* at "defect sites" on both of these surfaces of nickel. They also concluded that "the *observed* dissociative CO adsorption is not activated", and they emphasized that this result was inconsistent with the data of ROSEI et al. [43] who reported that $S_o^R \geq 10^{-4}$ for the dissociative adsorption of CO on Ni(110) with an associated activation energy of 23 kcal/mol. Furthermore, in comparing their results with those of GOODMAN et al. [44,45], who reported an activation energy of 22-25 kcal/mol for CO dissociation on Ni(100) in connection with their methanation studies, STEINRÜCK et al. [42] seemed to imply both that Goodman's barrier of 22-25 kcal/mol corresponded to a "perfect" Ni(100) surface on which all "defects" were passivated by carbon adatoms, and that their own barrier of zero corresponded to dissociative adsorption at defect sites on the Ni(100) surface.

This current situation is disconcerting for at least two reasons. First, there are apparently gross disagreements among the experimental measurements of a number of different research groups. Second, and even more distressing, the presence or absence of defect sites on a Ni(100) surface has been implied to lead to a difference in activation energies of approximately 25 kcal/mol for a reaction as mechanistically trivial as CO dissociation. At the measurement temperature of 500 K, this corresponds to a difference in the two rates of almost 11 orders of magnitude. This conclusion has a number of profound implications. For example, if it were true that defect sites of which the concentration were 1 part in 10^{11} could be equally important as the remaining "perfect" surface, then everyone engaged in attempting to clarify heterogeneous catalysis by studying reactions over single crystalline surfaces should recognize the futility of their efforts and terminate their research. Happily, however, it appears that this melancholy scenario will not need to be played out! By including the influence of molecularly chemisorbed CO as a precursor to dissociative adsorption, one can argue that all the experimental data discussed above are mutually consistent, and that the presence of defect sites enhances the rate of dissociation by a factor of no more than 20 rather than by 11 orders of magnitude.

If molecularly chemisorbed CO is considered to be a precursor to dissociation, then the initial probability of reaction (dissociation) is given by (29), where k_d and k_r are the rate coefficients of desorption and reaction of molecularly chemisorbed CO. For the experimental conditions of STEINRÜCK et al. [42], the lifetime of this molecular precursor on the surface is approximately 1s, and its coverage is always less than 4% of a monolayer. Since data are presented in [42] at only a single surface temperature (500 K), the detailed analysis presented in Sect. V.A cannot be carried out for this case. Nevertheless, plausibility arguments based on (reasonably) assumed rate coefficients can be presented. (Methods for testing these arguments experimentally will be delineated briefly.)

The initial heat of adsorption (equal to the activation energy of desorption) of molecular CO on the Ni(111), Ni(100) and Ni(110) surfaces is 26.5 kcal/mol, 30 kcal/mol and 30 kcal/mol, respectively [46], i.e. it is somewhat greater on the less "smooth" surfaces. Hence, it might be expected that the heat of adsorption of molecular CO would be slightly greater at defect sites on a surface compared to the perfect surface. Unfortunately, the ratio $k_d^{(0)}/k_r^{(0)}$ of (29) is not known in this case. Its value, however, would certainly be expected to be greater than one and very likely to lie between 10 and 1000. Calculated values of $E_d - E_r$ applying (29) are shown in Table 2 for $k_d^{(0)} > k_r^{(0)}$ equal to 10, 100 and 1000, making use of the experimental results of STEINRÜCK et al. [42], i.e. $S_o^R = 0.02$ for Ni(100) and $S_o^R = 0.40$ for sputter-damaged Ni(100), both at 500 K. A number of very important conclusions can be drawn from the results presented in Table 2. First, dissociative adsorption is found to be activated with respect to the gas phase energy zero only on the Ni(100) surface with $k_d^{(0)}/k_r^{(0)} = 10$. In all other cases, the barrier to dissociation lies below the zero of energy that has been adopted, by as much as 6.46 kcal/mol on the sputter-damaged Ni(100) surface for $k_d^{(0)}/k_r^{(0)} = 1000$. It should also be noted that the *difference* between the values of $E_d - E_r$ for the two surfaces is 3.46 kcal/mol for *every* assumed value of $k_d^{(0)}/k_r^{(0)}$. This implies that if E_d and E_r are shifted equally (statically) by virtue of the stronger bonding of CO at the defect sites, the increase in the binding energy of CO at the defect sites is only 1.73 kcal/mol. Alternatively, the maximum expected increase in the binding

Table 2. Values of $E_d - E_r$ in kcal/mol as a function of $k_d^{(0)} > k_r^{(0)}$ for precursor-mediated dissociative adsorption of CO. Experimental values of S_o^R at 500 K [42] have been used in deriving these results, as described in the text.

$k_d^{(0)}/k_r^{(0)}$	Ni(100)	Sputter-Damaged Ni(100)
10	-1.58	+1.88
100	+0.71	+4.17
1000	+3.00	+6.46

energy of CO at the defect sites compared to the perfect surface is 3.46 kcal/mol (assuming E_r is the same on the two surfaces). The intuitively appealing corollary to this conclusion is that the surface chemistry of the defect sites is subtly, rather than qualitatively, different from the perfect surface!

Furthermore, the activation energy of 23 kcal/mol reported by ROSEI et al. [43] for the dissociative chemisorption of CO on Ni(110) is that corresponding to the rate coefficient k_r above, which, based on the data of STEINRÜCK et al. [42] and CHRISTMANN et al. [46], would be expected to lie between approximately 27.0 and 31.6 kcal/mol (ignoring defects) or between approximately 25.3 and 29.8 kcal/mol (accounting for defects). Considering the fact that the analysis of ROSEI et al. [43] resulted in an unusually small value of $k_r^{(0)}$ ($\leq 7 \times 10^7$ s^{-1}), the actual agreement among [42], [43] and [46] is exceptionally good. Likewise, one would expect the activation energy of 22-25 kcal/mol reported by GOODMAN et al. [44,45] to be on the order of (or slightly less than) that corresponding to k_r in the analysis presented above (25.3-31.6 kcal/mol) as observed. Thus, the hypothesis that there is a molecularly chemisorbed precursor to the dissociation of CO on nickel surfaces is capable of reconciling all of the superficially inconsistent data discussed above.

The hypothesis of a molecularly adsorbed precursor state also is consistent with the observation by STEINRÜCK et al. [42] that the fractional surface coverage of dissociatively adsorbed CO exceeds 0.1 on Ni(100). If the probability of dissociative adsorption of CO on Ni(100) were essentially unity at defect sites and zero on the perfect surface, the observed value of S_o^R of 0.02 would imply that the maximum fractional surface coverage of dissociatively adsorbed CO would be 0.02. This inconsistency is removed if one adopts the point of view that the dissociative chemisorption is mediated by a molecularly chemisorbed precursor. The latter implies that S_o^R is 0.02 at 500 K on the perfect Ni(100) surface and 0.40 on defect sites on the Ni(100) surface at the same temperature. This difference in rates of a factor of 20 is considerably more appealing than the previously implied difference of 11 orders of magnitude!

The fact that the initial probability of dissociative adsorption of the CO was found to be independent of the translational energy of the CO for translational energies up to 20 kcal/mol [42] is also (only) consistent with precursor-mediated dissociative chemisorption. The apparent activation barrier to "direct" dissociative chemisorption would be expected to be much larger than 20 kcal/mol in view of the essential certainty of a multidimensional potential energy surface for dissociative adsorption. In keeping with this expectation, KANG et al. [47] have found that the threshold translational energy necessary for any measurable direct dissociation of CO on Ni(111) is 62.3 kcal/mol.

Finally, it should be noted that in order to test the proposition that CO dissociation on nickel is precursor-mediated, the rate of formation of $C^{12}O^{18}$ and $C^{13}O^{16}$ from a reactant mixture of $C^{12}O^{16}$ and $C^{13}O^{18}$ could be measured as a function of surface temperature. If the precursor mechanism is shown to apply, independent values of $k_d^{(0)}/k_r^{(0)}$ and $E_d - E_r$ could be deduced via (29) for different orientations of simple crystalline nickel surfaces.

VI. Synopsis

In this brief review of precursor states at solid surfaces, an attempt has been made to emphasize a number of different and important issues which include the following:

1. One can "think about" precursor intermediates in chemisorption and desorption reactions at surfaces as either *intrinsic* precursors which exist above unoccupied surface sites or *extrinsic* precursors which exist above surface sites occupied by a chemisorbed species. The extrinsic precursor should be regarded as physically adsorbed, whereas the intrinsic precursor could be regarded either as a physically adsorbed or a molecularly chemisorbed species. The molecularly chemisorbed intrinsic precursor would be expected only in the case of dissociative adsorption of a molecule that can also chemisorb molecularly (where a physically adsorbed intrinsic precursor to the molecularly chemisorbed intrinsic precursor could also be included).

2. In general, one should *expect* that precursor intermediates exist in chemisorption and desorption. A notable exception to this expectation, however, is the activated, dissociative chemisorption of molecules with sufficiently high incident translational energies, e.g. as can be achieved in molecular beam scattering measurements. Furthermore, one might expect the apparent activation energy deduced from beam scattering measurements of dissociative chemisorption to be greater than activation energies of precursor-mediated chemisorption for the same systems (due to the multidimensionality of the relevant potential energy surfaces).

3. Manifestations of the existence of precursor intermediates are probably most evident in either the variation of the probability of adsorption with coverage or the variation of the initial (low coverage) probability of adsorption with surface temperature. Whenever precursors are implicated in measured rates of adsorption, their existence should be considered also in the analysis of rates of desorption for the same systems.

4. In the mathematical modeling of reaction rates that include precursors, their concentration on the surface is assumed to be negligible. Although this approximation is eminently reasonable for those systems to which this modeling has been applied, it is not necessary in general. The stability of the precursor depends on the surface temperature (viz-a-viz the binding energy of the precursor and the activation energy separating it from the chemisorbed state) and on the gas pressure (flux). In fact, an intrinsic precursor to chemisorption has been isolated in the unfavorable case of molecular chemisorption of a "reactive" molecule (carbon monoxide) on a "reactive" surface [Ni(111)] when the surface temperature of adsorption was 6 K [27].

5. Assuming that the concentration of the precursor(s) is(are) negligible, expressions have been presented for rates of chemisorption and desorption for both first- and second-order reactions, and for a wide variety of assumptions regarding the detailed nature of the precursors. In all equivalent cases, the results of a kinetic (pseudo-steady-state) analysis are identical to those of a statistical (successive-site) analysis. Although useful, these models are limited, ultimately, by the fact that they involve macroscopic averages of rate coefficients and by the large number of parameters that can appear in their formulation.

6. Anticipating that a microscopic description of precursor-mediated surface reactions will supercede the macroscopic pictures, a molecular simulation based on combined Monte Carlo techniques and deterministic rate expressions (evaluated stochastically) has been presented for the case of molecular chemisorption and desorption. One might expect that this approach (or variations thereof) will provide both the basis of the "next generation" of theoretical descriptions of precursor-mediated surface reactions, as well as a more detailed understanding of the relevant "molecular chemistry". One should certainly hold out the hope, however, that sufficiently accurate potential energy surfaces will ultimately become available from theoreticians to permit a realistic analysis of precursor-mediated surface reactions by techniques involving molecular dynamics.

Acknowledgment. Financial support by the National Science Foundation via Grant No. CHE-8516615 is gratefully acknowledged.

References

1. J.A. Barker, D.J. Auerbach, Faraday Disc. Chem. Soc. *80*, 277 (1985)

2. J.C. Tully, Faraday Disc. Chem. Soc. *80*, 291 (1985)

3. I. Langmuir, Chem. Rev. *6*, 451 (1929)

4. J.B. Taylor, I. Langmuir, Phys. Rev. *44*, 423 (1933)

5. J.E. Lennard-Jones, Trans. Faraday Soc. *28*, 333 (1932)

6. J.A. Becker in *Structure and Properties of Solid Surfaces*, ed. by R. Gomer and C.S. Smith (Univ. of Chicago Press 1953) p. 459

7. J.A. Becker, C.D. Hartman, J. Phys. Chem. *57*, 157 (1933)

8. G. Ehrlich, J. Phys. Chem. *59*, 473 (1955)

9. P. Kisliuk, J. Phys. Chem. Solids *3*, 95 (1957); *5*, 78 (1958)

10. P.W. Tamm, L.D. Schmidt, J. Chem. Phys. *52*, 1150 (1970); *55*, 4253 (1971)

11. L.R. Clavenna, L.D. Schmidt, Surface Sci. *22*, 365 (1970)

12. C. Kohrt, R. Gomer, J. Chem. Phys. *52*, 3283 (1970)

13. D.A. King, M.G. Wells, Surface Sci. *23*, 120 (1971); Proc. Roy. Soc. (Lond.) A*339*, 2435 (1974)

14. M.R. Shannabarger, Solid State Commun. *14*, 1015 (1974); Surface Sci. *44*, 297 (1974); *52*, 689 (1975)

15. D.A. King, Surface Sci. *64*, 43 (1977)

16. R. Gorte, L.D. Schmidt, Surface Sci. *76*, 559 (1978)

17. A. Cassuto, D.A. King, Surface Sci. *102*, 388 (1981)

18. E.S. Hood, B.H. Toby, W.H. Weinberg, Phys. Rev. Letters *55*, 2437 (1985)

19. D.J. Auerbach, C. Rettner, these proceedings.

20. P.R. Norton, R.L. Tapping, J.W. Goodale, Surface Sci. *72*, 33 (1978)

21. P. Hofmann, K. Horn, A.M. Bradshaw, K. Jacobi, Surface Sci. *82*, L610 (1979)

22. D. Schmeisser, K. Jacobi, Surface Sci. *108*, 421 (1981)

23. D. Schmeisser, K. Jacobi, D.M. Kolb, Appl. Surface Sci. *11/12*, 164 (1982)

24. Y.-P. Hsu, K. Jacobi, H.H. Rotermund, Surface Sci. *117*, 581 (1982)

25. M.J. Grunze, J. Fuhler, M. Neumann, C.R. Brundle, D.J. Auerbach, J. Behm, Surface Sci. *139*, 109 (1984)

26. M. Shayegan, J.M. Cavallo, R.E. Glover, R.L. Park, Phys. Rev. Lett. *53*, 1578 (1984)

27. M. Shayegan, E.D. Williams, R.E. Glover, R.L. Park, Surface Sci. *154*, L239 (1985)

28. M. Shayegan, R.E. Glover, R.L. Park, J. Vacuum Sci. Technol. A*4*, 1333 (1986)

29. One would expect this preexponential factor to be smaller for dissociative chemisorption compared with either molecular chemisorption or molecular desorption when a value of 10^{13} s^{-1} is often observed (or assumed). If a value of 10^{13} s^{-1} rather than 10^{11} s^{-1} is assumed for this case of the dissociative adsorption of oxygen on Ni(111), the calculated activation barrier is approximately 1880 cal/mol rather than 1650 cal/mol.

30. M. Shayegan, R.E. Glover, R.L. Park (to be publishsed).

31. In these derivations, both the concentration and the time derivative of the concentration of the precursor state(s) are assumed to be small (i.e. zero).

32. E.S. Hood, B.H. Toby, W. Tsai, W.H. Weinberg, these proceedings.

33. K. Christmann, J. Demuth, Surface Sci. *120*, 291 (1982)

34. P. Feulner, D. Menzel, Phys. Rev. B*25*, 4295 (1982)

35. D. Menzel, H. Pfnür, P. Feulner, Surface Sci. *126*, 374 (1983)

36. A.B. Anton, N.R. Avery, B.H. Toby, W.H. Weinberg, J. Elect. Spect. Relat. Phenom. *29*, 181 (1983)

37. A.B. Anton, N.R. Avery, T.E. Madey, W.H. Weinberg, J. Chem. Phys. *85*, 507 (1986)

38. S.M. Gates, J.N. Russell, Jr., J.T Yates, Jr., Surface Sci. *146*, 199 (1984); J. Catal. *92*, 25 (1985)

39. W.H. Weinberg, R.P. Merrill, J. Vacuum Sci. Technol. *8*, 718 (1971)

40. J.E. Demuth, H. Ibach, Chem. Phys. Lett. *60*, 395 (1975)

41. J.T. Yates, Jr., private communication

42. H.P. Steinrück, M.P. d'Evelyn, R.J. Madix, Surface Sci. *172*, L561 (1986), and references therein

43. R. Rosei, F. Ciccacci, R. Memeo, C. Mariani, L.S. Caputi, L. Papagno, J. Catal. *83*, 19 (1983)

44. D.W. Goodman, J. Vacuum Sci. Tech. *20*, 522 (1982)

45. D.W. Goodman, R.D. Kelley, T.E. Madey, J.M. White, J. Catal. *64*, 479 (1980)

46. K. Christmann, O. Schober, G. Ertl, J. Chem. Phys. *60*, 4719 (1974)

47. H. Kang, T.R. Schuler, J.W. Rabalais, Chem. Phys. Lett. *128*, 348 (1986)

Precursor States, Myth or Reality:
A Perspective from Molecular Beam Studies

D.J. Auerbach and C.T. Rettner

IBM Almaden Research Center, K33/801, 650 Harry Road,
San Jose, CA95120, USA

In this short review we emphasize the perspective which can be gained with molecular beam techniques on the dynamical role of precursor states. By providing control over the initial state and measurement of the final state, molecular beams provide detailed information to help gain insight into the microscopic mechanism of chemisorption.

1. INTRODUCTION

It is widely accepted that precursor states play an important role in many cases of dissociative chemisorption. As discussed by Weinberg in the preceding paper, the concept of precursor states goes back to pioneering work of Langmuir [1] and Lennard-Jones [2] and has been further developed in subsequent years in an extensive literature. How then can the organizers of this workshop have the audacity to question this old and venerable concept by setting up a session entitled "Precursor States, Myth or Reality?"

To discuss this question sensibly, we must make an important distinction at the outset, namely the distinction between the question of the existence of a precursor state and the question of whether a precursor state plays an important dynamical role under a given set of conditions. Precursor state models are built on the notion that before chemisorbing, a molecule becomes trapped temporarily in a weakly bound, mobile, molecular state, which in the classical view arises because of long range physisorption (van der Walls) interactions. The existence of such a state is a very different question than that of whether incident molecules become trapped therein and then are free to migrate on the surface. Trivially, we could say every adsorption process involves passage through a precursor state, since all molecules have a long range attraction, and all molecules will pass through a region where this interaction is effective on their way to the surface. This however certainly is not what is meant by precursor states models. Even the existence of precursor states must involve not only a long range attraction, but also a barrier to the chemisorbed state as illustrated schematically in Fig. 1. Here we see two families of curves, of the type first described by Lennard-Jones. [2] At long range a family of curves is drawn to describe various possible weakly bound (physisorbed) states. These various states could be thought of as representing different systems, or different impact points or molecular orientations for a given system. At short range a curve is drawn to describe a strongly bound (chemisorbed) state. (Again we could have indicated multiple curves to denote multiple systems, impact parameters or orientations but to keep the illustration simple, only one is drawn.) In between the two states, there is a barrier, the height of which depends on which weakly bound curve we consider. This barrier is essential to allow molecules to become trapped on the surface in the weakly bound precursor state before chemisorbing. In some cases, direct spectroscopic identification of "precursor states" at low temperatures has been reported. However, this evidence does not settle, or even directly address, the question which underlies the title of this session, *viz.* what is the dynamical role of precursor states under a given set of conditions.

The notion of the dynamical role of precursor states is based largely on measurements of adsorption kinetics, in particular, on cases where measurements of sticking probability vs. coverage show a plateau region at low coverage. Very detailed microscopic models have been constructed to interpret this data. Although the models are able to fit measurements of sticking probability vs. coverage and surface temperature for an adsorption from an ambient gas, such data certainly do not provide sufficient information for building unique dynamical models.

In this short review we emphasize the perspective which can be gained with molecular beam techniques on questions relating to the dynamical role of precursor states. Molecular beams provide additional information to help gain insight into the microscopic mechanism of chemisorption. With beams we can control incidence energy, E_i, angle, θ_i, and surface temperatures, T_s; we can measure adsorption probabilities and the angular, velocity and residence time distributions of molecules which do not permanently stick. This additional information allows us to check the predictions of existing models and, perhaps more importantly, to provide a sound foundation for future attempts to build an accurate microscopic picture of the chemisorption processes.

This paper is organized as follows. We begin with a discussion of the potential hypersurfaces involved in interactions of diatomic molecules with surfaces. Although the question of precursor states is certainly not restricted to diatomics, for simplicity we restrict the discussion to dissociative chemisorption of diatomic molecules. Even for such systems, the potentials are known sketchily at best. Nonetheless, we can more precisely define the dynamical issues we wish to address by considering qualitative features of potential hypersurfaces and representations of such surfaces which go beyond the one-dimensional view illustrated in Fig. 1. The strategy which emerges is to examine energy transfer and trapping in systems without chemisorption, and then to apply these findings and techniques

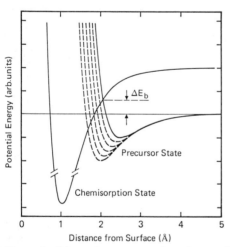

Figure 1. Schematic representation of potentials for dissociative chemisorption. The various curves for the molecular precursor state represent different adsorption sites, orientations, impact parameters, etc.

to reactive systems. Armed with a more precise definition of the problem, we turn briefly to a description of experimental techniques and capabilities, and then to a summary of several important aspects of inelastic scattering of atoms and molecules at surfaces. Studies of systems which show only weakly bound states (*i.e.* systems which do not have chemisorbed states or which do not have substantial sticking probabilities into chemisorbed states) provide models for understanding the dynamics of trapping in weakly bound "precursor" states. An important aspect of such studies is how molecular beam measurements allow the determination of trapping on a picosecond time scale from observations of angular and velocity distributions.

Next, we turn to the application of these ideas and techniques directly to chemisorption systems. We concentrate on two well studied systems which previous literature indicate are good examples of chemisorption via precursor states: N_2 on $W(100)$ and O_2 on $W(110)$. Molecular beam measurements allows us to probe the trapping probability into the precursor state and its dependence on T_S, E_i, θ_i, and coverage. Further, we are able to directly determine if a sticking probability of less than one is due to molecules which trap and then desorb. We also discuss the information which can be gained by variation of the energy and angle of incidence. We find in many cases a direct adsorption mechanism, ignored in precursor state models, can provide a parallel adsorption channel which becomes increasingly important at high energies.

1.1 Potential Surfaces for Dissociative Chemisorption

Figure 1 depicts a commonly used representation of potential energy surfaces for dissociative chemisorption of diatomic molecules. As mentioned above, two kinds of curves are drawn, one representing the weak (physisorption) interaction of the molecule before dissociation and the other representing the strong (chemisorption) interaction of the atoms resulting from dissociation. For the weakly bound state, multiple curves are drawn to denote variation of the repulsive potential with impact point on the lattice, variation with orientation of the molecule, or variation from system to system. Two important cases need to be distinguished: nonactivated adsorption and activated adsorption. Whether we are dealing with nonactivated or activated adsorption is determined by whether these curves cross below or above the zero of potential energy. If the crossing is below zero, molecules which trap in the weakly bound (precursor) molecular state have a smaller activation barrier to chemisorption than to desorption. Thus increasing the temperature will decrease the probability of chemisorption relative to desorption. In the case of a crossing above zero, the opposite is true and an increase in temperature results in an increase in the probability of dissociative chemisorption.

While curves of this type are widely used in the literature of dissociative chemisorption, they are not without difficulties. They are, after all, only a one dimensional representation of a much higher dimensional reality. The first problem which arises in the use of such curves is what meaning to attach to the ordinate. Many authors use this as distance from the surface; another interpretation is that of reaction coordinate. Either interpretation is inadequate if we want to ask dynamical questions about the path molecules follow in dissociative adsorption. Central to the question posed in the title to this session is the question of whether an incident molecule will trap in the weakly bound molecular state prior to dissociation. Curves of the type given in Fig. 1 are really unable to answer this question. As drawn there is really no reason why (for the nonactivated case of interest to precursor state models) a molecule could not follow the minimum energy path and go directly to the chemisorbed dissociated state without ever trapping in the precursor state.

127

To answer the question of whether or not a molecule will trap requires that we consider an important degree of freedom not depicted in Fig. 1, namely the bond distance in the diatomic molecule. The two families of curves of Fig. 1 for example are drawn for opposite extremes of this variable; the physisorbed state is drawn for the case of the diatom interatomic distance equal to the equilibrium bond distance for the molecule, while the dissociated chemisorbed state is drawn assuming a very large separation of the atoms. In passing from the physisorbed to the chemisorbed state at positive energy, no barrier would be encountered for the nonactivated case shown in Fig. 1, but in considering the actual path followed in a two dimensional space which includes the interatomic distance, a barrier might very well exist. Furthermore, whether or not we have to pass over a barrier might very well depend on the initial conditions.

The above remarks immediately suggest a better representation for the potential energy surfaces involved, a contour map of the potential as a function of two coordinates, the distance from the surface and the interatomic distance. Fig. 2 shows a potential energy surface for dissociative chemisorption in this representation taken from a calculation by Norskov et. al [3] for H_2 interactions with Mg(001). We do not wish to discuss the details of this particular system or method of calculation; rather Fig. 2 is presented as an illustration of qualitative features of potential hypersurfaces for dissociative adsorption. The contours shown are for the case of a molecule incident on an ATOP site with its axis parallel to the surface; the coordinates are distance orthogonal to the surface (z) and interatomic spacing parallel to the surface (R). Two wells are shown in the entrance channel. (We will return to the possibility of multiple entrance channel wells in later in the discussion.) The trajectory of a collision of a molecule with the surface can be followed as a path on this figure.

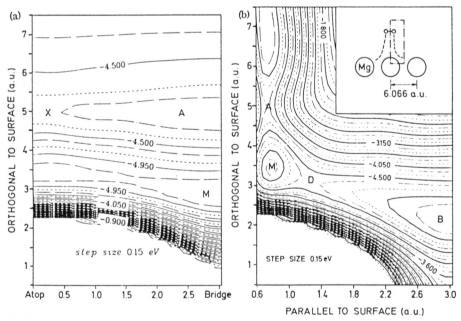

Figure 2. Contour representation of potentials calculated for H_2 on Mg from [3]; (a) H_2 parallel to the surface with fixed bond length of 1.5 a.u., and (b) H_2 dissociating over an atop site into bridge sites.

We are now able to address a question we could not address with curves of the type given in Fig. 1: how can a molecule with energy greater than the barriers between the physisorbed and chemisorbed states become trapped in the outer physisorbed state. We see that having an incident energy higher than the barriers (at A and D in Fig. 2) is not a sufficient condition to dissociate (in well B in Fig. 2). A path to dissociation must also be able to turn the corner. Thus even if the barriers are below the zero of potential energy, it is possible to trap (in state M) and whether or not this take place is a detailed dynamical question depending on the shape of the potential surface and the initial conditions.

The calculation of trajectories requires not only the information given in the representation of Fig. 2, but also the variation of potential with all the coordinates of the system. The variation of the potential with orientation may be important and can lead to steric restriction and to rotational energy transfer. Likewise, the variation with impact coordinates can lead also to steric restrictions and to deflection of scattered molecules away from the specular direction. The variation of potential with surface atom coordinates is also important, and is connected to transfer of energy to and from phonon channels.

We see from the above discussion that the question of whether a precursor state plays a dynamical role is very complicated. The answer will depend not only on the potential hypersurface but also on the detail of the dynamical motion on this surface and on the conditions of a given experiment or situation of interest.

1.2 Experimental Approach

The preceding discussion suggests an approach to understanding the dynamical role of precursor states. The approach involves study first of very simple model systems which isolate a given dynamical issue, and the moving step by step to add features by considering successively more complicated model systems. Clearly the question of trapping in weakly bound states is crucial. This requires that we understand the dynamics of inelastic collisions and the various channels of energy transfer available. To become trapped, an incident molecule must lose energy from the translational motion perpendicular to the surface and transfer this into phonons, electronic, rotational, or vibrational modes. Studies of weakly bound atomic systems (e.g. rare gases) let us probe phonon (and electronic) energy transfer. Variation of substrate allows us to probe the effects of surface corrugation. Moving to diatomic systems for which dissociation is negligible allows us to add the rotational and vibrational degrees of freedom. Finally we move to reactive systems and add the dissociative degree of freedom.

2. EXPERIMENTAL APPARATUS AND PROCEDURES

The basic concepts of the application of molecular beam techniques to the study of the scattering and reactions of atoms and molecules at a solid surface are very simple. A beam of atoms or molecules in a well defined initial state is incident on a well defined surface and the residence time, angle, velocity and internal state distributions of the particles leaving the surface are measured. The elements of an experiment thus include beam preparation, beam modulation, sample temperature and incidence angle control, and product detection. A wide range of techniques are available for each of the elements of a molecular beam experiment, and it is certainly beyond the scope of this short review to discuss, or even list them. Rather

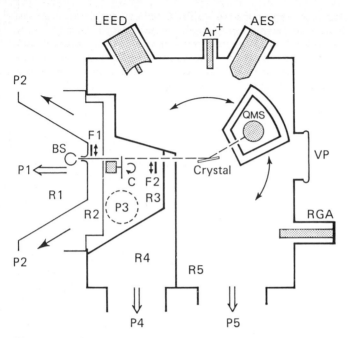

Figure 3. Molecular beam apparatus for determining scattering distributions and dissociative chemisorption probabilities. Key: R1-R5, vacuum regions; BS, beam source; C, high-speed chopper; F, beam flag; QMS, quadrupole mass-spectrometer detector; LEED, low energy electron diffraction screen; Ar+, argon ion sputtering gun; AES, Auger electron spectrometer; VP, 6 in. viewport; RGA, residual gas analyzer. The molecular beam system can detach at the R3 region and is non-bakable. The crystal is held on a manipulator which provides for heating and cooling and rotates about the same axis as the QMS.

we will briefly describe the molecular beam instrument in use in our laboratory for experiments of the type discussed in succeeding sections. Figure 3 shows this instrument in one of its configurations.

2.1 Definition of the Initial State

The beam source (BS), is of the supersonic type which yields a beam of high flux and narrow velocity distribution. This beam is collimated by a series of skimmers and apertures and pumped by four stages of differential pumping (via. pumps P1 .. P4) to reduce background loading on the scattering chamber to negligible values. A chopper (C) allows for the generation of narrow pulses, and flags (F1 and F2) allow the beam dose to be accurately controlled. By variation of source temperature and gas mixture (seeded beam technique), the energy of the incident beam can be controlled over a wide range. For molecules, some separate control over vibrational and translational energy of the beam is possible by manipulation of source conditions.

The two key angles in a scattering experiment are the initial and final polar scattering angles. These angles are controlled independently by mounting the target and detector on separate rotating platforms sealed with spring loaded teflon seals [4]. Sample manipulators are available which allow for variation of tilt and azimuthal angle. Liquid nitrogen cooling, electron bombardment heating, and a closed loop controller allow for accurate control setting and ramping of the sample temperature.

2.2 Final State Determination

The detector shown in Fig. 3 is a quadrupole mass spectrometer (QMS) with two stages of differential pumping. The output of this detector goes to pulse counting electronics and a digital multiscaler. Time resolutions to better than 1×10^{-6} secs. are possible. By measurement of the arrival time distribution relative to opening time of the chopper (C), velocity or residence time distributions can be measured. The detector can be rotated, as described above, to allow measurements at any desired final scattering angle or to allow the measurement of angular distributions.

For state-specific detection experiments, the QMS detector is replaced by a detection scheme based on laser spectroscopy. This can be laser fluorescence (LIF) or multi-photon ionization (MPI). The MPI set up is shown in Fig. 4. Here ions are produced by tuning a laser to a resonance in the molecule under study and then detected using simple repeller plate electron multiplier (EM) combination. Since a pulsed laser is used, time of flight discrimination of the ions allows separation of various background effects. Knowing the spectroscopy of the molecule, the populations of molecules in given rotational, vibrational, and electronic state can be deduced from measurements of intensity vs. laser wavelength. Information on polarization and orientation of scattered molecules is available from measurements as a function of the direction of polarization of the laser.

2.3 Surface Characterization

In addition to control of incidence conditions and measurements of scattered particle distributions, control, cleaning and and characterization of the surface is very important in these experiments. As shown in Fig. 3 and Fig. 4, some common surface science techniques are available in the molecular beam scattering instrument. These include: a set of hemispherical grids and phosphor screen for low energy electron diffraction (LEED), an ion source for sputtering (Ar^+), a cylindrical mirror electron energy analyzer for Auger spectroscopy (AES) and a residual gas analyzer (RGA). The base pressure in the scattering chamber is $\sim 2 \times 10^{-11}$ Torr.

2.4 Determination of Sticking Probabilities

Two methods are available for the measurement of sticking probabilities: (1) uptake methods and (2) beam reflection techniques. In uptake methods we measure coverage as a function of exposure; the sticking probability is obtained by differentiation of such curves. Coverage can be measured by AES with LEED calibrations or by thermal desorption techniques. Exposure can be controlled using the beam flags F1 and F2. If the sticking probability is high, it may be difficult to control the exposure in small enough increments. This problem can be overcome by spinning the chopper, C, with a chopper that has a small

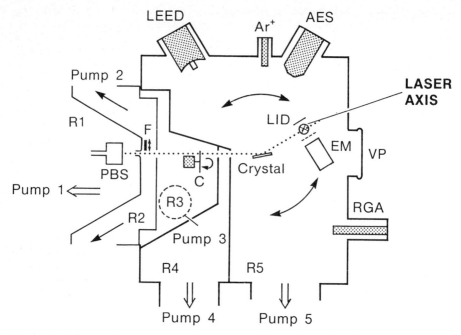

Figure 4. Molecular beam apparatus for state specific scattering distributions distributions and dissociative chemisorption probabilities. Key: LID, laser ionization detector; EM, electron multiplier; remainder as in Fig. 3.

(~ .01) fraction of open time. The uptake method is particularly well suited to measurements of small sticking probabilities since it is possible to effectively integrate by using large exposures. We have used this method for sticking probabilities in the range ~10^{-2} to 10^{-6}[5].

The beam reflection method we use for measuring the sticking probability, S, is a modification of the technique described by KING and WELLS [6]. The method is based on monitoring the fraction of the incident beam which is reflected from the surface, i.e. 1-S. The sample acts as a getter for the incident beam and the sticking probability is obtained by comparing the partial pressure increase vs time for the beam hitting the clean and saturated surface. Denoting these pressures by P_1 and P_2 respectively for constant pumping speed we have the simple relation

$$S(t) = \frac{P_2(t) - P_1(t)}{P_2(t)} . \tag{1}$$

Various small corrections necessary to take account of the vacuum time constant, changes in pumping speed, and effusive beam backgrounds are discussed by Rettner et. al [7].

3. INELASTIC SCATTERING AND TRAPPING

We give here a discussion of the inelastic scattering of atoms and molecules for systems with no (or negligible) dissociation or chemisorption channels.

These systems serve as useful models to understand the basic processes involved in trapping and inelastic scattering for later application to understanding results on more complicated systems with important dissociative chemisorption channels. A key parameter in any consideration of precursor state models is the trapping probability into the precursor state. We begin with a discussion of how molecular beam measurements can be used to gain information about trapping.

3.1 Techniques for Measuring Trapping Probabilities

By trapping we mean the process whereby a molecule incident on the surface looses kinetic energy and becomes bound in a minimum in the molecule-surface interaction potential; the molecule lives in this state for a long enough time to become equilibrated with the surface and eventually leaves the trapped state. (It can only return back to the gas phase for systems under consideration here; later we will consider systems where the molecule could leave the trapped state by dissociatively chemisorbing.) In a sense there is no real distinction between trapping and sticking except time scale. A molecule which sticks "permanently" will also eventually desorb only this desorption takes place on a time scale long compared to the experiment. Trapping involves sticking on the time scale of a fraction of a second to picoseconds. (Typical equilibration times are estimated to be 10-100 ps. for normal and parallel motion respectively. [8,9])

3.1.1 Residence Time Distributions

The most straight forward method of detecting trapping is to do an experiment on a faster time scale. With non-beam methods we are limited to time scales of a few seconds, while with chopped beams we can push this down to the microsecond regime. Fig. 5 shows the direct measurement of the residence time distribution for the system of NO on Ag(111). A beam of NO with $E_i = 0.1$ eV was incident on a Ag(111) crystal ($T_S = 100K$) at an angle $\theta_i = 60°$. These conditions were chosen to give a high trapping probability. Using a high speed rotating chopper, pulses of about 5 μs where formed and the intensity as a function of delay time from opening of the chopper was measured. The molecules were observed normal to the surface. (The measurement was actually made on a specific rotational state using LIF detection, but that is not important for the purposes of the present discussion.) As can be seen from Fig. 5, an exponential fall off of intensity is observed with a time constant of 2.1 ms. which can be identified with the residence time distribution for molecules on the surface.

While the direct measurement of residence time is straight forward in the ms or μs time regime, we do not have techniques for the direct measurement of residence time distributions on faster time scales. These cases are very important for weakly bound species (e.g. molecules in precursor states) so we need to consider other methods of detecting trapping. The basic idea of all these indirect methods is the same: we use the properties of the scattered molecules to give us a clock operating on the time scale of equilibration of molecules with the surface.

133

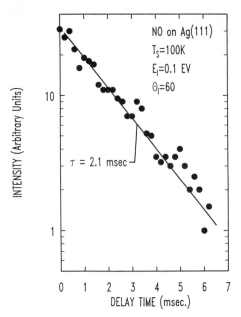

Figure 5. Residence time distributions for trapping desorption scattering of NO from Ag(111).

3.1.2 Angular Distribution Measurements

Angular distributions of scattered atoms or molecules have been widely used to determine the extent of trapping taking place. Molecules which undergo direct inelastic scattering retain a large fraction of their parallel momentum and have their normal momentum approximately reversed. Consequently, the direct inelastic scattering channel is characterized by an angular distribution peaked near the specular direction. Due to surface corrugation and energy transfer processes, the distribution may become shifted and broadened. An angular distribution of this type is illustrated in the upper panel of Fig. 6, [10] again for NO scattering from Ag(111). Here the incidence energy is high (.48 eV) and the incidence angle is 35°. If molecules trap on the surface a very different angular distribution will be observed. If the molecules live on the surface long enough to substantially equilibrate, any memory of the initial parallel momentum will be lost. Consequently, a distribution symmetric about the normal will be observed. In most case this will be a cosine or nearly cosine distribution. An angular distribution for NO undergoing trapping desorption scattering from Ag(111) is shown in the lower panel in Fig. 5 [11]. A cosine angular distribution is observed, which on a polar plot appears as a circle. Here the incidence energy was much lower (0.1 eV). Furthermore, the incidence angle was larger (60°) making the normal energy only 25 meV. Since to trap, a molecule must go into a state of negative normal energy, the normal energy of incidence is an important parameter in determining trapping probabilities.

Often, the determination of trapping from angular distributions is not as clear cut as in the examples of Fig. 6. We may observe a direct scattering lobe with a significant tail extending to normal. Is this tail to be associated with trapping or with the wings of the direct

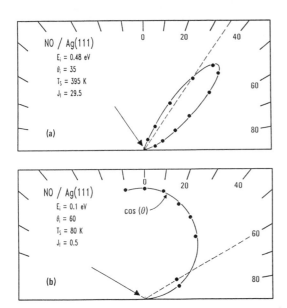

Figure 6. Angular distributions for NO scattering from Ag(111) for (a) direct inelastic scattering and (b) trapping desorption scattering.

inelastic scattering distribution ? This can be an impossible question to answer if (as is often the case) angular distribution data is only available over a limited range of in-plane angles.

3.1.3 Velocity Distribution Measurements

Velocity distribution data can also be used to measure trapping probabilities. The basic idea is similar to that which motivates the use of angular distributions; if a molecule is trapped on the surface its velocity distribution moves rapidly towards equilibration with the surface i.e. towards a Boltzmann distribution with a temperature of T_S. The velocity distribution of molecules which trap and then desorb will loose memory of the incident conditions, that is although the trapping probability and hence intensity of the trapping desorption scattering will change with the velocity of the incident particles but the form of the velocity distribution will remain unchanged. The surface temperature, on the other hand will have a strong effect on the velocity of molecules in the trapping desorption channel. We can separate the trapping desorption from direct inelastic scattering because the characteristics of the velocity distributions are very different in the two cases. For direct inelastic scattering the velocity distribution is generally a narrow distribution which is a sensitive function of the incidence conditions and a weaker function of the surface temperature.

The separation of scattering into direct inelastic and trapping desorption scattering on the basis of velocity distributions is illustrated in Fig. 7 taken from the work of Hurst et. al. [12] for Xe scattering from Pt(111). Fig 7a shows the time of flight distribution for the incident beam which had an energy of 0.14 eV and an angle of incidence, $\theta_i = 75°$. Time of flight distributions of scattered molecules at $\theta_f = 0°$, $45°$ and $75°$ (specular) are shown in

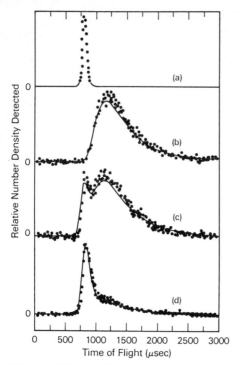

Figure 7. Time-of-flight spectra for Xe scattering from Pt(111) for $T_S=185$ K, $\theta_i=75°$, and $E_i=0.14$ eV; (a) incident Xe beam; (b) $\theta_f=0°$ (c) $\theta_f=45°$ (d) $\theta_f=75°$ (specular). From [12].

Figs. 7b, 7c, and 7d respectively. The curve measured at normal is a Boltzmann distribution at the surface temperature. Its peak position did not change if the incidence velocity was changed. The curve measured at 45° shows two peaks, one of which is the Boltzmann distribution and the other arising from direct inelastic scattering. The curve measured at specular shows a much stronger direct inelastic peak (because this channel peaks near specular) and a weak trapping desorption peak. These curves are not on the same scale; the trapping desorption peak falls off as $\cos(\theta_f)$.

For the case shown here, the separation of direct inelastic and trapping desorption scattering is very clear. However under different conditions it is possible for the velocity distributions from these two channels to overlap making a clear separation difficult. Nonetheless it has been possible to use the velocity distributions to learn about trapping even in these cases [13]. A series of measurements over a range of incidence conditions and surface temperatures is required. The basic idea is to learn about the behavior of the direct inelastic channel from measurements at high energy where trapping can be ignored. Then to use this knowledge to fit the data at low energy and extract the trapping behavior.

An important result of a number of studies [14] on the behavior of the direct inelastic scattering channel is that if both the incident and exiting kinetic energies are scaled by $2kT_S$ (the average kinetic energy of a molecule leaving the surface in equilibrium at T_S), all the data for mean final energy as a function of incident energy and surface temperature reduce to a single straight line. This correlation is represented by an equation in the form

$$< E_f > \; = \; a_1 < E_i > + \; a_2 (\; 2kT_s\;). \tag{2}$$

A correlation of the width of the velocity distribution with surface temperature is also found which is of the form

$$a^2 \; = \; b \left(\frac{2kT_S}{m} \right), \tag{3}$$

where α is related to the width of the velocity distribution by

$$I(v) \; = \; cv^3 \exp[\; - \; (v - v_0)^2 / \alpha^2]. \tag{4}$$

By using a nonlinear least squares procedure which constrains the direct inelastic scattering to (2)-(4), and fitting measured time-of-flight (TOF) curves to a form which includes a direct and Boltzmann component, the fraction trapped can be extracted from the TOF distributions.

As an illustration of the determination of trapping from velocity and angular distributions for a molecular system, consider the case of N_2 on Pt(111) as reported by Janda et al. [13]. Angular distributions for 0.09 eV N_2 incident at 45° on a Pt(111) surface at 200K are shown in Fig. 8, and corresponding time-of-flight distributions in Fig. 9. As one moves from specular to normal scattering angles, the peak of the TOF curves moves first to shorter times and then broaden and move towards longer times; at normal the TOF distribution is consistent with a Boltzmann distribution. Although the decomposition of these distributions into a direct inelastic and trapping desorption channel is not obvious to the eye, the

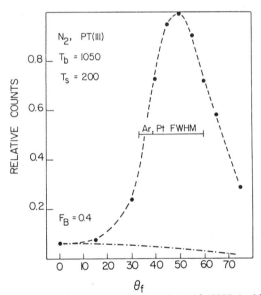

Figure 8. Angular distributions for 0.09 eV N_2 incident at 45° on a Pt(111) surface at 200K. The top line is the total intensity, while the bottom line is the intensity attributed to trapping-desorption. Also shown is the FWHM for Ar scattering from the same surface. From [13].

137

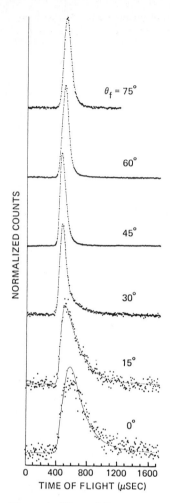

$\theta_f = 75°$

$60°$

$45°$

$30°$

$15°$

$0°$

NORMALIZED COUNTS

0   400   800   1200   1600
TIME OF FLIGHT (μSEC)

Figure 9. Time-of-flight distributions for 0.09 eV N_2 incident at $45°$ on a Pt(111) surface at 200K. From [13].

procedure outlined above gave a consistent result for the condensation coefficient of 0.4. For similar beam energy and surface temperature, the sticking coefficient of Xe would be nearly unity [12] while that for Ar would be < 0.15 [15].

3.2 Summary of Results on Inelastic Scattering and Trapping

It is worthwhile summarizing some of the major results obtained from the studies of inelastic scattering and trapping of atomic and weakly bound (non-dissociative) molecular systems which are pertinent to providing a perspective on precursor states.

1. The inelastic scattering can be divided into two channels: direct inelastic scattering and trapping desorption scattering. Velocity distributions for the former are characterized

by linear proportionality relations (2)-(4), a weak dependence on surface temperature and a strong dependence on incidence energy, while for the latter by a strong dependence on surface temperature.

2. These two channels can be experimentally separated under favorable conditions, and the trapping fraction determined.

3. This fraction is a strong function of the ratio of incidence energy to well depth, and also depends (more weakly) on the surface temperature.

4. Measurement of trapping by the velocity distribution method avoids some of the problems of interpretation of angular distributions and, for the few cases where measurements are available, gives results somewhat lower than older estimates based on angular distributions alone.

5. It is useful to have a measure of typical trapping probabilities. A rare gas system with similar van der Waals interactions to the molecular systems we consider below (O_2 and N_2) is Ar interacting with W or Pt surfaces. At room temperature and at energies characteristic of a room temperature gas, the fraction trapped is small, < 0.1-0.2. (We will return to this point later in the discussion).

4. DISSOCIATIVE CHEMISORPTION SYSTEMS

We now turn to the application of beam techniques to systems which exhibit dissociative chemisorption. We consider two systems, O_2 on W(110) and N_2 on W(100) which have been extensively studied before and concluded to exhibit adsorption mediated by precursor states.

4.1 Oxygen on W(110)

Many measurements have been reported for the sticking probability vs. coverage of O_2 on W(110),[16-23] using both effusive beam and ambient gas exposure. Data for this system is available over an unusually wide range of gas and surface temperatures. The general finding for gas temperatures around 300K is that the sticking probability, S, remains essentially constant until a critical coverage of about 0.35 ML, after which S decreases rapidly, falling to zero at a coverage of 0.5 ML. There is considerable variation in the value of the initial sticking probability which ranges from ~0.28 to 0.50 [16-23]. The sticking probability vs. coverage has been interpreted in terms of a precursor state model with a trapping probability into the precursor state equal to the initial sticking probability and a probability of unity for transmission from the precursor to the chemisorbed state.

Our initial interest in this system was sparked by Brundle [24] who pointed out a discrepancy in the literature between the interpretation just given for O_2 sticking probabilities on W(110) and the results for O_2 sticking probabilities on some stepped W surfaces with (110) terraces. On these stepped surfaces, a sticking probability of 1 is observed. Since one would not expect the trapping probability to vary very much between a flat W(110) surface and a stepped surface with (110) terraces, and since the sticking probability in a precursor model clearly can not exceed the trapping probability, the observation of unity sticking on the stepped surfaces clearly presents a problem for the model.

One possible explanation for this difficulty would be to account for a sticking probability of ~0.3 on the flat surface as arising from a trapping probability of ~1 and a probability of transmission from precursor to chemisorbed state of ~0.3. This hypothesis makes a prediction which is easily checked with molecular beam scattering measurements: all of the molecules which return to the gas phase would do so through a trapping desorption mechanism, which would be easily detectable from angular or velocity distribution measurements as outlined in section 3.

To test this hypothesis Rettner et al. [7] measured both angular and velocity distributions for O_2 scattering from W(110) for a wide range of incidence energies, angles and surface temperatures. All of the results are consistent with a direct inelastic scattering mechanism rather than the dominance of trapping desorption. An example of such data is given in Fig. 10 which shows the time-of-flight distribution of O_2 for $E_i=0.45$ eV, $\theta_i=45°$, and $\theta_f=55°$ The solid line shows the measured distribution which can be compared to the results of elastic scattering shown as a long dashed curve or trapping desorption shown by the short dashed curve. Clearly the measured results are quite different from either of the two comparison curves and no detectable trapping component is observed.

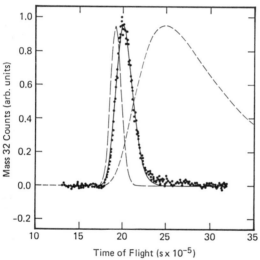

Figure 10. Time-of-flight distribution of oxygen molecules scattering from W(110) at an in-plane scattering angle of 55° for an incidence energy of 0.446 eV, an incidence angle of 45° and a surface temperature of 800 K. The surface coverage of atomic oxygen was kept below 0.08 ML. The dashed curve indicates the result expected for perfect elastic scattering and is the time-of-arrival distribution recorded for the beam when it is fired directly into the detector. The dotted curve displays the distribution expected for molecules which are accommodated to the surface temperature and desorb after a negligible delay with a Boltzmann velocity distribution at the surface temperature.

To further explore the mechanism for dissociative chemisorption in this system, the sticking probability was measured [7] as a function of the energy and angle of incidence. The results are shown in Fig. 11 for a surface temperature of 800K. Surprisingly, the initial sticking coefficient is found to rise quite rapidly with increasing translational energy at various incidence angles. The results scale with the normal "component" of the energy, $E_n = E_i \cos^2 \theta_i$. The increase of sticking coefficient with energy is opposite to the behavior

140

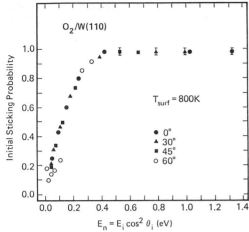

Figure 11. Initial dissociative chemisorption probability of O_2 on W(110) at a surface temperature of 800 K as a function of the kinetic energy of the incident molecules for angles of incidence between $0°$ and $60°$

expected for a classical precursor state model. As discussed in section 3, in order to trap, a molecule must lose excess energy into phonons and internal molecular motions. Trapping becomes less probable as the collision energy is increased. Careful examination of Fig. 11 shows that for $\theta_i = 60°$, there is a breakdown of strict normal energy scaling, and, more importantly, a minimum in the sticking probability at $E_n \sim 25$ meV in this case. It is possible that the increase below 25 meV represents a transition to precursor kinetics.

Although this data is in contradiction to previous interpretations of the mechanism for dissociative chemisorption for O_2 on W(110), it is important to point out that there is no direct contradiction of any previous measurements. By integrating the incidence energy distribution of a static gas or effusive beam over Fig. 11, Rettner et. al [7] were able to satisfactorily reproduce previous data if it was assumed that the increase observed below 25 meV continues as the energy is lowered. The mechanism for sticking apparently undergoes a transition from precursor-mediated to direct behavior at around this energy. In fact both channels contribute simultaneously to varying degrees as the energy is increased. For a Boltzmann distribution around 300K both channels are important.

The behavior of the sticking vs. coverage curves for this system is also very interesting. For a thermal energy beam or ambient gas, many authors observe a nearly constant sticking probability vs. coverage at low coverage. For a monoenergetic beam the results are dramatically different, and again reveal that the thermal results average over quite different behavior. Curves of sticking probability vs coverage for several incidence energies and angles [7] are shown in Fig. 12. In Fig. 12b we see the results for low energy. Here the sticking probability is seen to actually increase with coverage, reach a maximum and then decrease. In contrast, for 0.12 eV (Fig 12a lower curve) a roughly linear fall off of sticking probability with coverage is observed. At still higher energies, remarkably, the curves again start to resemble a "precursor" form. A hint of this is seen in Fig. 12a, and the trend continues as the energy is further increased. Certainly, a classical (weakly bound physisorption) precursor can not be expected to play a role at energies above 0.1 eV. This dramatically illustrates the danger in drawing conclusions from the coverage dependence of sticking probabilities alone.

141

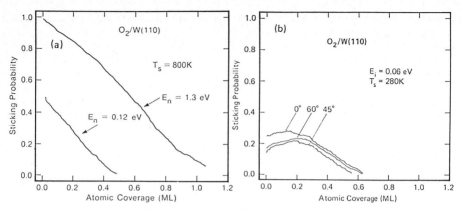

Figure 12. Sticking probability for O_2 on W(110) as a function of coverage for (a) E_n=0.12 eV and 1.3 eV and (b) E_i=0.0 6eV. From [7].

4.2 Nitrogen on W(100)

The dissociative adsorption of N_2 on W(100) has been widely studied, and has been used as a model system for precursor-mediated adsorption.[25-28] The results of sticking probability vs coverage measurements are accurately fit by precursor state kinetic models. We plan an extensive study of this system to carefully explore the assumptions of existing models and to provide a sound basis of microscopic understanding. A preliminary result of this study is shown in Fig. 13. [29] which shows the energy dependence of the initial sticking

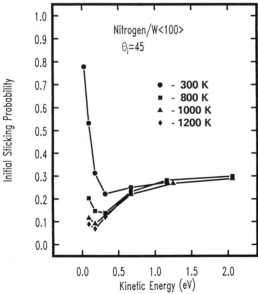

Figure 13. Initial dissociative adsorption probability for N_2 incident on W(100) as a function of the energy of incidence.

142

probability for energies up to 2.0 eV and surface temperatures from 300-1200K. For low surface temperature the sticking probability is seen to fall off dramatically with energy for energies up to about 0.4 eV (0.2 eV normal energy). This is exactly the behavior expected for precursor state models. These results are consistent with those of King and Wells [26] who varied the temperature of their effusive beam over a limited range. At higher energies, the initial sticking probability is seen to level off and increase slightly. For the higher temperature data, where the initial sticking probability is lower at low energy, the curves show an increase and merge at high energy with the low temperature curves. This behavior can again be understood in terms of a transition from precursor mediated to direct adsorption. Thus the paradigms of precursor mediated and direct (activated) adsorption are both manifest for this system (as for oxygen/W(110)) as the incidence energy is varied.

5. PRECURSORS: MYTH OR REALITY ?

Webster [30] defines myth as

> myth *n* 1: A story that is usually of unknown origin and at least partially tradi-
> tional, that ostensibly relates historical events usually of such character as to serve
> to explain some practice, belief, institution or natural phenomenon, and that is es-
> pecially associated with religious rites and beliefs.
> 2: A story invented as a veiled explanation of a truth.

In the sense of definition (2) at least, the popular view of precursor states is properly called a myth. While it is certainly true that precursor states play the role of dynamical interme-diates under some conditions, the situation is far more complicated than existing models allow. Many of the detailed assumptions of such models are contrary to what has been learned about energy accommodation and trapping in recent years using molecular beam techniques. As we have seen dramatically in the case of O_2 on W(110), even well studied systems which are generally accepted as proceeding by a precursor state mechanism, can show a mixture of behavior over the range of incidence energies and angles probed by a room temperature gas.

Molecular beam studies of the energy and angular dependence of sticking probabilities are already revealing a great deal about the dynamics of the dissociative chemisorption for a number of systems. However, even such detailed measurements can only probe the re-action dynamics at a rather superficial level and cannot be used to unambiguously determine the exact sequence of events leading to dissociation. Such shortcomings are often overcome by recourse to models that seem to permit gross extrapolation of available data to fully de-scribe the process at hand. We believe that such models should be introduced cautiously and their predictive capability should be carefully appraised. Clearly many more detailed experiments are required to provide a firm physical basis for understanding the dissociative chemisorption process.

6. ACKNOWLEDGMENTS

It is a pleasure to thank the many co-workers who have contributed to the results reported in this review. These include colleagues at the University of Chicago, notably C. A. Becker, J. P. Cowin, J. Hurst, K. C. Janda and L. Wharton, and those at IBM, notably A. W. Kleyn, A. C. Luntz, J. E. Schlaegel, J. A. Barker, C. R. Brundle, H. E. Pfnür, L. A. DeLouise, J. Kimman and H. Stein. We also thank co-workers R. J. Madix and J. Lee and J. C. Tully.

7. REFERENCES

1. I. Langmuir, *Chem. Rev.*, 6, 451 (1929).
2. J. E. Lennard-Jones, *Trans. Faraday Soc.* 28, 333 (1932).
3. J.K. Norskov, H. Houmoller, P.K. Johansson and B.I. Lundqvist, *Phys. Rev. Lettr.* 46, 257 (1981).
4. D.J. Auerbach, C.A. Becker, J.P. Cowin, and L. Wharton, *Rev. Sci. Instr.* 49, 1518 (1978).
5. C. T. Rettner, H. E. Pfnür and D. J. Auerbach, *Phys. Rev. Lett.* 54, 2716 (1985).
6. D. A. King and M. G. Wells, *Surface Sci.* 29, 454 (1971).
7. C.T. Rettner, L.A. DeLouise and D.J. Auerbach, *J. Chem. Phys.*, 85, 1131 (1986).
8. J.A. Barker and D.J. Auerbach, *Faraday Disc. Chem. Soc.* 80, 277 (1985).
9. J.C. Tully, *Faraday Disc. Chem. Soc.* 80, 291 (1985).

10. C.T. Rettner, J. Kimman and D.J. Auerbach, to be published.
11. A.W. Kleyn, A.C. Luntz and D.J. Auerbach, to be published.
12. J.E. Hurst, C.A. Becker, J.P. Cowin, K.C. Janda, L. Wharton, and D.J. Auerbach, *Phys. Rev. Lettr.* 43, 1175 (1979).
13. K.C. Janda, J.E. Hurst, J.P. Cowin, L. Wharton, and D.J. Auerbach, *Surface Science* 130, 395 (1980).
14. J.A. Barker and D.J. Auerbach, *Surface Science Reports* 4, 1 (1984).
15. J.E. Hurst, L. Wharton, K.C. Janda and D.J. Auerbach, *J. Chem. Phys.* 83, 1376 (1985).
16. C. Wang and R. Gomer, *Surface Sci.* 84, 329 (1979).
17. C. Kohrt and R. Gomer, *J. Chem. Phys.*, 52, 3283 (1969).
18. Besocke and S. Berger, *Proc. 7th Intern. Vacuum Congr. and 3rd Intern. Conf. on Solid Surfaces*, Vol. II Vienna, (1977) p. 893.
19. Butz and H. Wager, *Surface Sci.* 63, 448 (1977).
20. T. Engel, H. Niehus and E. Bauer, *Surface Sci.* 52, 237 (1975).
21. K. J. Rawlings, *Surface Sci.* 99, 507 (1980).
22. T. E. Madey, *Surface Sci.* 94, 483 (1980).
23. M. Browker and D. A. King *Surface Sci.* 94, 564 (1980).
24. C.R. Brundle, private communication (1985).
25. R. Clavenna and L.D. Schmidt, *Surface Science* 22, 365 (1970).
26. D.A. King and M.G. Wells, *Proc. Roy. Soc. (London)* A339, 245 (1974).
27. S.P. Singh-Boparai, M. Bowker and D.A. King, *Surface Science* 53, 55 (1975).
28. P. Alnot and D.A. King, *Surface Science* 126, 359 (1983).
29. C.T. Rettner, H. Stein and D.J. Auerbach, to be published.
30. Webster's Third International Dictionary, unabridged, D. & C. Merriam Co., Springfield Massachusetts (1961).

Molecular Beam Studies
of Dissociative Chemisorption on W(110)

C.T. Rettner and D.J. Auerbach

IBM Almaden Research Center, K33/801, 650 Harry Road,
San Jose, CA 95120, USA

We have employed molecular beam techniques to study the the dissociative chemisorption of a number of molecules on a W(110) surface. Chemisorption probabilities have been measured as a function of incidence angle, θ_i, and kinetic energy, E_i, and of surface coverage and temperature. The initial (zero coverage limit) sticking probabilities of CH_4, N_2 and O_2 on W(110) are all found to depend strongly on the incidence energy, scaling with $E_n = E_i \cos^2\theta_i$ in the case of CH_4 and O_2, but with the *total* energy, E_i in the N_2 case. This probability increases by factors of $\sim 10^5$, 10^2 and 10 for an increase of 1 eV in incidence energy for CH_4, N_2 and O_2, respectively. Similar behavior is observed at higher surface coverages, so that even the apparent saturation coverage is found to be energy dependent. In experiments with HD, we have examined the relationship between rotationally mediated selective adsorption, physisorption and dissociative chemisorption on this surface. We find that selective adsorption does not lead to chemisorption, but contributes to the diffuse scattering channel.

1. INTRODUCTION

The molecular dynamics of dissociative chemisorption remain poorly understood. This is partly because of the complex nature of condensed phase chemistry, and partly because there have been only relatively few dynamical studies of such reactions. Under these conditions it is useful to probe the form of even coarse features of the potential hypersurface of the system. Most previous work[1,2] has been discussed in terms of two mechanisms. In a number of cases there appears to be good evidence that dissociative chemisorption proceeds *via* a mobile-precursor mechanism,[3-9] while in others systems a direct mechanism seems more appropriate.[2,9-11] In the precursor model, the molecule loses sufficient incident translational energy upon collision to become trapped, or physisorbed, in a "precursor-state", becoming accommodated to the surface temperature. The molecule remains in this state until it reorients into a favorable configuration or diffuses to a favorable site for dissociation. Alternatively the molecule may desorb from the surface or remain in a molecular chemisorbed state. In the case of a direct mechanism, dissociation is pictured to occur upon essentially a single impact with the surface, provided the molecule possesses sufficient energy to surmount any activation barriers that may exist. We have employed molecular beam techniques to study various aspects of both such mechanisms for the chemisorption of N_2, O_2, CH_4 and HD on a W(110) surface. We report here a summary of these efforts.

A number of surprising results have been obtained. For example, in the case of N_2 chemisorption, we find[12,13] that translational energy is very effective in overcoming the barrier to dissociation, but that results at different angles of incidence scale approximately with the total incidence kinetic energy, rather than with the so-called "normal kinetic energy" predicted by a one-dimensional barrier model. In the case of O_2, we do observe normal energy scaling,[14] the surprise being that increasing energy causes an increase in the sticking probability. This system was previously believed to proceed *via* a precursor mechanism, for which one would expect the dissociation probability to decrease with increasing incidence energy. In fact the precursor channel seems to apply only to very low incidence energies. The chemisorption of methane on this surface is found to be dramatically activated by increasing translational energy,[15] and we find that our results for both CH_4 and CD_4 are consistent with a mechanism in which the hydrogen tunnels through the barrier to dissociation. We also find in this system that vibrational and translational energies are roughly equivalent in their ability to promote reaction.[16] Finally, in the case of HD chemisorption, we find[17] that accessing the physisorption potential well *via* rotationally mediated selective adsorption does not increase the probability of dissociative chemisorption.

2. EXPERIMENTAL

Figure 1 displays a schematic of the apparatus which has been described previously, as have our experimental techniques and procedures.[12-17] Supersonic beams are directed at a

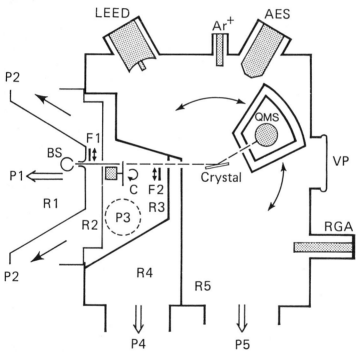

Figure 1. Molecular beam apparatus for determining scattering distributions and dissociative chemisorption probabilities. Key: R1-R5, vacuum regions; PBS, pulsed beam source; C, high-speed chopper; F, beam flag; QMS, quadrupole mass-spectrometer detector; LEED, low-energy electron diffraction screen; Ar+, argon ion sputtering gun; AES, Auger electron spectrometer; VP, 6 in. viewport; RGA, residual gas analyzer. The molecular beam system can detach at the R3 region and is non-bakable. The crystal is held on a manipulator which provides for heating and cooling and rotates about the same axis as the QMS.

crystal mounted in a UHV chamber on a manipulator which permits accurate control of the incidence angle and provides for accurate temperature control. Beam energies are varied by changing the nozzle temperature and seeding conditions. These energies are determined from flight times from a high-speed chopper to a differentially pumped quadruple mass spectrometer. Typical energy spreads of ~15% of the incidence energy (FWHM) are obtained with a reservoir pressure of 30 to 50 psi behind a 100 μm diameter nozzle.

Initial sticking probabilities, S_0, (corresponding to dissociation on the clean surface) are determined from the initial slopes of coverage *versus* exposure curves. Exposures are varied by controlling the dose time of a beam whose flux can be determined from its area and the partial pressure rise produced on the system. This method was cross-checked for high-energy measurements (with S_0 values $> 5 \times 10^{-2}$) by using a direct method involving measurement of the fraction of the beam reflected by the surface (following the approach of King and Wells[5]). In addition, Auger electron spectroscopy and LEED are also employed to determine surface coverages directly.

3. RESULTS AND DISCUSSION

3.1 Overview

We find that increasing translational energy generally causes a substantial increase in the dissociative chemisorption probability on this surface. This behavior is illustrated in Figure 2 which displays the variation of the dissociative chemisorption probability with incidence energy for N_2, O_2 and CH_4 on a clean W<110> surface at normal incidence and a surface temperature of 800 K. The dissociative chemisorption probabilities are seen to approach unity with increasing translational energy. Table I summarizes the observed sticking probabilities for these molecules for energies of 0.1 and 1.0 eV. It is clear that while great differences are observed in the sticking probabilities for energies of ~0.1 eV, the behavior above ~1.0 eV is much more uniform.

Table I.
Dissociative Chemisorption on W<110> at High and Low Incidence Kinetic Energies

Molecule	Energy (eV)	S_0	Θ_{sat}	Comments
N_2	0.09	3.0×10^{-3}	0.10	β state populated
	1.08	2.0×10^{-1}	0.45	β_1, β_2 and β_3 states populated.
O_2	0.06	1.0×10^{-1}	0.50	Possible involvement of precursor
	1.06	9.8×10^{-1}	1.00	No contribution by precursor
CH_4	0.10	1.0×10^{-5}		
	1.10	1.5×10^{-1}		Tunneling mechanism
CD_4	0.10	$< 10^{-7}$		
	1.08	8.0×10^{-2}		

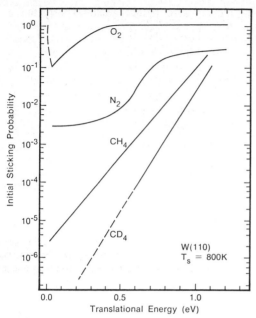

Figure 2. Effect of incidence energy on the dissociative chemisorption probability of O_2, N_2, CH_4 and CD_4 for a surface temperature of 800 K and at normal incidence. The dashed lines indicate theoretical curves (see text).

3.2 Dynamics of activation

The above behavior is consistent with the presence of a potential barrier to dissociation, one that can be overcome by the kinetic energy of the incident molecules. In this picture, the phenomenological barrier height distributions can be obtained by differentiating the curves of Fig. 2 for each system.[9,14] It will be apparent from Fig. 2 and Table I that the form of this distribution will vary greatly for different molecules. It is also worth noting that the height of any barrier for a direct single-collision process may be very different than the minimum activation energy for a system which is free to fully explore the available phasespace. Without knowing more about the nature of such barriers, we can say that the sensitivity to incidence energy implies that they must be encountered before the molecule becomes fully accommodated to the surface.

The nature of such barriers is central to the understanding of activated chemisorption. Most previous discussions of such systems have considered the barrier to be essentially one dimensional,[9,18-20] as first suggested by Lennard-Jones more than 50 years ago.[18] Here the potential energy of the system depends only on the distance of the molecule from the surface plane. In such a model the probability of overcoming the barrier to dissociation depends on the motion of the molecule parallel to the surface normal, and data recorded as a function of incidence energy, E_i, and angle, θ_i, would be expected to scale with the so-called "normal kinetic energy" $E_n = E_i \cos^2 \theta_i$.

Such a one-dimensional picture is clearly a gross over-simplification, and cannot be expected to accurately account for the dynamics of the true multidimensional

148

hyper-surface. Considering the general properties to be expected for such a potential surface, we may propose two possible (extreme) situations. In the simplest view, we might propose that the barrier lies in the approach coordinate, that is to say it lies in the *entrance channel* of the system. Here we might claim that dissociation is an essentially *direct* process, although the post-barrier dynamics would remain unknown. This could arise for example if the molecule passes into new electronic states close to the surface, where access to these states requires overcoming a potential barrier before attractive forces dominate. The negative ion intermediate of Gadzuk and Holloway[21] would be a candidate for such a state, as would states associated with molecular chemisorption. The requirement here would be that a significant proportion of molecules in such states should dissociate rather than desorb. The potential barrier would actually help here, by preventing desorption. Such an entrance-channel barrier would be efficiently overcome by initial translational energy, allowing barrier heights to be estimated from sticking vs. energy data. In the absence of significant surface corrugation, this picture would lead to normal energy scaling.

Alternatively, the barrier could lie predominantly in the molecular stretch coordinate, or in the *exit channel* of the system, preventing the atoms from separating. In this case it is necessary for the translational energy of the incident molecules to become channelled into this new coordinate. Since translation-to-vibrational energy transfer is unlikely to be highly efficient, this channeling will most likely require multiple encounters with the surface, as would occur if the molecules become temporarily trapped. This state would then act as a "dynamic" or "hot" precursor. Evidence for such a state has recently been found for the CO/Ni(100) system.[22] In this case the barrier height may be only a small fraction of that estimated from sticking vs. energy curves. It is unlikely that results would scale with the normal kinetic energy in this case, since such multiple encounters would most likely cause the momentum of the system in the directions parallel and perpendicular to the surface to become mixed. This might then lead to scaling with the *total* energy of the incident molecules.

Previous molecular beam studies of hydrogen dissociation on Cu(110), Cu(100) and Cu(310) surfaces found good agreement with the one-dimensional barrier model,[9] insofar as results were found to follow normal-energy scaling. Other aspects of the hydrogen/Cu system cannot be accounted for by such a picture, however. For example (in post-permeation studies) Comsa and David[23] find that the velocities of hydrogen molecules desorbing from Cu(100) and Cu(111) surfaces are constant for desorption angles out to 60°, and see no evidence for the increase in velocity with increasing angle predicted by the one-dimensional picture. Also Kubiak *et al.*[24] observe substantial vibrational excitation in hydrogen desorbing from Cu(110) and Cu (111), which cannot be accounted for by such a simple model. In the case of dissociation on W(110), we find the same kind of mixed behavior. While we find excellent normal energy scaling in the case of O_2 and CH_4 dissociation, this model is completely unable to account for our observations on the $N_2/W(110)$ system.

3.3 Nitrogen: a case of total energy scaling

In the case of dissociative chemisorption of nitrogen on this surface, we find that the initial sticking probability is almost independent of incidence angle. We believe that such a breakdown in the one-dimensional barrier model implies the presence of a strong chemical interaction prior to the activation barrier. For a barrier in the exit channel of the system, Gadzuk and Holloway have suggested that the molecule may become temporarily trapped as an intermediate N_2^- negative ion, [21] as discussed above. Alternatively, the chemical interaction may simply serve to produce a high degree of effective surface corrugation prior to a barrier in the entrance channel. A corrugated surface would indeed couple efficiently to motion parallel to the surface.

3.4 Methane: quantum tunneling and effect of vibrational energy

Again returning to Fig. 2, it will be apparent that the dissociation of methane on this surface is an extremely strong function of incidence energy, increasing exponentially up to about 1.0 eV. This behavior is consistent with a model in which a hydrogen atom tunnels through the activation barrier.[8] In fact the curves in Fig. 2 were actually obtained from a fit to the data points assuming such a model, where the dashed portion of the CD_4 curve has been extrapolated beyond the data points. Such a quantum tunneling model is also consistent with the pronounced isotope effect observed here, whereby the dissociation probabilities for CH_4 and CD_4 differ greatly at low energies but converge as the incidence kinetic energy is increased.[16] In the tunneling model, the curves for CH_4 and CD_4 converge as the incident energy exceeds the barrier height. This model can account for other reports[25,26] of strong isotope effects in methane chemisorption and our observations greatly strengthen the original proposal of Winters[25] that quantum tunneling may occur in methane chemisorption.

In separate experiments on the $CH_4/W(110)$ system, we have sought to probe the relative efficacy of translational and vibrational energies in promoting dissociation.[16] Here we have heated the nozzle beam source to produce controlled levels of vibrational excitation while adjusting the translational energy by varying the seeding ratio of CH_4 to a hydrogen carrier gas. We find that vibrational excitation enhances the initial chemisorption probability for this system but that, at least on average, this enhancement is not significantly larger than for an equivalent amount of energy placed into translation. Such behavior could arise if the system were to pass through a "hot precursor", as discussed above. In this case the intermediate would serve to mix translational and vibrational energy, however, this mixing would have to be relatively fast compared to the lifetime of the intermediate state and compared to the rate of mixing of parallel and perpendicular momenta.

3.5 Oxygen: precursor-mediated and "direct" dissociation

The increase in sticking probability with increasing E_n observed in the case of oxygen is somewhat surprising, considering previous measurements on this system which indicated that dissociation occurs by way of a mobile-precursor mechanism.[6] Trapping into such a precursor state would be expected to become increasingly inefficient with increasing translational energy: as it becomes increasingly difficult to lose sufficient energy for trapping to occur. The assignment of a precursor mechanism to this system is based on the observation that the sticking probability of oxygen gas at 300 K on this surface at ≤ 300 K remains roughly constant as the surface coverage increases up to about 0.2 atomic ML. This behavior is believed to arise because molecules striking a filled site can trap into a precursor state and diffuse to an empty site. However, we find that for energies of >0.1 eV the sticking probability falls linearly, or nearly so, with increasing coverage. Thus we believe that dissociative chemisorption can proceed *via* two different mechanisms in this system; the precursor mechanism dominating at energies of ≤ 0.03 eV, with direct dissociation at higher energies. It is also necessary to assume that the sticking probability is high for incidence energies close to zero in order to account for the sticking probability of about 0.5 to 0.3 measured[6,27] for oxygen gas on this surface at 300 K. The dashed part of the O_2 curve in Fig. 2 is attributed to a precursor channel and has been constructed to fit the known facts, as described elsewhere.[14]

3.6 Effect of E_i on saturation coverage

In the case of O_2 and N_2 we have systematically studied the effect of E_n on the saturation coverage of O and N on this surface. We find that in both cases increasing translational energy increases this coverage considerably. In the N_2 case, the saturation coverage increases from ~0.25 atomic ML at 0.1 eV to 0.5 at ~1.0 eV.[13,27] Here the increased coverage at high energies is accompanied by the observation of new "states" in temperature programmed desorption spectra.[28] In the O_2 case, the saturation coverage increases from 0.5 ML at $E_i<0.25$ eV to 1.0 ML at $E_i>0.8$ eV. Such findings can be understood if we propose that the effect of increased surface coverage is simply to increase the average activation barrier for dissociation. The saturation coverage is then reached when the entire distribution is shifted above the available energy. The sticking vs. coverage curves then simply reflect the ability of an atom adsorbed on one site to increase the activation barrier at adjacent sites.

3.7 Selective adsorption and chemisorption of HD

In studies of HD sticking and scattering at this same surface we have sought to answer the question "can rotationally mediated selective adsorption lead to dissociative chemisorption?". In rotationally mediated selective adsorption (RMSA) a molecule makes a rotationally inelastic collision with the surface such that a fraction of its kinetic energy is converted to rotation, allowing a "vertical" descent into a selective adsorption resonance. Such resonances can be witnessed by the concomitant loss (or occasionally gain) in scattering from diffractive channels. The resonant molecules are temporarily bound into discrete vibrational (or librational) states appropriate to the laterally averaged molecule-surface potential. Since the dissociative chemisorption probability on this surface was known to be of the order of 0.1, we postulated that molecules trapped in RMSA resonances might go on to chemisorb with a high probability. For example, in the simplest picture, resonant molecules make multiple "hops" on the surface, and if at each hop the probability of dissociative chemisorption is p, then for small p the net dissociation probability will be greatly enhanced. By measuring S_0 as a function of E_n (by varying θ_i) across RMSA resonances, we have been able to show that there is no correlation between chemisorption and selective adsorption resonances in this system. Nor do these molecules contribute significantly to incoherent zero-phonon scattering, insofar as they do not appear in the wings of the coherent scattering peaks. We believe that molecules which undergo selective adsorption may trap into the physisorption well before leaving the surface, but that less than 1 in 15 go on to dissociate. It is possible that this behavior is due to the presence of an energetic barrier between the physisorbed and chemisorbed states. However we think it is more likely that this result is due to differences in the pre-exponential or A-factors for dissociation compared to desorption. Alnot and King[29] find that for N_2 on W(100) the ratio of these factors is about 40 to 1 in favor of desorption. Thus even in the absence of energetic barriers, dissociation may be unfavorable simply because the volume of phasespace leading to dissociation is small compared to that for desorption.

4. Acknowledgements

It is a pleasure to thank the many co-workers who have assisted us at different stages of this research program. These include J. E. Schlaegel, J. A. Barker, R. J. Madix, J. Lee, H. E. Pfnür, L. A. DeLouise, J. P. Cowin, and H. Stein. We also thank P. Alnot for valuable contributions to these experiments, and helpful discussions with H. F. Winters, J. W. Gadzuk and P. S. Bagus are gratefully acknowledged.

1. M. P. D'Evelyn and R. J. Madix, Surface Science Reports, 3, 413 (1984).
2. J. A. Barker and D. J. Auerbach, Surf. Sci. Reports, 4, 1 (1984).
3. G. Ehrlich, J. Phys. Chem. Solids, 1, 1 (1956).
4. I. Langmuir, Chem. Rev., 6, 451 (1929).
5. D. A. King and M. G. Wells, Proc. R. Soc. Lond. A., 339 245 (1974).
6. C. Wang and R. Gomer, Surface Science, 84, 329 (1979).
7. D. Menzel, H. E. Pfnür and P. Feulner, Surf. Science, 126, 374 (1983).
8. E. S. Hood, B. H. Toby and W. H. Weinberg, Phys. Rev. Lett., 55, 2437 (1985).
9. M. Balooch, M. J. Cardillo, D. R. Miller and R. E. Stickney, Surface Science, 46, 358 (1974).
10. A. Gleb and M. Cardillo, Surface Science, 59, 128 (1976).
11. K. Christmann, O. Schober, G. Ertl and M. Neumann, J. Chem. Phys., 60, 4528 (1974).
12. D. J. Auerbach, H. E. Pfnür, C. T. Rettner, J. E. Schlaegel, J. Lee and R. J. Madix, J. Chem. Phys., 81, 2515 (1984).
13. H. E. Pfnür, C. T. Rettner, D. J. Auerbach, R. J. Madix and J. Lee, J. Chem. Phys., in press.
14. C. T. Rettner, L. A. DeLousie and D. J. Auerbach, J. Vac. Sci. Tech., 4, 1491 (1986); C. T. Rettner, L. A. DeLousie D. J. Auerbach, J. Chem. Phys., 85, 1131 (1986).
15. C. T. Rettner, H. E. Pfnür and D. J. Auerbach, Phys. Rev. Lett., 54, 2716 (1985).
16. C. T. Rettner, H. E. Pfnür and D. J. Auerbach, J. Chem. Phys., 84, 4163 (1986).
17. C. T. Rettner, L. A. DeLousie, J. P. Cowin and D. J. Auerbach, Chem. Phys. Lett., 118, 355 (1985); C. T. Rettner, L. A. DeLousie, J. P. Cowin and D. J. Auerbach, Faraday Discuss. Chem. Soc., 80, 127 (1985).
18. J. E. Lennard-Jones, Trans. Faraday Soc., 28, 333 (1932).
19. W. van Willigen, Phys. Lett., 28A, 80 (1968).
20. M. Balooch and R. E. Stickney, Surface Science, 44, 310 (1974).
21. J. W. Gadzuk and S. Holloway, Chem. Phys. Lett., 114, 314 (1985); S. Holloway and J. W. Gadzuk, J. Chem. Phys., 82, 5203 (1985).
22. M. P. D'Evelyn, H. P. Steinrück, A. Winkler and R. J. Madix, Surface Science, to be published.
23. G. Comsa and R. David, Surface Science, 117, 77 (1982).
24. G. D. Kubiak, G. O. Sitz and R. N. Zare, J. Chem. Phys., 81, 6397 (1984); J. Chem. Phys., 83, 2538 (1985)
25. H. F. Winters, J. Chem. Phys., 62, 2454 (1975); J. Chem. Phys., 64, 3495 (1976).
26. C. N. Stewart and G. Ehrlich, J. Chem. Phys., 62, 4672 (1975).
27. K. Besocke and S. Berger, Proc. 7th Intern. Vacuum Congr. and 3rd Intern. Conf. on Solid Surfaces, Vol. II Vienna, (1977) p. 893; R. Butz and H. Wager, Surface Science, 63, 448 (1977); T. Engel, H. Niehus and E. Bauer, Surface Science, 52, 237 (1975); K. J. Rawlings, Surface Science, 99, 507 (1980); T. E. Madey, Surface Science, 94, 483 (1980); M. Browker and D. A. King, Surface Science, 94, 564 (1980).
28. J. Lee, R. J. Madix, J. E. Schlaegel and D. J. Auerbach, Surface Science, 143, 626 (1984).
29. P. Alnot and D. A. King, Surface Science, 126, 359 (1983).

The Role of Precursors in Molecular Chemisorption and Thermal Desorption

E.S. Hood[1], B.H. Toby[2], W. Tsai[2], and W.H. Weinberg[2]

[1]Department of Chemistry, Montana State University,
  Bozeman, MT 59715, USA
[2]Division of Chemistry and Chemical Engineering,
  California Institute of Technology, Pasadena, CA 91125, USA

The concept of a "mobile precursor" to molecular chemisorption and thermal desorption is not new /1/; the theoretical foundation was established in 1932 with the discovery of physical adsorption. Seminal calculations performed by Lennard-Jones /2/ revealed that a molecule incident on a metal surface passes through a shallow potential energy minimum resulting from an attractive Van der Waals interaction. Indirect experimental evidence for the existence of a mobile precursor or intermediate state in chemisorption was presented in 1933. The adsorption or sticking coefficient for cesium atoms on a tungsten substrate was determined to remain close to unity even near saturation coverage. Taylor and Langmuir /1/ invoked a mobile intermediate state to interpret this functional dependence of the adsorption coefficient on adsorbate coverage.

Much recent experimental /3-10/ and theoretical /11-18/ effort has been directed toward understanding the phenomena of precursor-mediated dissociative and molecular chemisorption. Since 1933 experimental measurements of the coverge-dependent adsorption coefficients for the following systems of diatomic molecules on metal surfaces have provided additional indirect evidence for precursor-mediated chemisorption: carbon monoxide on Ru(001) /3/ and Ni(111) /4,5/ and Ni(100) /6/; and nitrogen on polycrystalline tungsten /7/, /W(100) /8/, and Ni(110) /9/. Most recently, the spectroscopic techniques of XPS /10/ and HREELS /6/ have been employed in an attempt to verify directly the existence of molecular precursor states. The spectroscopic search for an intrinsic precursor to the chemisorption of nitrogen on Ni(100), Re(001) and W(100) /10/ and carbon monoxide on Ni(111) /5/ has thus far proven unsuccessful.

The influence of mobile precursor states on the kinetics of molecular chemisorption and thermal desorption has been examined theoretically by both a reaction kinetics approach /11-13/ and statistical modeling /14-17/. The chemical reaction kinetics approach has been employed to analyze the influence of intermediate states on chemisorption kinetics. Complex rate expressions have been derived utilizing the stationary state approximation. Kinetic expressions for molecular chemisorption via a mobile precursor state have also been derived by Kisliuk /14/ using a successive site statistical model. Lateral interactions between the precursor and chemisorbed particles have been ignored. A random distribution of filled sites has been assumed at all adsorbate coverages; thus precluding examination of ordering phenomena such as island formation. An adaptation of this successive site statistical model has been utilized by King /15-16/ to examine the kinetics of precursor state mediated thermal desorption. Both theoretical methods have achieved some success in reproducing experimentally measured adsorption coefficients and thermal desorption rates; but neither approach is capable of probing the interrelationship between adsorbate lattice structure and the kinetics of adsorption and desorption.

We now employ a novel theoretical modeling scheme for studying precursor-mediated molecular chemisorption and thermal desorption. Our stochastic formulation utilizes a combination of time-dependent Monte Carlo (TDMC) simulation and deterministic rate equations /18/. The microscopic detail inherent in our numerical approach explores directly the interrelationships among adsorption and desorption kinetics, energetics and the development of structure within the adsorbate overlayer.

We examine nondissociative molecular chemisorption from a mobile precursor state within the constraints of a lattice gas model. The time-dependent Monte Carlo (TDMC) simulation is initiated with the trapping of gas phase molecules onto the solid substrate. Molecules are placed sequentially on randomly chosen surface sites to simulate physical adsorption into the mobile precursor state. Each precursor may migrate across the substrate in single jumps to nearest-neighbor surface sites. Physically adsorbed molecules occupy this mobile intermediate state until either desorption or chemisorption occurs. The energy of physisorption binding the precursor to the substrate surface is a function of the local molecular environment. Lateral interactions between a precursor state molecule and previously chemisorbed molecules are assumed to be pairwise additive and limited in range to nearest-neighbor and next nearest-neighbor adsorbates. The binding energy (ε_p) of the intrinsic precursor state, a molecule physically adsorbed over an unoccupied lattice site, is given by:

$$\varepsilon_p = \varepsilon_o - \sum_j n_j \varepsilon_r + \sum_k n_k \varepsilon_a , \qquad (1)$$

where ε_o is the binding energy of an isolated precursor to the metal surface, ε_r and ε_a are, respectively, the repulsive and attractive energies of interaction between a precursor molecule and a neighboring chemisorbed molecule. The sums in Eq. (1) are performed over nearest-neighbor (j) and the next nearest-neighbor (k) lattice sites where n_j and n_k are occupation numbers. The binding energy (ε_n) of the extrinsic precursor state, a molecule physically adsorbed over a lattice site occupied by a chemisorbed molecule, is assumed fixed at eighty-five percent of the intrinsic precursor state binding energy /19/.

Each precursor state molecule is temporarily confined to a particular lattice gas site. The migration of the precursor is restricted by a periodic potential parallel to the substrate surface. For the purpose of computing migration probabilities, the parallel adsorbate-surface potential is represented as a series of adjacent, intersecting, harmonic potential wells. The relative depth of each potential energy minima is determined by the sum of lateral interactions specific to that surface site. The activation barrier (ε_m) to precursor state migration is defined as the difference between the energy at the intersection point between adjacent harmonic potentials and the ground state minimum at the occupation site:

$$\varepsilon_m = \varepsilon_m^o + \delta\varepsilon_p(i,j)/2 + [\delta\varepsilon_p(i,j)]^2/16\varepsilon_m^o . \qquad (2)$$

In (2) ε_m^o is the activation barrier to migration for an isolated intrinsic precursor state molecule in the absence of lateral interactions with chemisorbed molecules and $\delta\varepsilon_p(i,j)$ is the difference in precursor state binding energies or harmonic potential well depths between the ith and j^{th} sites given by [$\varepsilon_p(i,j) = \varepsilon_p(j) - \varepsilon_p(i)$]. Thus the probability for precursor migration ($P_m(i,j)$) from the ith to the neighboring j^{th} lattice gas site is given by:

$$P_m(i,j) = N\exp(-\varepsilon_m/k_BT) , \qquad (3)$$

where ε_m is the activation barrier to migration, k_B is Boltzmann's constant, and T is the surface temperature. The constant N is introduced to ensure unitarity. The probability for desorption (P_d) from the precursor state is given by:

$$P_d = NX\exp(-\varepsilon_p/k_BT) \,, \tag{4}$$

where X is a dynamical factor representing the ratio between desorption and migration prefactors or "attempt frequencies". The probability for precursor-mediated chemisorption (P_c) is determined from the initial, zero coverage, adsorption coefficient (S_0) and the binding energy of an isolated intrinsic precursor state molecule and is given by:

$$P_c = NX \, S_0\exp (-\varepsilon_0/k_BT)/(1-S_0) . \tag{5}$$

Equation (5) is derived from a geometric series expansion describing molecular chemisorption from the mobile precursor state at infinite adsorbate dilution. (Neither direct chemisorption from the gas phase nor chemisorption from the extrinsic precursor state are considered.)

The relative probabilities for desorption from the intrinsic precursor state, migration to adjacent sites, and molecular chemisorption are all functions of the extrinsic precursor binding energy. Because the extrinsic precursor binding energy depends directly upon neighboring chemisorption occupation numbers, these microscopic probability functions are also functions of the local molecular environment. Maps displaying the locations of chemisorbed molecules are constructed as the TDMC simulation proceeds. These adsorption maps detail the development of adsorbate lattice structure as a function of coverage, thus providing a clear demonstration of the profound effects of a mobile intermediate to molecular chemisorption on island formation and growth.

Thermal desorption spectra are computed from adsorbate maps using a Monte Carlo formalism in conjunction with deterministic rate equations. Chemisorbed molecules populate a number of distinct binding states despite the assumptions of a single binding site and a single binding geometry; the degeneracy of the chemisorption state is broken by lateral interactions among the chemisorbed molecules. Each chemisorption state is characterized by a unique activation barrier to desorption $(\varepsilon_b(i))$ which is determined partially by the number and magnitude of lateral interactions between the chemisorbed molecules:

$$\varepsilon_b(i) = \varepsilon_b^0 - \sum_j n_j \, \tilde{\varepsilon}_r + \sum_k n_k \, \tilde{\varepsilon}_a \,, \tag{6}$$

where ε_b^0 is the binding energy of an isolated chemisorbed molecule in the low coverage limit. The repulsive energy of interaction between nearest-neighbor chemisorbed molecules is given as ε_r; while the attractive energy of interaction between next nearest-neighbor chemisorbed molecules is denoted as $\tilde{\varepsilon}_a$. The sums appearing in (6) are defined in the text following (1). The rate of desorption from each chemisorption state is computed using a modified first-order Polanyi-Wigner equation /20/

$$\frac{d\sigma_i}{dt} = -\nu(\theta) \, \sigma_i \exp (-\varepsilon_b(i)/k_BT) \tag{7a}$$

with a coverage-dependent preexponential term

$$\nu(\theta) = \nu_o \exp (\alpha \theta) \cdot \tag{7b}$$

In (7a,b) σ_i is the occupation number of the ith chemisorption state and θ is the fractional coverage of chemisorbed molecules with respect to the total number of available lattice gas sites. The system of coupled differential equations describing the simultaneous desorption from all chemisorption states is solved numerically by the method of finite differences. The change in population of each chemisorption state is computed over a finite temperature interval (usually smaller than 0.25K) using the second-order Euler-Cauchy method. During each computational cycle the appropriate number of molecules is selected to desorb from each chemisorption state according to the prescription given in (7). To avoid introduction of artificial correlation effects in desorption modeling, these molecules are removed from randomly chosen lattice gas sites. The desorption calculation is repeated until all chemisorbed molecules have been removed from the substrate surface and a thermal desorption spectrum has been constructed. Thermal desorption spectra are computed for a wide range of initial adsorbate coverages for comparison to experiment.

Precursor-mediated thermal desorption has also been analyzed by a similar theoretical modeling scheme. This second, more indirect, mechanism of desorption proceeds via excitation of a chemisorbed molecule into the intrinsic precursor state with subsequent "desorption" into the gas phase. The structure of the equation describing the excitation from the chemisorption to precursor state is identical to (7), except that $\varepsilon_b(i)$ is replaced by the activation barrier to precursor excitation $\varepsilon_x(i)$. The kinetic prefactors to excitation and direct desorption are assumed equal. Once excited, the previously chemisorbed molecule is indistinguishable from a molecule entering the intrinsic precursor directly from the gas phase. The same kinetic processes are operative; the now mobile molecule migrates across the substrate surface until "desorption" or chemisorption occurs. Excitation and subsequent precursor migration may precipitate readsorption at a "new" lattice gas site. If this "new" adsorption site is more energetically favorable than the original, excitation followed by readsorption may result in thermal annealing of the adsorbate lattice.

The interaction of molecular carbon monoxide with the Ru(001) surface has been examined using the previously described methodology for TDMC simulations of adsorption and thermal desorption modeling. This chemisorption system has been well-characterized by numerous experimental studies /21,22/. Carbon monoxide adsorbs non-dissociatively in a vertical geometric configuration with the carbon atom bonded to the metal. Due to the hexagonal geometry of the Ru (001) surface and the nature and magnitude of adsorbate interactions, a well-developed LEED pattern possessing ($\sqrt{3}$ x $\sqrt{3}$) R30° symmetry is observed at low adsorbate coverages, reaching maximum intensity at $\theta = 0.33$. Coverage-dependent adsorption coefficients for molecular carbon monoxide are displayed in Fig. 1. The results of TDMC simulations are in excellent agreement with experimental measurements at surface temperatures of 100K and 200K at adsorbate coverages below 0.40. (Recent experimental studies /24/ have revealed that the carbon monoxide overlayer assumes more complicated structures at higher adsorbate coverages; thus application of the lattice gas model is suspect at coverages greater than $\theta = 0.35$.) Computer simulated and experimentally measured spectra for the desorption of carbon monoxide from Ru(001) are compared in Fig. 2. Comparison with experimental studies provides numerical estimates of the kinetic and energetic parameters characterizing this chemisorption system. (Please see Table 1.)

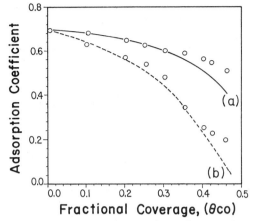

Figure 1. The coverage-dependent molecular adsorption coefficient for molecular carbon monoxide on Ru(001) at (a) 100K and (b) 200K. The lines represent calculational averages from TDMC simulations; circles are experimental results from Pfnur et al. /21/

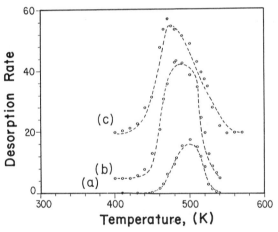

Figure 2. Thermal desorption spectra for molecular carbon monoxide from Ru(001) at fractional CO coverages of (a) $\theta = 0.10$, (b) $\theta = 0.28$ and $\theta = 0.34$. The dotted lines represent calculational averages from thermal desorption modeling; the symbols are experimental data from Pfnur et al. /22/

The interaction of molecular nitrogen with the Ru(001) surface has also been analyzed. Recent kinetic measurements performed at surface temperatures 77K and 90K reveal an unusual functional dependence of the probability of adsorption on adsorbate coverage /23-25/; a maximum in the coverage-dependent adsorption coefficients is observed at adsorbate coverages of approximately $\theta = 0.2$. Saturation coverage is a sensitive function of substrate temperature, occurring at approximately $\theta = 0.53$ at 78K and $\theta = 0.37$ at 90K. The results of TDMC simulations compare favorably with the experimentally measured adsorption coefficients. (Please see Fig. 3.)

Table 1. Kinetic and energetic parameters characterizing the chemisorption systems of molecular nitrogen and carbon monoxide CO on Ru(001)

		Value /kcal/	
		N_2/Ru(001)	CO/Ru(001)
ε_p :	Intrinsic precursor binding energy	1.6	2.0
ε_n :	Extrinsic precursor binding energy	1.4	1.5
ε_b^0 :	Chemisorption state binding energy	6.2-6.4	39
ε_m^0 :	Precursor migration barrier	0.3	0.7
ε_a :	Attractive next-nearest-neighbor interaction energy (precursor-chemisorbed molecule)	0.45	1.1
ε_r :	Repulsive nearest-neighbor interaction energy (precursor-chemisorbed molecule)	0.275	1.6
$\tilde{\varepsilon}_a$:	Attractive next-nearest-neighbor interaction energy (chemisorbed-chemisorbed molecule)	0.45	1.3
$\tilde{\varepsilon}_r$:	Repulsive nearest-neighbor interaction energy (chemisorbed-chemisorbed molecule)	0.275	4.0
χ :	Ratio of precursor preexponential factors of desorption to migration	500-1000	500
ν_o :	Preexponential factor for desorption from chemisorption states (sec^{-1})	10^{12}	10^{16}
α :	Desorption preexponential coverage dependence	17	21

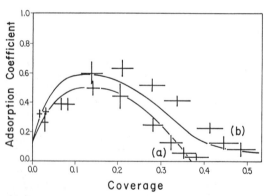

Figure 3. The coverage-dependent molecular adsorption coefficient for molecular nitrogen on Ru(001) at the following surface temperatures: (a) 90K and (b) 78K. The solid lines represent calculational averages from TDMC simulations; the bars are experimental results.

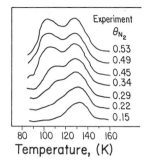
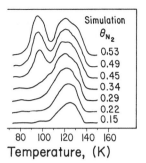

Figure 4. Thermal desorption spectra for molecular nitrogen from Ru(001) at fractional adsorbate coverages ranging from θ = 0.15 to θ = 0.53

Experimentally measured spectra for the desorption of molecular nitrogen from Ru(001) are displayed in Fig. 4. The appearance of two distinct peaks at adsorbate coverages greater than 0.25 monolayer suggests the presence of two independent chemisorption states; yet electron energy loss spectra indicate chemisorption of molecular nitrogen occurs only at ruthenium atop sites with the molecular axis oriented perpendicular to the surface plane /25/. Fig. 4 shows simulated thermal desorption spectra computed under the assumption of a single binding geometry at atop sites. The results of the TDMC simulations are in good agreement with experimental thermal desorption spectra obtained over adsorbate coverages ranging from θ = 0.15 to θ = 0.53 following adsorption at 77K. TDMC simulations of direct desorption and precursor-mediated desorption generate nearly indistinguishable spectra for the chemisorption system of molecular nitrogen on Ru(001). The only significant, recurring difference between these two approaches to thermal desorption analysis involves the full-width at half-maxima of the simulated spectra: precursor-mediated chemisorption simulations yield discernably broadened spectra. Because the precursor-mediated desorption calculations are an order of magnitude more costly, our discussions will focus on results from direct desorption calculations and the associated analyses.

The chemisorption of molecular nitrogen on the Ru(001) surface is examined using the TDMC methodology. The adsorption simulations are performed at a variety of surface temperatures between 78K and 95K on a substrate lattice composed of approximately 10,000 surface ruthenium atoms. The chemisorption maps displayed in Fig. 5 allow examination of the growth of adsorbate lattice structure as a function of coverage. The order in the adsorbate overlayer results from the hexagonal geometry of the Ru(001) surface, the nature of the chemisorption site, and the nature and magnitude of the precursor-chemisorbate and chemisorbate-chemisorbate lateral interactions. The energetically favored overlayer structure (as indicated by a well-developed LEED pattern at 95K) possesses ($\sqrt{3}$ x $\sqrt{3}$) R30° symmetry. However, three independent, degenerate adsorbate phases can exist on this hexagonal surface. These phases are distinguishable one from another by 120° rotations about a three-fold axis centered within the triangle formed by three adjacent ruthenium surface atoms. Distinct adsorbate domains can be observed on a surface where two or more of these phases coexist.

A detailed analysis of the molecular adsorption process is afforded through concurrent examination of the coverage-dependent adsorption coeffi-

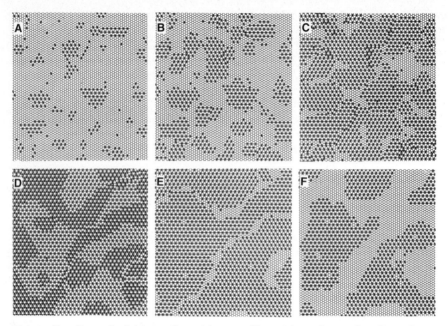

Figure 5. Maps depicting adsorption configurations for molecular nitrogen on a Ru(001) surface. The open circles represent ruthenium atoms at the single crystal surface ; the dots represent nitrogen molecules chemisorbed atop ruthenium surface atoms. Figs. (a-d) are generated from TDMC simulations of adsorption; Figs. (e-f) are generated from thermal desorption modeling. Solid lines join nitrogen molecules contributing to a single phase. The nitrogen coverages are (a) 0.06, (b,f) 0.15, (c,e) 0.29, and (d) 0.53

cient and maps of the chemisorbate overlayer. In the system of molecular nitrogen on Ru(001), chemisorption at 78K occurs in four distinct stages: (I) island nucleation, (II) isolated island domain growth, (III) concurrent growth of isolated island domains and the development of antiphase domain boundaries, and (IV) antiphase domain growth. Adsorbate overlayer formation begins with the chemisorption of isolated molecules which act as nucleation centers. A map detailing the structure of the adsorbate layer representative of growth stage I is shown in Fig. 5a. During growth stage II, cooperative behavior dominates the adsorption process as nucleation centers initiate the growth of numerous, small island domains. The adsorption coefficient $S(\theta)$ undergoes a three-fold increase from an approximate initial value of 0.2 to a maximum value between 0.6 and 0.7 at a coverage of about 0.20. Due to the attractive nature of the next nearest-neighbor interaction between coadsorbed molecules, the edges of island domains serve as the most probable site for chemisorption. Development of structure in the chemisorbed overlayer is evidenced by increased correlation in adsorbate positions as displayed in Fig. 5b. Adsorption stage II continues with the growth of isolated island domains up to a fractional coverage of approximately 0.22. The onset of growth stage III is revealed in Fig. 5c as "crowding" between neighboring islands is observed. Adjacent domains of similar phase coalesce to form larger islands, while adjacent domains of differing phase develop grain boundaries. Molecules adsorbed within the grain boundaries experience nearest-neighbor repulsions not encountered by

160

molecules adsorbed within the islands proper. During stage III appreciable molecular chemisorption occurs within grain boundary regions in addition to the energetically more favorable island domains. In the coverage regime between 0.2 and 0.4 monolayer, the adsorption coefficient suffers a three-fold decrease in magnitude from 0.6 to 0.2. This dramatic decrease in the probability of adsorption is attributable to appreciable, concurrent thermal desorption from the domain boundary regions and to a decrease in available adsorption sites. The onset of stage IV is accompanied by an abrupt change in slope of the coverage-dependent adsorption coefficient at 0.4 monolayer. At adsorbate coverages exceeding 0.4 monolayer, virtually none of the most favorable adsorption sites within single island domains remain vacant; thus chemisorption in stage IV occurs almost exclusively in regions of antiphase domain. Saturation coverage is achieved at approximately 0.55 monolayer due to the establishment of a steady state balance between the rate of precursor-mediated adsorption and the rate of desorption from the chemisorption states. An adsorbate configuration representative of saturation coverage is displayed in Fig. 5d.

A detailed analysis of the molecular adsorption of molecular carbon monoxide on Ru(001) with the concurrent evaluation of adsorbate lattice maps has also been performed. This adsorption process follows closely the first three (3) phases of development observed for the system of molecular nitrogen on Ru(001). (At adsorbate coverages grater than $\theta = 0.40$ where stage IV development is observed in the system of molecular nitrogen on Ru(001), the system of molecular carbon monoxide on Ru(001) is no longer reasonably described by the simple, lattice gas model.) Computer simulations of low-energy electron diffraction (LEED) studies are now in progress for both chemisorption systems (molecular carbon monoxide and nitrogen on Ru(001)) in an attempt to quantify the ordering of the adsorbate layer during island formation and development.

The influence of lateral interactions between coadsorbates on the process of molecular thermal desorption can be evaluated by concurrent examination of adsorbate lattice maps and the coverage-dependent temperature-programmed desorption spectra (TPDS). For coverages below $\theta = 0.25$, the TPDS displayed in Fig. 4 consist of a single peak with a maximum intensity at a surface temperature of approximately 125K. This "high-temperature" feature is associated with desorption from the perimeters of isolated islands. A nitrogen molecule chemisorbed within an isolated island experiences attractive lateral interactions not experienced by an isolated adsorbate; thus the activation barrier to desorption for a molecule within an island or a molecule adsorbed at an island edge is increased relative to isolated molecules. At adsorbate coverages greater than $\theta = 0.3$, a second peak appears in the TPDS. This "low-temperature" peak with a maximum in intensity at surface temperatures less than 100K is associated with desorption from antiphase domains. The activation barrier to thermal desorption for nitrogen molecules chemisorbed within these antiphase domain regions is diminished by repulsive lateral interactions; thus the characteristic desorption temperature for molecules chemisorbed within the antiphase domains is reduced relative to those molecules chemisorbed within the islands proper. At saturation coverage the number of molecules desorbing from antiphase domains and the number desorbing from the isolated island domains are nearly equal, as are the intensities of the "low-temperature" and "high-temperature" thermal desorption peaks.

The microscopic insight provided by TDMC simulations facilitates the interpretation of the complex TPD spectra for nitrogen thermal desorption from Ru (001). Comparing directly the development of adsorbate structure with kinetic measurements reveals clearly the extreme value of this type of numerical simulation technique. The two independent peaks appearing in the TPD spectra at high adsorbate coverages are shown to result solely from the lateral interactions between chemisorbed nitrogen molecules. Desorption from two distinct chemisorption states of differing coordination number or binding geometry need not be invoked.

Numerical values for the energetic parameters characterizing the systems of molecular nitrogen and molecular carbon monoxide on Ru(001) are obtained by minimizing the differences between the calculated and experimental coverage-dependent adsorption coefficients and thermal desorption spectra over adsorbate coverages ranging from 0.05 to 0.55 and 0.10 to 0.40 monolayers, respectively. (Please see Table 1). Precursor-chemisorbate and chemisorbate-chemisorbate lateral interaction energies are treated as separately variable parameters. Adsorption and desorption simulations for the system of carbon monoxide on Ru(001) yield numerical values for the energies of the chemisorbate-chemisorbate interaction nearly three times greater in magnitude than for the precursor-chemisorbate interaction. In contrast, the system of molecular nitrogen on Ru(001) is characterized by nearly identical values for both interactions. This similarity in lateral interaction energetics is due primarily to the weak binding energy of chemisorbed molecular nitrogen on this ruthenium surface. The success of the description of both the "weak" chemisorption system of molecular nitrogen on Ru(001) and the "strong" chemisorption system of molecular carbon monoxide on Ru(001) provided by TDMC simulations attests to the extreme verstility and wide range of applicability of this methodolgy.

Examination of the maps detailing adsorbate positions as a function of coverage exposes obvious differences between the processes of precursor-mediated molecular chemisorption and thermal desorption from the chemisorbed state. TDMC simulations reveal an adsorption mechanism dominated by kinetic trapping and a thermal desorption mechanism dominated by thermal relaxation or annealing. Precursor-mediated chemisorption results in the formation of numerous, small island domains; while thermal desorption results in the formation of fewer, but much larger, island domains. Thus, as displayed in Fig. 5, the results of the thermal desorption simulations provide a sharp contrast to the results of the adsorption simulations. The extreme value of this integrated approach to molecular chemisorption and thermal desorption simulations is demonstrated through the ability to disclose in microscopic detail the inter-relationships among kinetics, energetics, mechanism, and adsorbate overlayer structure.

References:
1. J.B. Taylor and I. Langmuir: Phys. Rev. 44, 423 (1933)
2. J.E. Lennard-Jones: Trans. Faraday Soc. 28, 333 (1932)
3. D. Menzel, H. Pfnur and P. Feulner: Surf. Sci. 126, 374 (1983)
4. M. Shayegan, E.D. Williams, R.E. Glover III and R.L. Park: Surf. Sci. 154, L239 (1985)
5. S.L. Tang, J.D. Beckerle, M.B. Lee and S.T. Ceyer: J. Chem. Phys. 84, 6488 (1986)
6. M.P. D'Evelyn, H.P. Steinruck and R.J. Madix: Surf. Sci. (to appear)
7. K. Janda, J.E. Hurst, C.A. Becker, J.P. Cowin, L. Wharton and D.J. Auerbach, Surf. Sci. 93, 270 (1980)
8. D.A. King and M.G. Wells: Proc. Royal Soc A339, 245 (1976)
9. M. Golze, M. Grunze, R.K. Driscoll and W. Hirsch: Surf. Sci. 6, 464 (1981); M. Grunze, W.N. Unertl and M. Golze: J. Vac. Sci. Technol. A2(2), 896 (1984)

10. C.R. Brundle, J. Behm, D.J. Auerbach, and M. Grunze: J. Vac. Sci. Technol. A$\underline{2}$, 1014 (1984); M.J. Grunze, J. Fuhler, M. Neumann, C.R. Brundle, D.J. Auerbach and J. Behn: Surf. Sci. $\underline{139}$, 109 (1984)
11. D.A. King and M. Wells: Proc. Roy. Soc. London, A$\underline{339}$, 245 (1974)
12. R. Gorte and L.D. Schmidt: Surface Sci.$\underline{102}$, 559 (1978)
13. D.L. Freeman and J.D. Doll: J. Chem. Phys. $\underline{78}$, 6002; $\underline{79}$, 2343 (1983)
14. P. Kisliuk: J. Phys. Chem. Solids $\underline{3}$, 95 (1957); $\underline{5}$, 78 (1958)
15. D.A. King: Surf. Sci. $\underline{102}$, 388 (1981)
16. A. Cassuto and D.A. King: Surf. Sci. $\underline{102}$,(1981)
17. J.E. Adams and J.D. Doll: Surf. Sci. $\underline{103}$, 472; $\underline{111}$, 492 (1981)
18. E.S. Hood, B.H. Toby and W.H. Weinberg: Phys. Rev. Lett. $\underline{55}$, 2437 (1985)
19. K. Christmann and J.E. Demuth: Surf. Sci. $\underline{120}$, 291 (1982)
20. P.A. Redhead: Vacuum $\underline{12}$, 203 (1962); D.L. Adams, Surf. Sci. $\underline{42}$, 12 (1974)
21. H. Pfnur and D. Menzel: J. Chem. Phys. $\underline{79}$, 2400 (1983)
22. H. Pfnur, P. Feulner and D. Menzel: J. Chem. Phys. $\underline{79}$, 4613 (1983)
23. P. Feulner and D. Menzel: Phys. Rev. B$\underline{25}$, 4295 (1982)
24. D. Menzel, H. Pfnur and P. Feulner: Surf. Sci. $\underline{126}$, 374 (1983)
25. A.B. Anton, N.R. Avery, B.H. Toby and W.H. Weinberg: J. Electron Spectrosc. Relat. Phenom. $\underline{29}$, 181 (1983)

Adsorption and Reaction of CO_2 on Metal Surfaces. Detection of an Intrinsic Precursor to Dissociation

B. Bartos[1], *H.-J. Freund*[1], *H. Kuhlenbeck*[2], *and M. Neumann*[2]

[1]Institut für Physikalische und Theoretische Chemie der
  Universität Erlangen-Nürnberg, Egerlandstr. 3,
  D-8520 Erlangen, Fed. Rep. of Germany
[2]Fachbereich Physik der Universität Osnabrück, Barbarastr. 7,
  D-4500 Osnabrück, Fed. Rep. of Germany

1. Introduction

In the late seventies much progress was made in understanding the mechanism of the CO oxidation reaction using the spectroscopic machinery of surface science /1-2/. ERTL and his group /3-5/ unambiguously showed by molecular beam and other experiments that CO oxidation on transition metal surfaces proceeds via a Langmuir-Hinshelwood mechanism rather than an Eley-Rideal mechanism. This implies that both reactants, CO and oxygen are adsorbed on the surface when CO_2 is formed. The latter is readily desorbed at the temperatures used in the above-mentioned studies. Therefore, adsorbed CO_2 has not been observed in those studies but only detected after desorption in the gas-phase /3-5/. In light of the fact that in addition to the importance of CO oxidation, CO_2 dissociation /6-9/, the reverse reaction, is of considerable - even technical /10/ - interest, several groups have started to investigate the interaction and reactivity of CO_2 with and on metal surfaces /7-9, 12-16/. In particular, from molecular beam experiments on CO_2 reaction dynamics performed on different surfaces a picture arises that can *schematically* be represented in a simplified manner by the twodimensional potential energy diagram shown in Fig. 1 /17-18/. CO_2 approaches the surface along the entrance channel and may, after pas-

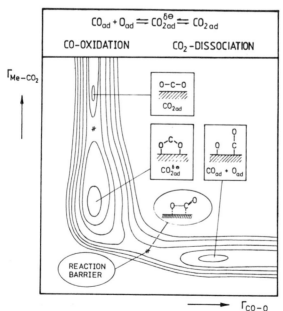

Fig. 1: Schematic two-dimensional potential energy diagram for CO_2 metal interaction (vertical axis) and CO-O-dissociation (horizontal axis) from ref. /16/

sing through some kind of physisorbed (van der Waals) state,be trapped into an intermediate state which then dissociates along the exit channel into adsorbed CO and adsorbed oxygen. In favour of the existence of an intermediate state, which from the standpoint of CO_2 adsorption can be called a precursor to dissociation, and whose electronic and geometric structure is not yet known, is the observed cosine-like angular distribution of scattered CO_2 for certain surfaces, e.g. Ni(100) /13/ and Pd(111) /1,3,4/.

Most of the studies referenced above proposed some kind of bent CO_2 species as an intermediate, transient species. Many authors consider the bent species to represent the transition state of the reaction rather than a stable intermediate. In a recent theoretical study /15/ we have tried to evaluate possible interaction and reaction channels between CO_2 and a metal via cluster calculations. Upon interaction between CO_2 and metal the CO_2 molecule bends due to partial electron transfer from the metal to the molecule. Bonding to the metal can be established either by pure carbon, bidentate CO_2 oxygen coordination or by the C-O bond. i.e. mixed carbon-oxygen coordination. The total binding energy is rather small, approximately of the order of 10 kcal/mol.

In the present paper we review in some detail experimental results for the system CO_2/Ni(110) and CO_2/Fe(111) and mention results for other metal surfaces, i.e. the face specificity of CO_2 adsorption observed recently by WEDLER and coworkers /9/.

2. CO_2 Adsorption

Figure 2 shows angle-resolved photoelectron spectra (normal emission) of the CO_2 covered Fe(111) and Ni(110) surfaces in comparison with the clean

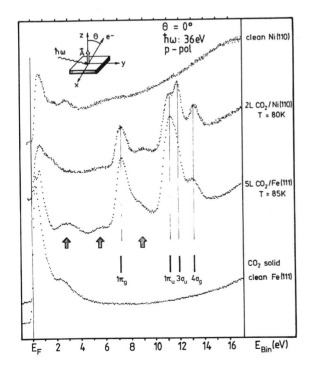

Fig. 2: Photoelectron spectra in normal emission (p-polarization) of CO_2/Ni(110) and CO_2/Fe(111). For comparison the corresponding solid phase /21/ energy levels (aligned at the $4\sigma g$ level) are indicated in the form of a bar diagram.

165

surfaces at T = 85 K and T = 80 K, respectively. CO_2 exposure leads to three (four) strong peaks marked with lines and some weaker bands marked with arrows. The assignment of the strong peaks to the $1\pi_g$, $1\pi_u$, $3\sigma_u$ and $4\sigma_g$ ion states of CO_2 is obvious when comparing the measured spectra with those in the gas phase /19,20/ or in the solid phase /21/. Clearly, CO_2 adsorbs molecularly on both surfaces without a strong influence on the electronic structure of the adsorbed species. However, the binding energies of adsorbed CO_2 are lowered slightly as a function of increasing coverage /9,14,16,22/ indicating some interaction between the absorbed species. To explain these findings the changes in work function of the metal surfaces for increasing exposure to CO_2 at 80-85 K was measured /9,16/. Up to an exposure of about 1L the work function increases, passes through a maximum, drops by a smaller amount and saturates. This is explainable in terms of the competition of two different adsorbed species, one causing a strong work function increase, the other one a smaller work function decrease. On this basis we interpret the peaks marked by arrows in fig. 2 as due to the species that increases the work function, while CO_2, which dominates the spectrum, decreases the work function.

Even though for none of the CO_2 adsorbates a sharp LEED pattern has been observed, E vs $k_{\parallel}$ dispersions observed in angle-resolved photoelectron spectra indicate some long-range order /23/, i.e. on the Ni(110) surface. Some of the ionization bands (independent of the indication of long-range order) show pronounced intensity variations as a function of polar angle. The intensity variations contain information on the local geometry of the adsorbate. The strongest intensity variation was observed for the totally symmetric $4\sigma_g$ ion state within the (110) azimuth for s-polarization. The intensity of the $4\sigma_g$ state varies with respect to the $1\pi_g$-state, whose intensity is almost independent of polar angle. We have plotted in Fig. 3 the intensity ratio $4\sigma_g/1\pi_g$ along the azimuth (100) and (110). We observe a rather pronounced peaking of the $4\sigma_g$ along the (110) azimuth. The physics is very simple. The $4\sigma_g$ state is mostly localized on the oxygen atoms. It represents one of the oxygen lone pairs

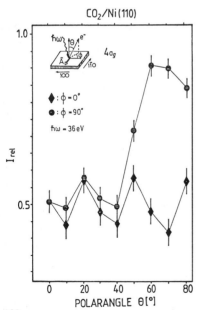

Fig. 3: Experimentally determined relative intensities (referring to the $1\pi_g$ level intensity) of the $4\sigma_g$ ion state as a function of polar angle θ for $CO_2/Ni(110)$; 2L exposure.

with strong O2p_z character (z is along the molecular axis). By symmetry the main emission direction from this orbital has to be pointing along the molecular axis. Symmetry also asks for vanishing 4σ_g intensity in normal emission (see Fig. 3) if the linear CO_2 molecules were all oriented with their axis parallel to the (110) azimuth. This is at variance with observations.

The contribution to the cross-section in normal emission is mainly due to the carbon 2s-character in the 4σ_g-orbital. In other words, a polarization component perpendicular to the molecular axis could explain the observed emission perpendicular to the surface. There are several possibilities how such a situation could be achieved. A structure model /15/ that accounts for many of the observations is shown in Fig. 4: Half of the molecules is oriented along the (110), the other half along the (100) azimuth. Such a structure is closely related to the (001) plane of solid CO_2 /24/ and represents a state of lower energy for an arrangement of linear quadrupoles /25/. As far as the orientation of the molecular axis is concerned, a fit of calculated ionization intensities to the observed values (Fig. 3) as function of polar angle suggests a tilt of the molecules from perfect parallel orientation within ± 20°. A detailed discussion including band structure will be published elsewhere /26/. For the Fe(111)/CO_2 system, which does not show dispersion, a similar intensitiy analysis indicates a larger tilt angle, which is consistent with the more open structure of the bcc-(111)-surface.

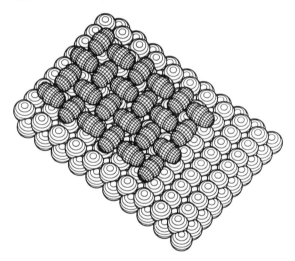

Fig. 4: Tentative structure model for the relative orientation of linear CO_2 molecules within a two-dimensional layer on the Ni(110) surface. The relative internuclear distances were taken from a (100) cut through a CO_2 single crystal as reported in ref. /24/.

3. Reaction of Adsorbed CO

Figure 5 shows a set of photoelectron spectra of clean and adsorbate covered Ni(110) (Fig. 5a) and Fe(111) (Fig. 5b) surfaces at various temperatures and exposure conditions. Firstly, we discuss the Ni(110)/CO_2 system. Spectrum (a) (taken from Fig. 2) represents molecular CO_2 that is adsorbed without strong distortion of its molecular electronic structure. Only adsorbate-induced features are shown. Upon heating this layer to 114 K (spectrum (b)) and 140 K (spectrum (c)) the spectrum changes considerably as compared to 80 K: Firstly, the adsorbate-induced features lose intensity relative to the substrate emission close to the Fermi edge indicating partial desorption of the adsorbate layer. Secondly, there are

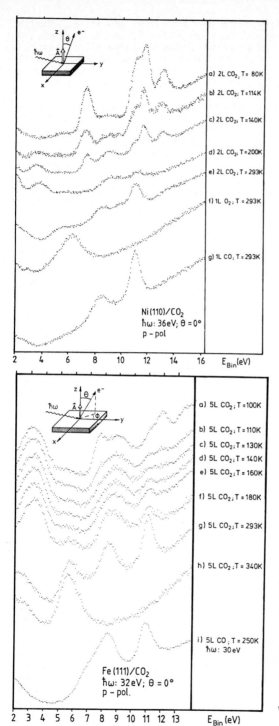

Fig. 5: a) left panel: Photo-electron spectra of a Ni(110)/CO$_2$ adsorbate at various temperatures (a) - (e) in comparison with a CO covered (g), and an oxygen covered surface (f).
b) right panel: Photoelectron spectra of a Fe(111)/CO$_2$ adsorbate at various temperatures (a) - (h) in comparison with a CO covered surface (i).

extra spectral features observed at 3.5 eV and 8.7 eV, and the CO_2 band at 10.8 eV binding energy loses intensity upon increasing the temperature. At 140 K it is only visible as a shoulder of less than half the intensity at 80 K. It appears that the lower the intensity of the 10.8 eV band is, the more intense are the bands at 3.5 eV and 8.7 eV binding energy. Up to about 200 K the spectrum remains basically unchanged except that more and more undisturbed CO_2 is desorbed. At 140 K the work function increases up to 0.95 eV due to the loss of coadsorbed undisturbed CO_2. The observed work function change can be reached either by admitting CO_2 at 140 K surface temperature or by heating a saturated CO_2 layer produced at 80 K to 140 K. Further heating of the adsorbate decreases the CO_2 induced features until they vanish at 200 K (spectrum (d)). This intensity decrease is accompanied by a slight further increase in work function up to 1 eV. Upon heating up to room temperature (spectrum (e)) a distinctly different photoemission spectrum is observed. Also, the work function decreases considerably to 0.5 eV indicating a change in the magnitude of surface polarization. The photoelectron spectrum at room temperature consists of three bands at 5 eV, 8.1 eV and 10.8 eV binding energy. By using spectrum (g) in Fig. 5a which shows the photoelectron spectrum of a disordered CO/Ni(110) adsorbate at room temperature, the peaks at 10.8 eV and 8.1 eV binding energy can be identified as being due to CO. The band at 5 eV binding energy in spectrum (e) is situated within the tpyical "oxygen range". Even though a spectrum of an oxygen overlayer, created by exposing the surface to 1L of O_2 at room temperature, spectrum (f) shows a binding energy about 1 eV larger than in the CO/O coadsorbate the assignment of the peak at 5 eV in spectrum (f) to adsorbed oxygen is reasonable. The Ni(110) surface is known to undergo rather complex reconstructions upon oxygen adsorption /27/ and the oxygen peak is expected to shift in binding energy depending on the surface structure ,i.e. coordination number etc. /3/. The position of the oxygen peak in a spectrum taken for an oxygen CO adsorbate prepared at 80 K is much closer to the position of the oxygen peak in spectrum (e). Summarizing so far, we are led to the conclusion that CO_2 undergoes dissociation when adsorbed molecularly at low temperature and subsequently heated to room temperature.

Figure 5b shows a set of spectra for CO_2/Fe(111) in normal emission with p-polarized excitation ($h\omega$ = 32 eV) taken at different surface temperatures. Again only the adsorbate – induced features are shown. Clearly, upon raising the temperature to 100 K some CO_2 induced peaks (marked with arrows in Fig. 2) lose intensity, while those that must be attributed to a second species, show an intensity increase. Upon heating to 140 K the peaks caused by linear, molecular CO_2 disappear, and the intermediate species dominate the spectrum. Further heating of the substrate leads at room temperatures to spectrum (g). By comparison of spectrum (g) with spectrum (h) which is a pure Fe(111)/O spectrum and a pure Fe(111)/CO spectrum (spectrum (i)), we can unambiguously assign the peaks to be due to coadsorbed CO and O. Upon raising the temperature further to 340 K the molecularly adsorbed CO dissociates.

Both sets of spectra (Fig. 5a and b) indicate the existence of an intermediate molecular species that seems to be the precursor for dissociation of CO_2.

In a previous theoretical paper /15/ we have proposed a bent,anionic CO_2 species coordinated to the metal as a possible candidate for such a precursor. Such a bent CO_2 species can be formed by transferring electronic charge from the metal to the CO_2 molecule. The presence of charge transfer onto the adsorbate is corroborated by the strong work function increase observed at 140 K, where much of the coadsorbed, linear

CO_2 has been desorbed. In the language of molecular orbital theory, the extra charge transferred from the metal occupies a previously unoccupied orbital ($2\pi_u$) of linear CO_2, which consequently becomes a 17-electron system. Such a system tends, according to the Walsh rules /28/, to avoid a linear geometry. On the basis of one-electron orbital energies we show in Fig. 6 a correlation diagram for CO_2 and a bent CO_2^- species as calculated in ref. /15/. On the left, the comparison is made with the experimental Ni(110) data, on the right with the Fe(111) data set. The energy scale for the anion has been shifted so as to align the orbital at highest binding energies ($4\sigma_g$, $4a_1$). Clearly, the first observation is that while the CO_2 spectrum spreads over about 5-6 eV binding energy, the CO_2^- spectrum extends over more than 10 eV /29/. Let us try to assign spectrum (d) in Fig. 5a shown in comparison to theory /15/ in Fig. 6a. It is most obvious to assign the peak at 3.5 eV binding energy to the $6a_1$ orbital. The $6a_1$ orbital is mainly localized on the carbon atom /15/. It is therefore appropriate to compare its binding energy with carbon-induced bands in Ni/C adsorbates. Recently, BRADSHAW and coworkers /30/ have taken angle resolved spectra of such systems. On Ni(110) carbon adsorption leads to a peak at 4.1 eV binding energy relative to E_f. Our $6a_1$ orbital therefore appears to be in the correct energy range. However, formation of atomic carbon in the course of our experiments is very unlikely at the low temperatures used. Note, that for preparation of the carbon adlayers the metal surface was exposed to CO around 200°C /30/. Coming back to our

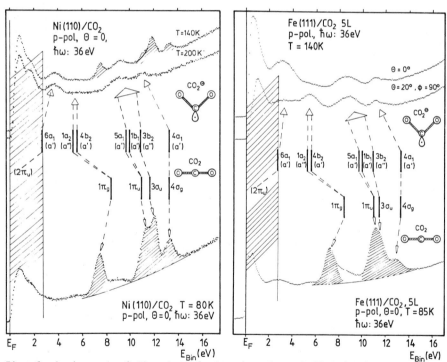

Fig. 6: Assignment of the photoelectron spectra of CO_2 adsorbates on the basis of ab initio calculations.
a) left panel: Ni(110)/CO_2 at 80 K, 140 K and 200 K.
b) right panel: Fe(111)/CO_2 at 85 K and 140 K.
Undisturbed linear CO_2 peaks are shaded.

assignment in Fig. 6a the next set of orbitals is expected between 5-6 eV binding energy. The following one should be situated between 9-10 eV and the final one at approximately 12 eV. Based on the observed energies the assignment proposed by the arrows in Fig. 6 is obvious, except that there is no peak observed between 5-6 eV. However, off normal a peak in the energy region in question clearly shows up which is demonstrated for the $CO_2/Fe(111)$ system in Fig. 6b. The observation that the peaks at 3.5 eV, 8.7 eV and 11.4 eV show considerable intensity in normal emission, while the peak at 5.5 eV is absent in normal emission is consistent with the symmetry of the states given for CO_2^- in Fig. 6: Those regions of binding energies that are assigned to groups of orbitals containing totally symmetric (a_1) orbitals should contribute to electron emission normal to the sample surface. On the other hand the orbitals of non totally symmetric character should lead to emission that peaks off normal. The orbital assignment given in Fig. 6 for the bent species refers to the C_{2v} point group. Clearly, the local geometry of the precursor state is such that there is still a mirror plane present in the system, since otherwise there would be no selection rule. Possible coordination sites with mirror plane symmetry are shown in Fig. 7. While the sites on the left and right hand side of Fig. 7 have C_{2v} symmetry, the one shown in the middle has C_s symmetry. The symmetry of the CO_2^- orbitals in C_s symmetry are given in Fig. 6 in parentheses. Such a symmetry corresponds to coordination of one of the C-O bonds to the surface. Obviously, the orbital with $4b_2$ character in C_{2v} symmetry becomes totally symmetric in C_s symmetry, while the $1a_2$ orbital in C_{2v} remains non totally symmetric in C_s. Thus, were the symmetry of the adsorbed molecule C_s we would expect some intensity in the energy range around 6 eV binding energy. However, the absence of appreciable intensity in the corresponding energy region suggests that the adsorbed molecule has C_{2v} symmetry.

$CO_2^{\delta\ominus}$ - POSSIBLE COORDINATION SITES

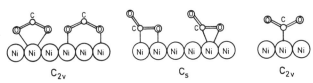

C_{2v} C_s C_{2v}

Fig. 7: Possible coordination sites of CO_2^- together with the corresponding point group symmetries.

The results of an electron energy loss study /16/ parallel to a large extent the results of the photoemission studies, and it appears that a C_{2v} coordination site with direct oxygen coordination is more likely than a carbon coordinated site. This conclusion is in line with the results of our theoretical study /15/. If the assignment given in Fig. 6 and discussed above is correct the shaded peaks in the upper curve of Fig. 5a are not caused by the *bent* CO_2 species but rather by neutral, linear CO_2 as can be clearly seen by comparison with the spectrum shown at the bottom of Fig. 6a and b. Here the assignment is unambiguous as discussed in section 3 and indicated by the calculated ionization energies which have been shifted rigidly so as to line up with the experimental $4\sigma_g$ ionization energy. It appears that some CO_2^- is formed at low temperature and coexists with neutral CO_2 on the surface. We do not know at present whether the CO_2 and the bent CO_2 species are adsorbed in separate islands or are coadsorbed forming a homogeneous phase. It is known from gas phase experiments on CO_2 clusters that a bent CO_2-anion can be stabilized by solvation with neutral CO_2 molecules /31,32/.

4. Synopsis

We have reviewed in the present work mainly photoemission results on adsorption and reaction of CO_2 on Ni(110) and Fe(111). Exposed at 80-85 K CO_2 adsorbs molecularly on the clean surfaces. The electronic structure of the main fraction of molecules is, if at all, only slightly distorted in comparison to the gas phase and a thick solid film. However, even at low temperature part of the molecules adsorb into a state that is identified as a precursor to dissociation ($CO_2^{ad} \rightarrow CO^{ad} + O^{ad}$) which occurs at elevated temperatures. The precursor is suggested via the photoemission results /14,16/ the HREELS study /16/, and supported by a theoretical study /15/ to be a bent "anionic" CO_2^- species. All studies favour a coordination site with C_{2v} symmetry. Wether the species is oxygen or carbon coordinated (see Fig. 7) is not yet clear, although oxygen coordination is more plausible. Recently, the structure of the formate ion (HCO_2^-) coordinated to a Cu surface has been determined by SEXAFS /33,34/. This species, which differs from our proposed CO_2^- species only by the hydrogen atom attached to the carbon atom, has been shown to have C_{2v} symmetry with a similar coordination site as proposed in the present work for CO_2. Interestingly, the stability of the precursor towards dissociation depends on the particular system, as is evident from a comparison of the Ni(110) with the Fe(111) system. Decomposition of CO_2^- starts to occur above 200 K on Ni(110) while it starts at lower temperature $T \geq 180$ K on the Fe(111) surface. Parallel to the lower dissociation temperature for CO_2 on Fe(111) the formation of the dissociation precursor seems to be facilitated on Fe(111) indicated by the larger CO_2^- signals at low adsorption temperature (compare Fig. 2a and Fig. 2b). At present, we can only speculate on the reason for these observations. A key quantity that determines the formation of the dissociation precursor could be the work function. WEDLER and coworkers /9/ e.g. observed a strong face specificity for CO_2 adsorption on different Fe surfaces; Table 1 collects work function values in addition to those of other single crystal surfaces for

Table 1: CO_2^- formation as a function of work function on various adsorbate systems

Metal	Orientation	Workfunction	CO_2^- formation	ref.
Fe	(110)	5.0	no*	/9/
Fe	(110) stepped	4.5	yes	/9/
Fe	(111)	4.3	yes	/9,14/
Ni	(110)	4.65	yes	/16/
Pd	(111)	5.95	no	/35/
Pd	(100)	5.65	no*	/12/
Al	(111)	4.53	yes	/36/
Al/Na	(111)	2.8	yes	/36/
Rh	(111)	5.5	no*	/7/
Pd/K	(100)	2.0	yes	/37/

* = CO_2 condensation

which CO_2 adsorption has been studied. It appears that large work functions inhibit the formation of CO_2^-. This is reasonable, since we know from our theoretical work that the bonding of CO_2 to transition metals is dominated by Coulomb interactions: The more energy it takes to transfer an electron from the metal to the CO_2 molecule the more unstable becomes the CO_2 coordination complex. If one therefore changes the work function, i.e. decreases it via alkali coadsorption or increases it via oxygen coadsorption, one can influence the stability of the CO_2^- moiety (see table 1). Also, the presence of the CO_2^- anion may be key for subsequent CO_2 adsorption due to possible CO_2^- - CO_2 interactions.

5. Acknowledgements

We would like to thank the "Bundesministerium für Forschung und Technologie", the "Deutsche Forschungsgemeinschaft" and the "Fonds der Chemischen Industrie" for their financial support.

6. References

1. T. Engel and G. Ertl "The Chemical Physics of Solid Surfaces and Heterogeneous Catalysis" (Ed. D.A. King, D.P. Woodruff) Vol. 4, p. 92 (Elsevier, Amsterdam (1982)) and references therein
2. M.P. D'Evelyn, R.J. Madix, Surf. Sci. Rep. 3 (1984) and ref. therein
3. T. Engel and G. Ertl, Adv. Catalysis 28, 1 (1979)
4. T. Engel and G. Ertl, J. Chem. Phys. 69, 1267 (1978)
5. H. Conrad, G. Ertl and J. Küppers, Surf. Sci. 76, 323 (1978)
6. D.W. Goodman, D.E. Peebles, J.M. White, Surf. Sci. 140, L 239 (1984)
7. W.H. Weinberg, Surf. Sci. 128, L 224 (1983)
8. L.H. Dubois, G.A. Somorjai, Surf. Sci. 128, L 231 (1983)
9. H. Behner, W. Spieß, D. Borgmann, G. Wedler, Surf. Sci. 175, 276 (1986) and
    H. Behner, W. Spieß, G. Wedler, D. Borgmann, H.-J. Freund, Surf. Sci. submitted
10. D.E. Peebles, D.W. Goodman, J.M. White, J. Phys. Chem. 87, 4378 (1983)
11. D.C. Grenoble, M.M. Estadt, D.F. Ollis, J. Catal. 67, 90 (1981)
12. F. Solymosi, J. Kiss, Surf. Sci. 149, 17 (1985)
13. M.P. D'Evelyn, A.V. Hamza, G.E. Gidowski, R.J. Madix, Surf.Sci. 167, 451 (1986)
14. H.-J. Freund, B. Bartos, H. Behner, G. Wedler, H. Kuhlenbeck and M. Neumann, Surf. Sci, in press
15. H.-J. Freund, and R.P. Messmer, Surf. Sci. 172, 1 (1986)
16. B. Bartos, H.-J. Freund, H. Kuhlenbeck, M. Neumann , H. Lindner and K. Müller, Surf. Sci. in press
17. G. Ertl, Ber. Bunsengesellschaft Phys. Chem. 86, 425 (1982)
18. J.C. Tully, Adv. Chem. Phys. 42, 63 (1980)
19. D.W. Turner, A.D. Baker, C. Baker, C.R. Brundle "Molecular Photoelectron Spectroscopy" (Wiley, New York, 1970)
20. H.-J. Freund, H. Kossmann, V. Schmidt, Chem. Phys. Lett. 123, 463 (1986)
21. J.-H. Fock, H.-J. Lau, E.E. Koch, Chem. Phys. 83, 377 (1984) and J.-H. Fock, Dissertation, Universität Hamburg (1983)
22. W. Spieß, Dissertation, Universität Erlangen-Nürnberg (1984)
23. E.W. Plummer and W. Eberhardt, Adv. Chem. Phys. 49, 533 (1982)
24. R.W. Wyckhoff, Crystal Structures 2nd ed. Vol. 1 (Wiley, New York 1963)
25. M.A. Morrison, P.J. Hay, J. Phys. B10, 647 (1977)

26. G. Odörfer and H.-J. Freund, unpublished results
27. see e.g. H. Niehus, G. Comsa, Surf. Sci. 151, L 171 (1985)
28. A.D. Walsh, J. Chem. Soc. 2266 (1953)
29. see also: J. Pacanski, U. Wahlgren, P.S. Bagus, J. Chem. Phys. 62, 2740 (1985); W.B. England, Chem. Phys. Lett. 78, 607 (1981)
30. K.C. Prince, M. Surmann, Th. Lindner, A.M. Bradshaw, Solid State Comm. 59, 71 (1986), and G. Paolucci, R. Rosei, K.C. Prince, A.M. Bradshaw, Appl. Surf. 22/23, 582 (1983)
31. A. Stamatovic, K. Leiter, W. Ritter, K. Stephan, T.D. Märk, J. Chem. Phys. 83, 2942 (1985)
32. K.-H. Bowen, G.W. Liesegang, R.A. Sanders, D.R. Hershbach, J. Phys. Chem. 87, 557 (1983)
33. J. Stöhr, D.A. Outka, R.J. Madix, and U. Döbler, Phys. Rev. Lett 54, 1256 (1985)
34. T.H. Upton, J. Chem. Phys. 83, 5084 (1985)
35. G. Odörfer, B. Bartos, H.-J. Freund, H. Kuhlenbeck, M. Neumann, to be published
36. A. Baddorf, D. Heskett, E.W. Plummer, unpublished
37. F. Solymosi and A. Berkó, to be published

Influence of Electronic and Geometric Structure on Desorption Kinetics of Isoelectronic Polar Molecules: NH_3 and H_2O

T.E. Madey, C. Benndorf[1], and S. Semancik[2]

Surface Science Division, National Bureau of Standards, Gaithersburg, MD 20899, USA

Consideration of the thermal desorption kinetics for isoelectronic NH_3 and H_2O from certain metal surfaces reveals an interesting contrast, which relates to geometrical and electronic structural effects in the adsorbed layers. Both NH_3 and H_2O are polar molecules bonded to the surface via lone pair orbitals on N and O, respectively. Hydrogen – bonding <u>attractive</u> interactions between neighboring H_2O molecules lead to formation of 2-d and 3-d clusters; thermal desorption kinetics of H_2O are characterized by sharp desorption peaks over narrow temperature ranges [the full width at half maximum (fwhm) is $\Delta T < 10K$ in some cases]. In distinction, lateral interactions between neighboring NH_3 molecules are largely <u>repulsive</u> (dipole-dipole interactions) and the thermal desorption spectra are considerably broader in temperature than for H_2O (fwhm $\Delta T \sim$ 70K to 150K, depending on substrate). In the following paragraphs, we summarize results obtained in our laboratory during the last few years which illustrate these points for desorption of NH_3 and H_2O from metal surfaces, including Ni(111), Ru(0001) and Ag(110). In each of these cases, the adsorption of NH_3 and H_2O is molecular, and desorption proceeds without dissociation.

First, consider the bonding geometry and adsorbate lateral interactions for adsorbed H_2O and NH_3. In the free H_2O molecule, there are two lone pair orbitals associated primarily with the O atom; in liquid and solid H_2O, hydrogen-bonding interactions involving the H-atoms and the lone pair orbitals result in tetrahedral coordination of the O atoms. There is a great deal of evidence from high resolution electron energy loss spectroscopy (EELS), work function measurements, ultraviolet photoemission spectroscopy (UPS), and electron-stimulated desorption ion angular distribution (ESDIAD) [1-3] that bonding of H_2O to metal surfaces occurs via one of the lone pair orbitals, with the other lone pair orbital available for participation in attractive hydrogen-bonding interactions with neighboring molecules. This leads to the formation of two and three-dimensional clusters on the surface even at cryogenic temperatures, where surface

---

[1] Permanent address: Inst. for Physical Chemistry, U. of Hamburg, W. Germany

[2] Chemical Process Metrology Division, NBS

mobility of adsorbed molecules is reduced. This configuration is indicated schematically here:

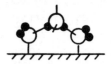

In the free NH_3 molecule, there is only one lone pair orbital, $3a_1$, which is associated primarily with the N atom. When NH_3 is adsorbed on metal surfaces the evidence from many surface measurements and calculations [4-7] indicates that bonding occurs via the lone pair orbital on the N atom, with H atoms pointed away from the surface. The dipole moment of an adsorbed NH_3 molecule can be rather high (< 2 Debye) and there is a net dipole-dipole repulsive interaction between neighboring molecules, as illustrated schematically:

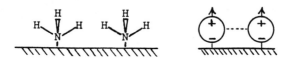

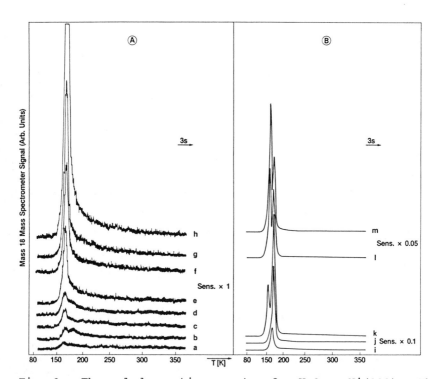

Fig. 1. Thermal desorption spectra for H_2O on Ni(111). The relative H_2O coverage corresponding to each spectrum is: (a) 0.02; (b) 0.04; (c) 0.06; (d) 0.07; (e) 0.18 (f) 0.22; (g) 0.33; (h) 0.50; (i) 0.30; (j) 0.70; (k),(l),(m)>1. from Ref. [8]

A series of results using thermal desorption spectroscopy (TDS) is shown in Fig. 1A and 1B for H_2O on Ni(111)[8].

From the lowest coverages in Fig. 1A, each spectrum is dominated by a single narrow peak (fwhm ~ 9K) which shifts from 165K in curves a-d to ~ 170K at higher coverages. At higher H_2O doses, the peak at 170K reaches a saturation value, and a second peak appears at ~150K (curves k, l, m) that does not saturate with higher exposures. The peak at 165-170K is due to desorption of the H_2O monolayer bound to the Ni substrate, whereas the peak at ~150K is due to desorption from an ice multilayer. The small peak at ~180K may be related to surface defects. Similar results for H_2O on Ni(111) were reported by STULEN AND THIEL [9]; other surfaces for which two TDS peaks are seen due to monolayer and multilayer H_2O include Pt(111) [10], Cu(110) [11], and Rh(111) [12].

The sharpness of the monolayer desorption peak and its coverage-dependent shift in peak temperatures are not consistent with simple first order desorption kinetics with a "normal" pre-exponential of 10^{13} s^{-1}. However, these data can be fit remarkably well by a model proposed by GOLZE, GRUNZE, AND HIRSCHWALD [13] based on first order desorption and coverage-dependent attractive lateral interactions. Following GOLZE et al. we assume

$$-\frac{d\Theta}{dt} = \nu\Theta \exp\left(- \frac{E_D^0 + W\Theta}{RT} \right).$$

Here ν is the preexponential factor, Θ is fractional coverage, E_D^0 is the desorption energy for a single admolecule at zero coverage and W is the pairwise interaction energy (positive for attractive interactions). If we choose $\nu = 10^{13}$ s^{-1}, $E_D^0 = 42$ kJ/mole, and W= 1400 J/mole, we calculate using the above equation that $T_p \sim 170K$ at $\Theta = 1$, the fwhm= 9K, and $\Delta T_p \sim 5K$ over the range of measurements; this is in excellent agreement with the experimental measurements shown in Fig. 1.

This interpretation has some interesting implications [1,8,9]. The coverage - dependent attractive lateral interaction term is significantly smaller than single hydrogen - bond energies in H_2O (~18 kJ/mole). In this model, the TDS behavior is not due simply to the existence of attractive lateral interactions, but to the increase in the strength of those interactions (i.e., an increase in binding energy) as coverage increases. This may occur via dipole reorientation in hydrogen - bonded H_2O surface clusters as coverage increases [1].

Another interpretation of these data is similar to that proposed previously for H_2O on Pt(111) [14] and discussed below for H_2O on Ag(110). In this model, H_2O diffuses rapidly at low coverages to form hydrogen-bonded clusters on the surface. Desorption of H_2O from clusters may proceed with a desorption order less than one [e.g., free sublimation from an ice multilayer proceeds via zero order kinetics]. This is consistent with the shift of T_p as coverage increases. BAUER et al [15] have shown similar results for desorption of metals (Ag, Cu) from islands on W(110).

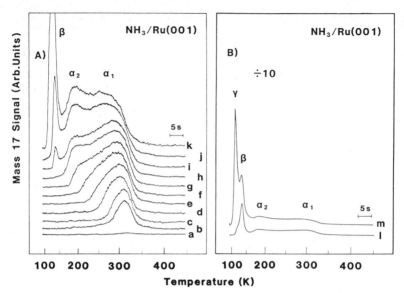

Fig. 2. NH_3 thermal desorption spectra for NH_3 on Ru(0001): (A) TDS for low NH_3 coverages (α_1, α_2 and β state); (B) TDS for high NH_3 coverages including the γ state. The NH_3 coverage corresponding to each spectrum is: (a) 0.00; (b) 0.04 (c) 0.05; (d) 0.07; (e) 0.09 (f) 0.12; (g) 0.15; (h) 0.18; (i) 0.2; (j) 0.28; (k) 0.35; (l) 0.52; (m) 0.76. (1 monolayer= 1.58×10^{15} molecules/cm^2).

Consider now the evidence for repulsive interactions in desorption of NH_3. A series of thermal desorption spectra (TDS) which illustrate the desorption of NH_3 from the hexagonal Ru(0001) surface is given in Fig. 2A, B [16]. At low NH_3 coverages, only one desorption peak is detected at 315K. With increasing NH_3 coverage, this desorption state (designated α_1) broadens and the peak temperature shifts toward lower desorption temperatures.

This behavior is consistent with repulsive interactions in which the desorption energy decreases with increasing coverage. For coverages > 0.15, a second desorption peak (designated $_2$) is observed. The saturation coverage of $\alpha_1 + \alpha_2$ is 0.25 monolayers (4×10^{14} molecules/cm^2). The β and γ peaks are due to 2$\underline{nd}$ layer and multilayer NH_3, respectively.

Assuming first order desorption kinetics for NH_3 in the α_1 state and a "normal" pre-exponential factor 10^{13} s^{-1}, we calculate the coverage dependence of desorption energy. The data were fit well by an expression of the form

$$E_D = E_D^0 - W\Theta^{3/2},$$

where E_D^0= 87 kJ/mole and W= 650 kJ/mole. Using a modified Topping model, we also calculated the coverage dependence of the dipole-dipole repulsive energy; E_D was found to vary as $\Theta^{3/2}$,

178

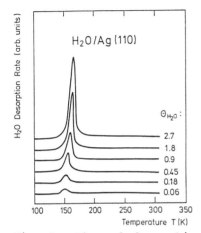

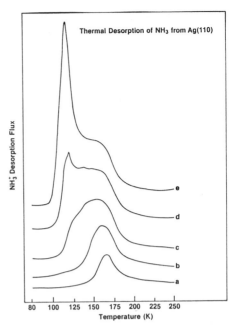

Fig. 3. Thermal desorption
spectra for desorption of
H_2O from Ag(110). The H_2O
coverage is indicated for
each curve. From BANGE
et al. [17].

Fig. 4. Thermal desorption
spectra for desorption of
NH_3 from Ag(110). From
SEMANCIK and MADEY [18].

although the calculated value of W was ~3 times smaller than that
observed experimentally [16].

Two other series of desorption spectra are shown in Figures
3 and 4; TDS of H_2O and NH_3 from Ag(110) are compared and
contrasted. For TDS of H_2O from Ag(110)[17], a single peak
desorbing with zero order kinetics is seen at all coverages, from
fractional monolayer to multilayer. The presence of only one
peak indicates that the binding energies for H_2O in the monolayer
and multilayers are similar, and that the binding energy for H_2O
on Ag(110) cannot exceed the sublimation energy of ice. Based on
EELS, UPS and ESDIAD studies, coupled with the TDS data of Fig.
3, it is believed that H_2O is adsorbed in poorly ordered hydrogen
bonded clusters (attractive lateral interactions) for T>80K, and
that the rate-limiting step in desorption involves evaporation of
H_2O from clusters, even for fractional monolayers [17].

In contrast, TDS of NH_3 from Ag(110) (Fig. 4) provides clear
evidence for repulsive lateral interactions: the desorption
spectra shift and broaden to lower temperature as coverage
increases [18]. As in the case of NH_3/Ru(0001) discussed above,
the data are consistent with a dipole-dipole repulsion model. In
further support of this model, ESDIAD data for NH_3 on Ag(110)
yield a "halo" pattern indicating that NH_3 is bonded (as on Ni,

Ru, etc.) via the N atom with the H atoms pointed away from the surface.

The cases discussed above (TDS of H_2O from Ag(110) and Ni(111)) have been relatively simple, with at most two TDS peaks corresponding to monolayer and multilayer, respectively. There are more complex situations (H_2O on Ru(0001)[3] and Ni(110)[19]) where three or even four TDS peaks are seen. In these cases, there appears to be a direct correlation between the more complex TDS spectra and the formation of highly ordered hydrogen-bonded H_2O layers: on both surfaces, both long-range (LEED) and short-range (ESDIAD) order is seen.

In summary, the thermal desorption kinetics for H_2O and NH_3 desorbing from metal surfaces provide evidence for attractive (H_2O) and repulsive (NH_3) lateral interactions in the adsorbed layers.

Acknowledgement.

We acknowledge, with pleasure, the contribution of several colleagues, including D.L. Doering, F.P. Netzer, K. Bange, J.K. Sass and E. Stuve. This work was supported, in part, by the Division of Basic Energy Sciences of the U.S. Department of Energy.

References

[1]   P.A. Thiel and T.E. Madey, Surface Science Reports, in press.

[2]   P.A. Thiel, F.M. Hoffmann and W.H. Weinberg, J. Chem. Phys. 75 (1981) 5556.

[3]   D.L. Doering and T.E. Madey, Surface Science 123 (1982) 305.

[4]   M. Grunze, in: The Chemical Physics of Solid Surfaces and Heterogeneous Catalysis, Vol. 4, Eds. D.A King and D.P. Woodruff (Elsevier, Amsterdam, 1982) Ch. 5 and references therein.

[5]   C.W. Seabury, T.N. Rhodin, R.J. Purtell and R.P. Merrill, Surface Sci. 93 (1980) 117.

[6]   B.A. Sexton and G.E. Mitchell, Surface Science 99 (1980) 523.

[7]   R.C. Baetzold, Phys. Rev. B29 (1984) 4211.

[8]   T.E. Madey and F.P. Netzer, Surface Science 117 (1982) 549.

[9]   R.H. Stulen and P.A. Thiel, Surface Science 157 (1985) 99.

[10]   G.B. Fisher and J.L. Gland, Surface Science 94 (1980) 446.

[11]   K. Bange, D.E. Grider, T.E. Madey and J.K. Sass,
         Surface Science 137 (1984) 38.

[12]   J.J. Zinck and W.H. Weinberg, J. Vac. Sci. Technol. 17
         (1980) 188.

[13]   M. Golze, M. Grunze, and W. Hirschwald, Vacuum 31
         (1981) 697.

[14]   G.B. Fisher and J.L. Gland, Surface Sci. 94 (1980) 446.

[15]   E. Bauer, F. Bonczek, H. Poppa and G. Todd, Surface
         Sci. 53 (1975) 87.

[16]   C. Benndorf and T.E. Madey, Surface Sci. 135 (1983)
         164.

[17]   K. Bange, T.E. Madey, J.K. Sass and E. Stuve, Surface
         Sci., in press.

[18]   S. Semancik and T.E. Madey, unpublished.

[19]   C. Benndorf and T.E. Madey, Surface Science, in press.

A Molecular Orbital Analysis of Chemisorption Precursor States

E.L. Garfunkel and Xing-hong Feng

Department of Chemistry, Rutgers, The State University of New Jersey, New Brunswick, NJ 08903, USA

I. Introduction

In molecular adsorption two kinds of precursor states are thought to occur [1]. An <u>intrinsic</u> precursor state refers to a weakly bound state of a molecule over a "clean" surface, while an <u>extrinsic</u> precursor state refers to weak adsorption on top of other adsorbates. Both kinds of precursor states are more mobile than chemisorbed states, and both can give rise to complex adsorption and desorption behavior. In this report, we focus on the intrinsic precursor state, presenting a simple molecular orbital analysis of precursor state potential wells and the activation barrier to chemisorption (see Fig. 1). Results of semi-empirical calculations for carbon monoxide adsorbing on nickel and other first row transition metals are described [2].

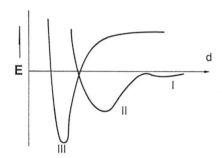

Fig. 1. A general potential energy diagram for a diatomic molecule approaching a surface showing: I- precursor, II associatively chemisorbed, and III dissociately chemisorbed states.

Recently, UHV surface spectroscopic methods have been used to study intrinsic precursor states and the chemisorption barrier [3-8]. For CO adsorbed on the nickel (111) surface, several experimental observations of precursor-like states have been reported [3,4]. Theoretical studies of Ni/CO systems have focused on the stable chemisorption state [9], rather than precursor states.

II. Model of Interaction

There are several contributions to the interaction between an adsorbate and a metal surface. In this work we focus our attention on the molecular orbital (MO) interaction between CO and the transition metal surface. Small metal clusters have been used as simplified models of surfaces for our calculations. Semi-empirical MO calculations [10] neglect some of the weak attractive interactions (induced dipole, quadrupole, etc.) [1c,11]. The inclusion of these terms either explicitly or implicitly (ab initio CI calculations) would lower the energies of both the wells and the barriers, but they should not alter the basic arguments presented below.

The basic argument we wish to present is that the adsorbate-substrate system undergoes several electronic state changes as it approaches the surface [12]. We begin the explanation by reviewing simple filled-shell repulsion. When two atoms or molecules approach one another, their orbitals will overlap and create bonding and antibonding combinations. The solution of the variational equations for a homonuclear diatomic molecule yields:

$$E_1 = H_{aa} - \alpha, \quad E_2 = H_{aa} + \beta,$$

where:

$$\beta = (SH_{aa} - H_{ab})/(1 - S_{ab}), \quad \alpha = (SH_{aa} - H_{ab})/(1 + S_{ab}),$$

as represented in Fig. 2.

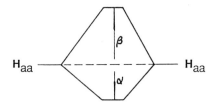

H_{aa} β H_{aa}

α

Fig. 2 Simplified energy diagram for 2 interacting orbitals.

Normally, β is bigger than α [10b]. Therefore, if there are 4 electrons in this system, then all levels will be filled and the net interaction between the two species will be repulsive (except at long ranges where dispersion forces lead to weak clustering).

When the system changes from a diatomic molecule, with a well-defined number of valence electrons, to a metal-adsorbate system, the picture becomes more complicated. For a solid, we first assume a Fermi level below which all orbitals are occupied. For the molecular orbitals of an adsorbate, the simplest case to consider is one where the highest occupied molecular orbital lies below the Fermi level of the metal. (For Ni/CO, the 5σ orbital of CO is implied.) We now let the molecule approach the surface and follow the orbital energies as a function of distance. The occupied valence molecular orbitals of the adsorbate will begin to couple with a metal orbital (or a group of them) and create bonding and antibonding combinations as the adsorbate approaches closer. As noted above, the absolute value of the repulsive antibonding combination is larger than that of the attractive bonding one, hence if both orbitals are occupied, the overall interaction is initially repulsive (i.e., closed-shell repulsion).

At a well-defined distance d_c, however, the antibonding combination should pass above the Fermi level (see Fig. 3). For distances closer

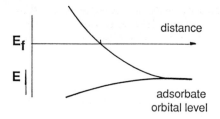

Fig. 3. Schematic splitting of occupied adsorbate level into bond and antibonding pair.

than this critical distance, the anti-bonding combination will become empty, donating its electrons to an unoccupied metal orbital at the Fermi level. We believe that this "state" change occurs mostly as an electronic rearrangement within the substrate orbitals. The Fermi level changes only infinitesimally as a molecule adsorbs on the surface, hence the main interaction term changing for distances closer than d_c will be the attractive bonding one. The energy then becomes attractive (see Fig. 4) until core repulsion causes the next minimum to be reached. A similar type of adsorption barrier has also been observed for hydrogen [13], and we believe that this MO analysis of the barrier to chemisorption holds for many molecule-surface systems. (The upper curve corresponds to the electronically excited adiabatic curve for a molecular system. On the surface, however, there are many other intermediate energy potential curves which correspond to excitations in the metal.)

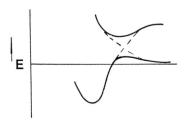

Fig. 4. Potential energy diagram corresponding to the electron orbital system shown in Fig. 3.

An adsorption barrier must exist between the precursor and chemisorption wells for the precursor state to exist. This holds for precursor states due to either dispersion-type attractions or to an actual molecular orbital attraction as we describe below.

A second molecular orbital term results from the interaction of initially unoccupied orbitals of the adsorbate (for CO, the 2π orbitals) with occupied metal ones. As the orbitals interact, unoccupied molecular ones will become occupied when they pass below the Fermi energy. A more accurate picture is that a substrate orbital of the correct symmetry will become stabilized by mixing with the unoccupied orbital. In this situation there is only an attractive component to the interaction (from the stabilization of the substrate orbital), as the antibonding combination is always empty. If the originally unoccupied adsorbate orbital is diffuse, then the attractive interaction may occur before a filled shell barrier is reached. This

184

can help explain the existence of a molecular orbital-induced precursor well. As the molecule approaches closer, other orbitals will also mix and can give rise to additional energy barriers and wells.

III. Results of Calculations.

Below we present the results of EHMO [10] semi-empirical calculations on clusters [2]. The clusters are used as approximate models of real adsorbate surface systems. As the EHMO method is not accurate, the energies and distances of the results should only be taken as approximate values.

In Fig. 5 we show the orbital energies of a CO/Ni_4 cluster as CO approaches the surface in a perpendicular configuration. As can be seen, the M-CO(5σ) antibonding combination passes above Ef (~9eV) near 2.5 Å.

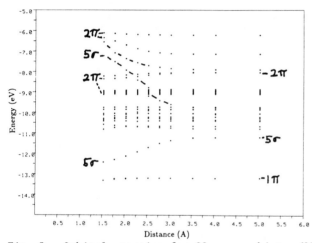

Fig. 5. Orbital energies for CO approaching a Ni_4 cluster bridge site, carbon end first.

The sum of the occupied one-electron orbital energies is the potential energy of the system (at the EHMO level of approximation). The potential energy diagram for this system is similar to that shown schematically in Figure 4. The energy and distance of the potential barriers are listed in Table 1 for top, bridge, and threefold site approaches [2].

Also listed in Table 1 are the results for CO approaching nickel clusters oriented parallel to the nickel cluster surface. For this orientation we do observe an initial precursor well before the barrier is reached, as depicted in Fig. 1. Our interpretation of the appearance of a precursor well for the parallel orientation (and not the perpendicular orientation) follows from the argument presented above. The 2π (CO) levels are oriented away from the C-O axis, interact most strongly (attractively) from the side of CO, and hence

Table 1

CO site	Orientation	barrier		precursor well	
		distance	energy	distance	energy
bridge	vertical	2.6 Å	0.34 eV		
	parallel	2.45	0.04	2.9	0.03
threefold	vertical	2.3	0.57		
	parallel	2.2	0.11	3.1	0.01
top	vertical	2.4	0.37		
	parallel	1.0	3.44	3.0	0.03

favor a side-on configuration. The 5σ (CO) level is oriented along the C-O axis and interacts most strongly for a carbon end first approach. The 5σ interaction is at first repulsive, hence the best attractive long distance interaction occurs for the side-on parallel orientation.

We have also performed a series of calculations for CO on the other first row transition metal series (2). The main results of these calculations are that 1) the adsorption barrier heights were similar for most of the metals, and 2) the precursor wells get much weaker as one moves from nickel to the left on the periodic table (or to copper). We can draw correlations between the energy and position of the CO precursor well, and the ionization potential (Fermi energy) and metal orbital diffuseness of the cluster atoms. Although the molecular orbital contribution to the precursor state appears to be small for the metals other than nickel, the attractive weak interactions (dispersion, etc.) combined with the filled shell barrier may result in a precursor state.

Ab initio Hartree-Fock calculations with nickel pseudo-potential cores have recently been performed on these systems. The initial results confirm the general idea of state crossings and adsorption barriers as predicted by the EHMO calculations. Complete CI calculations are in progress [14], as they are needed to quantify the problem of the relative contribution of MO overlap and weak interactions for the precursor state, confirm the nature of the barrier, and better understand the various hybridizations of substrate and overlayer orbitals.

References

1. a) P. Kisliuk: J. Phys. Chem. Solids 5, 78 (1958); b) R. Gorte and L. Schmidt: Surface Sci. 76 (1978) 559; c) J.E. Lennard-Jones, Trans. Faraday Soc. 28, 333 (1932); d) D.A. King; Surf. Sci. 64, 43 (1977)

2. a) E.L. Garfunkel and X.H. Feng: Surf. Sci. 176, (1986); b) X.H. Feng and E.L. Garfunkel: Langmuir, submitted for publication

3 a) S.L. Tang, M.B. Lee, J.D. Beckerle, M.A. Hines and S.T. Ceyer: J. Chem. Phys. 82, 2826 (1985); b) S.L. Tang, J.D. Beckerle, M.B. Lee and S.T. Ceyer: J. Chem. Phys. 84, 6488 (1986)

4. M. Shayegan, E.D. Williams, R.E. Glover III and R.L. Park: Surface Sci. 154, L239 (1985)

5. a) J. Eickmans, A. Otto and A. Goldmann: Surface Sci. 149, 293 (1985); K.C. Prince, G. Paolucci, A.M. Bradshaw, K. Horn and C. Mariani: Vacuum 33, 867 (1983)

6. T. Engel: J. Chem. Phys. 69(1), 373 (1978)

7. P.R. Norton, R.L. Tapping and J.W. Goodale: Surface Sci. 72, 33 (1978)

8. A. Cassuto and D.A. King: Surface Sci. 102, 388 (1981)

9. a) S. Sung and R. Hoffmann: J. Am. Chem. Soc. 107, 578 (1985); b) C.M. Kao and R.P. Messmer: Phys. Rev. B31, 4335 (1985); c) J.N. Allison and W.A. Goddard, III: Surf. Sci. 110, L615 (1981); d) L.M. Rohlfing and P. Hay: J. Chem. Phys. 83 (9), 4641 (1985)

10. a) R. Hoffmann: J. Chem. Phys. 39, 1397 (1963); b) T.A. Albright, J.K. Burdett and M.H. Whangbo, Orbital Interactions in Chemistry (Wiley, New York, 1985)

11. a) L.W. Bruch: Surf. Sci. 125, 194 (1983), and refs. therein; b) G. Vidali, M.W. Cole, and J.R. Klein: Phys. Rev. B28, 3064 (1983)

12. S. Holloway and J.W. Gadzuk: J. Chem. Phys. 82 (11), 5203 (1985), and refs. therein

13. a) B.I. Lundqvist, O. Gunarsson, H. Hjelmberg and J.K. Norskov, Surface Sci. 89, 196 (1979); b) J.K. Norskov, A. Houmoller, P.K. Johansson and B.I. Lundqvist: Phys. Rev. Letters 46, 257 (1981); c) J.K. Norskov, S. Holloway and N.D. Lang: Surface Sci. 137, 65 (1984)

14. M. Marino, W. Ermler, X.H. Feng and E.L. Garfunkel: to be published

Thermal and Non-thermal Activation of Si(100) Surface Nitridation

F. Bozso and Ph. Avouris

IBM T.J. Watson Research Center, P.O. Box 218, Yorktown Heights, NY 10598, USA

INTRODUCTION

There is currently great interest in understanding the surface chemistry and thin film growth mechanisms of technologically important semiconductors. One intriguing finding of many studies in this area is that unusually high temperatures are required for sustained reactivity and film growth. Thus, for example, silicon nitride film growth using the reaction of Si with NH_3 is carried out at temperatures of 1000-1300K.[1,2] Intuitively one would expect that the presence of surface dangling bonds should give to semiconductor surfaces a free-radical-like reactivity characterized by low activation barriers. The high process temperatures are often undesirable because they can also enhance the rates of unwanted processes such as dopant diffusion. For these reasons non-thermal means of reaction are actively being sought.[3] Essential for the success of such efforts is to understand the mechanism of the reaction, particularly the nature of the rate-limiting step.

Here we present results from our studies of the reactions of Si(100) with NO and NH_3 [4] to give technologically important silicon nitride films. In these studies, using a variety of surface spectroscopic and analytical techniques, we have been able to both identify the reaction step requiring high thermal activation and to implement a non-thermal activation scheme which allows controlled Si_3N_4 thin film growth even at 90K.

RESULTS AND DISCUSSION

A. The reaction of Si(100) with NO

In Fig. 1 we show N(1s) x-ray photoelectron spectra (XPS) obtained after exposing a Si(100) 2x1 surface at 90K to NO. Two main peaks are observed: a low binding energy peak at 397.7 eV which is characteristic of N in silicon nitride[4] and a high binding energy peak at 401.9 eV. Upon gentle annealing of the sample to 300K we observe the disappearance of the 401.9 eV peak with a simultaneous increase of the intensity of the 397.7 eV peak. Electron stimulated desorption (ESD) from the NO-exposed surface at 90K shows strong NO^+ desorption which however, is completely eliminated when the surface temperature is raised to 300K. We thus conclude that the NO exists on the Si(100) surface at 90K in both a dissociated and in a weakly adsorbed molecular form. Heating leads to the partial dissociation and partial desorption of the molecular form. Cooling the annealed surface to 90K and re-exposing it to NO does not lead to any further increase of the XPS peaks corresponding to atomically or molecularly adsorbed NO. The surface shows a passivation towards any further reaction.

Ion scattering spectroscopy (ISS) provides information about the composition of the reacted surface (Fib. 1b). Upon NO exposure the Si ISS signal decreases with the simultaneous appearance of a strong O ISS peak. No N could be detected by ISS except for a weak shoulder on the O ISS peak which was eliminated upon annealing. No further change of the ISS spectrum

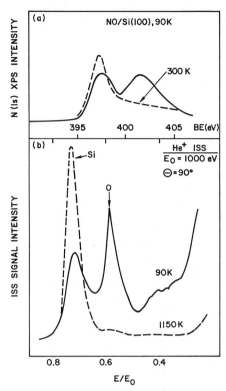

Fig. 1.   (a) N(1s) x-ray photoemission spectrum (XPS) of the Si(100)-(2x1) surface after exposure to 10L NO at 90K (solid line) and after annealing to 300K (dashed line). (b) He⁺-ion scattering spectra (ISS) of the Si(100)-(2x1) surface after exposure to 10L NO at 90K (solid line) and after annealing to 1150K (dashed line).

was detected with increasing temperature until about 900K where thermal desorption spectroscopy (TDS) showed the desorption of SiO molecules (Fig. 3). Above this temperature the O ISS peak decreases sharply while the Si peak increases. At ~1150K the ISS spectrum shows no evidence of oxygen or nitrogen while the Si signal is as strong as that of the initial clean Si(100) surface. XPS studies, on the other hand, show no decrease in the N(1s) signal as a result of the heat treatment.

From these studies we conclude that when NO dissociates on the Si(100) surface the O atom stays on the Si surface and ties up surface dangling bonds, while the N atom occupies a sub-surface site and is thus shielded from the incident He⁺ ions in ISS.

B. The Reaction of Si(100) with NH_3

The behavior of the Si(100) + NH_3 reaction is in many respects similar to that of the Si(100) + NO reaction. Exposure of the Si(100) 2x1 surface to 10L of NH_3 at 90K yields the N(1s) XPS spectrum shown in Fig. 2a. Again, two main N(1s) peaks are observed: a peak at a binding energy of 399.7 eV and a second one at 397.7 eV. The 399.7 eV peak is ascribed to molecularly adsorbed NH_3 . Heating the sample to 300K leads to NH_3 desorption (Fig. 3) and to the dis-

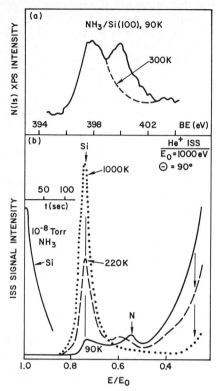

Fig. 2.   (a) N(1s) x-ray photoemission spectrum (XPS) of the Si(100)-(2x1) surface after exposure to 10L NH_3 at 90K (solid line), and after annealing the NH_3-exposed surface to 300K (dashed line).   (b) He^+ ion scattering spectra (ISS) of the Si(100)-(2x1) surface after exposure to 10L NH_3 at 90K (solid line), after subsequent annealing to 200K (dashed line), and after annealing to 1000K (dotted line). Inset: Si ISS peak intensity decrease in the course of a 10L NH_3 exposure at 90K. (Laboratory scattering angle 90°.)

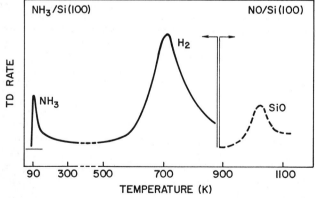

Fig. 3.   Thermal desorption spectra of NH_3, H_2 and SiO from a Si(100)-(2x1) surface after exposure at 90K to 10L NH_3 and 10L NO respectively.

appearance of the 399.7 eV peak. Cooling the surface back down to 90K and re-exposing to NH_3 regenerates the XPS peak characteristic of molecularly adsorbed NH_3. Again however, no further NH_3 dissociation takes place, the surface having been rendered inactive. He^+ ISS provides important clues regarding the nature of the surface passivation. As Fig. 2b shows, exposure of the Si(100) to NH_3 at 90K leads to a fast decrease of the Si ISS signal indicating a high sticking probability (s>0.5). The ISS spectrum of the mixed nitride-molecularly adsorbed NH_3 phase at 90K shows weak Si and N signals and a strong background at low energies. Mild annealing of this phase to 220K and desorption of the molecularly adsorbed NH_3 (Fig. 3) leads to the increase of the Si ISS signal, which now has recovered approximately half of its original intensity, and the total elimination of the N ISS signal. (A small peak at $\sim E/E_0 = 0.6$ is due to a small oxygen contamination during the ISS experiment.) Simultaneously, the low energy background is greatly reduced. No further significant change of the ISS spectrum is observed until the temperature is raised above ~700K. At 1000K (Fig. 2b) the Si ISS signal is as strong as that from the clean surface. The background at low energies has been eliminated, but, as in the case of the reaction with NO, no surface nitrogen is detected. Parallel XPS measurements, on the other hand, show no decrease in the N(1s) signal as a result of heating to 1000K. We conclude that the nitrogen resulting from the dissociation of NH_3 resides in sub-surface sites and is thus shielded from the incident He^+ ions. The only surface nitrogen detected is that from molecularly adsorbed NH_3. The hydrogen from NH_3 could shield the Si atoms and give rise to the background at small E/E_0. However, the fact that after NH_3 desorption the Si ISS signal has only about half of its original intensity and that there is still significant background indicates the presence of surface hydrogen directly bonded to silicon. Since reactions between closed-shell species have high activation energies,[5] this surface hydrogen, by tying up the dangling bonds of the silicon surface, may inhibit further NH_3 dissociation and, thus, lead to the observed self-limiting behavior. Strong support for this hypothesis came from XPS studies of nitrogen uptake by the Si(100) surface exposed to NH_3 at increasing temperatures. These studies showed that there is a strong correlation between the extent of nitridization and the thermal desorption of H_2 from the surface (Fig. 3). Thus it was found[4] that the surface passivation persists till about 700K where the hydrogen thermal desorption rate is sufficiently high. Above that temperature sustained silicon nitride growth was observed.

From the above experiments it can be concluded that the Si(100) 2x1 surface is highly reactive and can dissociate both NO and NH_3 even at 90K. This high reactivity is ascribed to the existence of surface dangling bonds, up to one per surface Si atom for symmetric dimers on the 2x1 surface or two per Si atom in the bulk-like terminated surface structure, giving the surface free-radical-like reactivity which is characterized by very low activation barriers. Surface oxygen and particularly surface hydrogen tie up the dangling bonds and lead to surface passivation. Nitrogen, on the other hand, tends to reside in sub-surface sites and does not appear to affect significantly the reactivity of the surface. Having understood the cause of the surface passivation we sought to enhance the thin film growth processes by non-thermal means. Here, we present results showing that silicon nitride thin film growth by the reaction of Si(100) with NH_3 can be achieved even at 90K, provided that the surface hydrogen has continuously been removed by electron stimulated desorption (ESD). Oxynitride thin film growth by the reaction of Si(100) with NO can also be enhanced by electron beam irradiation. The more complex behavior of the silicon oxynitride films under electron irradiation will be described elsewhere.[6]

In Fig. 4 we show the dependence of the H^+ ESD yield from the NH_3-exposed Si(100) surface on incident electron energy. The absolute energy threshold is observed at ~21 eV in agreement with previous studies of H^+ ESD from Si(111) surface.[7] No enhancement of the ESD yield is observed above the N(1s) ionization energy but an increase is observed above the Si(1s) ionization energy.

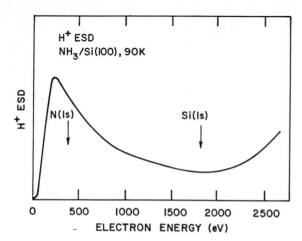

Fig. 4.   Electron energy dependence of H$^+$ electron stimulated desorption (ESD) yield from a Si(100)-(2x1) surface exposed to 10L NH$_3$ at 90K. (Electron current ~100nA.)

In Fig. 5 we show the influence of electron irradiation on the reaction of Si(100) and NH$_3$ using as probes the Si LVV and N KLL Auger spectra excited by 1 keV electrons. In these studies the same electron beam is used to induce the reaction and also excite the Auger process. In Fig. 5 curve (a) shows the Auger spectra of the surface after exposure to NH$_3$ at 90K. As can be seen the N Auger signal is very small, corresponding to sub-monolayer nitride, while the Si Auger spectrum is the same as that of clean Si. Curve (b) shows the Si LVV spectra obtained after raising the sample temperature to 900K and exposing it to 10^{-7} torr of NH$_3$ for 5 min. The spectra show the thermal growth of silicon nitride which gives the characteristic Si Auger peak at ~ 84 eV.[8] In addition, satellites displaced by 11 eV and 21 eV from this peak are observed. These satellites are ascribed to electron energy losses resulting from the excitation of silicon surface plasmons[9] and silicon nitride bulk plasmons,[10,11] respectively. Curve C shows the Si LVV spectra obtained when the sample from step (b) is cooled back down to 90K and exposed for 10 min simultaneously to 10^{-7} torr NH$_3$ and the 20μA, 1 keV electron beam. We see that even at 90K the electron beam converts essentially all silicon that can be probed by the Auger process to silicon nitride. In addition, the intensity of the N KLL Auger is approximately 10 times stronger than the corresponding signal from the surface not irradiated by the electron beam. Assuming a layer-by-layer growth mechanism for the silicon nitride film we estimate its thickness to be 20±5Å.

Thus, we see that through the electronic excitation induced by the electron beam we have achieved controlled silicon nitride film growth even at 90K. Our studies suggest that the main role of the electron beam is to induce hydrogen desorption and thus help maintain a hydrogen-free Si surface which can then dissociate NH$_3$. Although, in principle, electron beam induced dissociation of adsorbed NH$_3$ and the possible introduction of Si defects could further enhance the film growth process, neither seems to be essential. This is deduced from the fact that no measurable enhancement of the growth rate is observed on going from incident-electron energies below the N(1s) threshold to energies above this threshold (see also Fig. 4). Also, after heating the irradiated surface to ~1000K and cooling it back down to 90K we find that it can readily dissociate NH$_3$, although this process should have annealed most of the possible radiation-induced defects. The H-free surface can dissociate NH$_3$ even when N(1s) XPS studies indicate the presence of several silicon nitride layers.

192

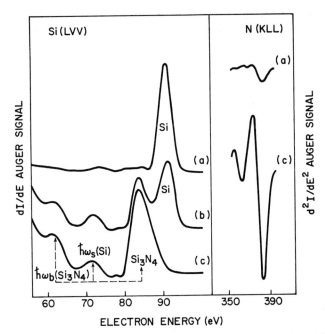

Fig. 5.   Si(LVV) and N(KLL) Auger spectra of the Si(100(-(2x1) surface, (a) after exposure to NH_3 at 90K, (b) after partial thermal nitridation at 900K by exposure to
10^{-7} torr NH_3 for 5 min, and (c) after simultaneous exposure to NH_3 and an
electron beam of 20μA, 1000 eV at 90K. Note that the Si(LVV) and N(KLL) Auger
spectra were recorded in dI/dE and d^2I/dE^2 mode respectively.

The nature of the surface layer of the thin silicon nitride film presents an interesting
problem. The Si Auger spectra of Fig. 5 indicate the presence of Si surface plasmon excitations
in both the thermally grown and e-beam grown silicon nitride films. Note also that while the
nitride bulk plasmon is observed, no corresponding surface plasmon is detected. The above result coupled with the observed full recovery of the Si ISS signal after the hydrogen thermal
desorption (Fig. 2b) and also depth profiling results using SIMS of the Si_2N^+ ion (not shown)
strongly suggest that the Si segregates on top of the silicon nitride thin film. This Si surface layer
appears to have very similar reactivity to that of the clean Si(100) 2x1 surface and should play
a key role in the film growth process. The detailed electronic structure of this surface layer is
under study.

REFERENCES

1.   M. M. Moslehi and K. C. Saraswat, IEEE J. Solid State Circ. **20**, 26 (1985).

2.   J. A. Nemetz and R. E. Tressler, Solid State Tech.. 79 (February 1983).

3.   For reviews see: Nanometer Structure Electronics: An Investigation of the Future of
     Microelectronics, edited by Y. Yamamura, T. Fujisava and S. Nambs (North-Holland,
     Amsterdam 1985); Laser and Electron Beam Processing of Materials, edited by C. W.
     White and P. S. Peercy (Academic Press, New York 1980).

4.   F. Bozso and Ph. Avouris, Phys. Rev. Letters **57**, 1185 (1986).

5.   S. W. Benson, "Thermochemical Kinetics," J. Wiley, New York, 1976.

6.   Ph. Avouris and F. Bozso, to be published.

7.   H. H. Madden, D. R. Jennison, M. M. Traum, G. Margaritondo and N. G. Stoffel, Phys. Rev. B **26**, 896 (1982).

8.   J. F. Delord, A. G. Schrott and S. C. Fain, J. Vac. Sci. Technol. **17**, 517 (1980).

9.   H. Raether, "Excitation of Plasmons and Interband Transitions by Electrons" (Springer-Verlag, Berlin-Heidelberg-New York 1980).

10. R. Kärcher, L. Ley and R. L. Johnson, Phys. Rev. B **30**, 1896 (1984).

11. N. Lieske and R. Hezel, Thin Solid Films **61**, 217 (1979).

Precursor Mediated and Direct Adsorption: Two Channels Leading to Chemisorption

T. Engel

Department of Chemistry, University of Washington,
Seattle, WA 98195, USA

The preceding papers by WEINBERG and by AUERBACH and RETTNER have addressed the question of how important precursors are in kinetic processes on surfaces from different perspectives. Both papers have considered the evidence for the existence of precursors and all authors agree that precursors to adsorption have been clearly identified under certain conditions. The clearest proof of existence of the precursor is its spectroscopic identification, and numerous studies [1-4] carried out at very low temperatures have obtained "fingerprint" spectra or other electronic features which can unambiguously be identified with extrinsic and intrinsic precursors. However, as WEINBERG points out, molecularly chemisorbed species can also be regarded as precursors to dissociatively adsorbed species. They can be isolated at much higher temperatures since the energies of adsorption, desorption and migration are scaled to higher values.

It is important to consider the concept of a precursor in a historical perspective since essentially all the important kinetic and statistical models describing precursor adsorption were formulated before 1960. At that time various methods had been developed to measure surface coverages and sticking coefficients on relatively well defined surfaces. The key experimental feature which researchers were attempting to understand was the way in which the sticking coefficient, S, decreased with adsorbate coverage. As WEINBERG has discussed, precursor models contain a great deal of microscopic information which could not be determined independently. The only test of the models at the time they were proposed was the variation of $S(\theta)$ with θ which is of course insufficient to determine the various input parameters unambiguously. Recent computer simulation studies by HOOD et al. [5] have used a microscopic model containing all relevant parameters to calculate adsorption and desorption rates. Comparisons with experimental results show good agreement. These studies provide the best justification for the microscopic models underlying precursor mediated kinetics. Not until it became straightforward to vary the surface temperature and impinging molecule energy independently was it possible to explore through experiments whether predictions which follow if precursor mediated adsorption occurs are indeed satisfied. For instance, the surface temperature should affect the adsorption probability more critically than the impinging particle energy for precursor mediated adsorption, whereas the opposite can be true for a direct adsorption process.

Molecular beam techniques allow a detailed exploration of the individual processes such as accommodation, trapping, and the transition from a physisorbed to a chemisorbed state which are the atomistic steps leading to adsorption. Time scales down to the microsecond range can be measured directly using chopping techniques. However, shorter (but less well defined) time scales can also be investigated by measuring angular resolved scattering intensity distributions or the corresponding distributions in velocity. By carrying out scattering experiments at surface temperatures for which the chemisorbed species does not desorb, it can be clearly established whether those molecules which do not adsorb have accommodated to the surface temperature. Since desorption from the precursor state is assumed to follow equilibration to the surface temperature, a cosine angular distribution and a Maxwell-Boltzmann distribution at the surface temperature is expected. Molecules or atoms which undergo direct inelastic scattering should scatter in a lobular pattern with a velocity distribution which is centered near that of the incident beam. Note that since these experiments described sample only those atoms or molecules which do *not* go on to chemisorb, there are limitations on the information which can be obtained. For instance, if 90% of the molecules trapped in a precursor state went on to chemisorb, a measurement of the backscattered intensity as a function of exit angle might not be able to distinguish between precursor mediated and direct adsorption channels under realistic experimental conditions. An additional experiment which can test for precursors is to examine the dependence of S on E_i. Here the scattered flux is not measured, and the beam is used as a variable energy source with various uptake measurements leading to a determination of S.

Precursor mediated adsorption relies on the initial step of trapping in the weakly bound state. This occurs most readily for incident species of low energy since the probability that an incident molecule will lose enough energy to be trapped falls with increasing incident energy. This can be seen simply by considering a binary collision between an incoming gas atom and an effective mass representing a surface cluster involved in the collision. This result has been verified experimentally in rare gas trapping, for which the complicating feature of a transition to a chemisorbed species is absent [6]. Therefore, the sticking coefficient in a precursor mediated process should *decrease* as the incident energy *increases*.

However, molecular beam studies have shown that the opposite often occurs. This can be seen in figure 11 of the paper by AUERBACH and RETTNER which shows results for the $O_2/W(110)$ system. The increase in initial coefficient sticking with energy is clearly incompatible with precursor mediated adsorption. This seems particularly puzzling since this adsorption system is one in which precursor kinetics were thought to be well established [7,8]. This leaves us with an apparent contradiction between two different types of measurements on the same adsorption system.

Let us consider the two types of experiments in detail. We characterize classical measurements as those for which adsorption of a gas in equilibrium with

196

the walls of a chamber at 300K on a surface of variable temperature are studied. The energy spread of the impinging gas particles is large and is described by the Maxwell-Boltzmann distribution. All angles of incidence are represented. In the molecular beam experiment, the energy spread is small and the spread in angles of incidence is typically much less than one degree. It is difficult to compare the consequences of these very different conditions on adsorption kinetics in general. However, they can be compared for the case in which only the normal component of the incident atom velocity, v_z, determines its adsorption probability. This is for instance the behavior observed for O_2 adsorption on W(110) [9]. It is convenient to define a normal energy E_n given by $1/2\ mv_z^2$.

To compare the two types of experiments one must first determine what fraction of the molecules or atoms incident upon the surface from a gas in equilibrium with a chamber at temperature T have a normal energy greater than a given value E_n. This fraction is $\varrho^{-E_n/kT}$. By substitution in this formula for $T = 300K$ we obtain values of 0.38, 0.14 and 0.02 for $E_n = 0.025eV$, $0.050eV$ and $0.10eV$ respectively. We note that the lowest energy values for which the sticking coefficient was determined by AUERBACH, RETTNER et al. for the O_2/W(110) and N_2/W(100) adsorption systems is 0.05eV [9]. For this value of E_n, 86% of the incident molecules or atoms in the classical experiment have normal energies which are below the *lowest value* which have been used in molecular beam adsorption experiments to date. This shows that rather than being contradictory, the two methods are complementary in that they span different energy ranges with essentially no overlap.

Viewed in this way, the most significant contribution of the molecular beam measurements is that they have shown that a direct adsorption channel is available to molecules with high incident energies. The sticking coefficient for this channel can be high when compared to adsorption at low energies. At this point there is a good deal of evidence which suggests that adsorption at low incident energies is influenced to a significant extent by precursor states. It is not clear to what extent direct adsorption processes also play a role under these conditions. Although the contribution of direct adsorption processes with high values of E_n is negligible in classical experiments at low pressures, they may still be of importance in catalytic processes at high pressures. The fraction of molecules with $E_n > 0.5eV$ in a gas at 300K is 3.8×10^{-9}. However, if the impingement rate is high enough the contribution from such high energy species may be significant [10].

It is instructive to view precursor mediated and direct adsorption in terms of scattering from a potential energy surface. Multidimensional potential energy surfaces have been discussed by AUERBACH and RETTNER in the preceding article. Clearly, whether an adsorption system is precursor or direct adsorption dominated, scattering takes place from the same potential energy surface. However in the two different cases, different regions of the energy surface are being sampled. Since dynamical effects are important in surface scattering, trajectories may sample quite different regions in energy space for different initial conditions such as E_i. For this reason, there is in general no simple relationship between the rise in the sticking coefficient with energy for the direct adsorption channel and an effective activation energy for adsorption. This is illustrated by recent calculations for the scattering of H_2 from nickel surfaces [11]. These studies show that features such as normal energy scaling do not unambiguously show whether the barrier is in the entrance or exit channel. In molecular beam experiments, one is tracing out complex trajectories in energy space for well-defined initial conditions but has no way to determine what the path is in terms of molecular and surface coordinates. Similarly, for the classical adsorption experiment, the interpretation becomes complex since the range of incident parameters due to the Maxwell-Boltzmann distribution in velocities and the distribution in incident angles is enormous. So much averaging over initial conditions takes place that it is difficult to describe such an experiment in terms of scattering from a potential energy surface. These considerations make it imperative to focus on those aspects of the experiment which are easily interpretable.

In molecular beam experiments, this suggests that measurements of S for lower values of E_i should be carried out to give a larger overlap with classical studies. For classical studies, several useful tests for establishing whether a precursor mediated mechanism is dominant have been suggested by WEINBERG. These tests should be applied. Unfortunately, most classical studies have used the dependence of $S(\theta)$ on θ as the sole criteria as to whether precursor mediated adsorption occurs. This is less easily interpreted than measurement of S as θ approaches zero for different surface temperatures due to possible heterogeneities induced by already adsorbed species. A number of other mechanisms could also be proposed to explain why $S(\theta)$ falls off more slowly than $(1-\theta)^2$ or $(1-\theta)$ depending on whether adsorption is dissociative or not. For instance, if an incident molecule loses enough normal energy to be trapped perpendicular to the surface but insufficient energy to be trapped parallel to the surface, it can sample a number of adsorption sites before becoming strongly bound. This does not imply the importance of a precursor since we are not dealing with a well-defined state, but rather a non-equilibrium effect. Similarly, at very high incident energies, $S(\theta)$ falls off slowly with θ for O_2 adsorption on $W(110)$ although precursor dominated adsorption seems very unlikely. This behavior has been attributed to the ability of the O_2 molecule to penetrate into energetically less favorable positions on the surface due to its high kinetic energy [12]. Under these conditions, the concept of well-defined sites has little meaning until equilibration has occurred.

In summary, the articles by WEINBERG and by AUERBACH and RETTNER have eloquently summarized the current state of research concerning the elementary steps leading to adsorption. There is little question that adsorption studies carried out under classical conditions in which adsorption takes place from a gas in thermal equilibrium with a chamber at 300K are often precursor mediated. There is also little question that at high incident energies, a second channel involving direct adsorption is dominant. These two mechanisms should not be viewed as alternatives, but rather as extremes in the scattering behavior from the same potential energy surface.

References

1. P.R. Norton, R.L. Tapping, J.W. Goodale, *Surface Science* **72**, 33 (1978).
2. P. Hofmann, K. Horn, A.M. Bradshaw, K. Jacobi, *Surface Science* **82**, L610 (1979).
3. M.J. Grunze, J. Fihler, M. Neumann, C.R. Brundle, D.J. Auerbach, J. Behm, *Surface Science* **139**, 109 (1984).
4. M. Shayegan, E.D. Williams, R.E. Glover, R.L. Park, *Surface Science* **154**, L239 (1985).
5. E.S. Hood, B.H. Toby, W. Tsai and W.H. Weinberg, this volume.
6. K.C. Janda, J.E. Hurst, J. Cowin, L. Wharton and D.J. Auerbach, *Surface Science* **130**, 395 (1983).
7. C. Wang and R. Gomer, *Surface Science* **84**, 329 (1979).
8. T. Engel, H. Niehus and E. Bauer, *Surface Science* **52**, 237 (1975).
9. See figures 11 and 13 of the article by D.J. Auerbach and C.T. Rettner in this volume. Although somewhat lower values of E_n are used for the O_2/W(110) system, the data points correspond to rather grazing angles of incidence for which normal energy scaling breaks down.
10. M.B. Lee, Q.Y. Yang, S.L. Tang and S.T. Ceyer, *J. Chem. Phys.* **85**, 1693 (1986).
11. C.-Y. Lee and A.E. DePristo, *J. Chem. Phys.* **84**, 485 (1986).
12. C.T. Rettner, L.A. DeLouise and D.J. Auerbach, *J. Chem. Phys.* **85**, 1131 (1986).

Part III

Kinetics of Phase Transitions
at Surfaces

Experiments on Kinetic Effects at Phase Transitions

K. Heinz

Institut für Angewandte Physik, Lehrstuhl für Festkörperphysik
der Universität Erlangen-Nürnberg, Erwin-Rommel-Str. 1,
D-8520 Erlangen, Fed. Rep. of Germany

1. Introduction

The physics of phase transitions has experienced substantial progress and
an increase in experimental and theoretical activity during the past decade.
Due not least to recognition of the importance of dimensionality, this
development also spread over into surface physics, where 2-dimensional or
quasi-2-dimensional systems are available for investigation, allowing a
check of the fundamental theoretical concepts. Considerable efforts were
made, both theoretical and experimental, and today a variety of phase diagrams of adsorbate systems are known /e.g. 1-5/. In some cases even critical
parameters of phase transitions have been experimentally determined /6-8/.

The situation concerning the kinetics of phase transitions, however, is
quite different. The considerable theoretical efforts undertaken in the past
(see J. D. Gunton, this volume) are not quantitatively balanced by experiment. Only about a dozen investigations of the kinetics of surface phase
transitions exist, and it was only very recently that the experimental activity in this field began to increase. It is the purpose of this paper to
review the present state of this experimental work.

Analogous to the bulk transition, kinetics in the surface of a solid display the path along which the atoms of the first or first few layers transform
from one state of order to another. Therefore, experimental techniques to
observe this process must be both surface sensitive and sensitive to atomic
order. Both requirements are easily met by diffraction experiments, which
display the autocorrelation function of surface atoms only. In the case of
atom diffraction, only the first-layer atoms are probed, however, with the
disadvantage that their correlation to the next-layer atoms is hidden. This
information is additionally available when low-energy electron diffraction
(LEED) is used with an electron penetration depth of a few atomic layers.
In addition, the LEED experiment is comparatively inexpensive and in principle
easy to perform, though sophisticated techniques must be used when high-
accuracy measurements and/or large transfer widths are required (for recent
reviews see /9-11/). Therefore, most of the experimental investigations
of surface transition kinetics done so far and which are presented here use
low-energy electrons. In some cases other tools were also used, e.g. work
function measurements, Rutherford backscattering, imaging techniques such
as scanning tunneling microscopy and the recently developed low-energy electron microscopy, electron spin resonance and X-ray diffraction. Comparing
the latter with LEED it must be emphasized that LEED is highly dominated by
multiple scattering which is the price to be paid for its high surface sensitivity. As a consequence, and in contrast to X-ray scattering, a full dynamical theory has to be used when integral intensities are interpreted to
give structural data /12-14/. Moreover, the dynamics of scattering can also
modify spatial intensity profiles which display the surface order which is

of central interest when dealing with phase transitions. Due to multiple scattering, LEED in principle also probes multisite correlations, in addition to the autocorrelation, and it was only recently that consequences arising from this fact were pointed out /15/. Fortunately, however, kinematic scattering seems to dominate intensity profiles in many cases, especially for normal incidence of the primary beam, and many successful investigations based on kinematic data evaluation are known (for recent reviews see e.g. /16-17/). Therefore, in this paper it is assumed that kinematic interpretation of spot profiles is allowed. However, it should be pointed out that surface-sensitive X-ray diffraction with grazing incidence /18/ could be of importance in the future, as dominance of kinematic scattering is out of the question there.

Kinetics of phase transitions are involved whenever time dependences of processes are under investigation rather than their final equilibrium states. Unfortunately, in many cases different processes can mix. On the surface this may happen in the case of adsorption, when the adsorption kinetics interferes with the ordering kinetics, starting with the arrival of the first few atoms on the surface. Or, when the temperature of a surface is changed, reordering of the surface atoms can take place, mixing with the time dependence of the temperature or even with that of partial desorption, especially in the case of weak binding to the substrate, i.e. physisorption. Separation of the different processes is established when their time constants belong or can be made to belong to different orders of magnitude. Then the process with the longest time constant can be observed in its pure form, which is important when it is still not fully understood. Therefore, in experiments dealing with the kinetics of surface phase transitions care must be taken that no other processes become time limiting. In adsorption experiments this could be achieved by a sudden change in coverage (which is held constant in the following) allowing one to trace the kinetics of ordering. In practice, however, this is difficult to realize because in many cases adsorption is made to take place from the surrounding gas phase, which cannot be reduced with sufficient speed when a certain coverage is reached. Additionally, to allow for a rapid adsorption process itself, the sample must be cooled in many cases, which in turn may inhibit ordering processes, especially when activation barriers are involved. It might be owing to these difficulties that at present no experiments with a step-like change of coverage are known to the author.

Therefore, all experiments performed to record the kinetics of a pure ordering process try to produce a step-like temperature change which is quick enough to leave the surface for subsequent ordering unaffected by the time dependence of the temperature change. Of course, this situation is far from being ideally realized, and in fact most experiments cannot completely avoid a mixing of the kinetics with the temperature time dependence, at least at the very beginning of the transition. This is especially true for very fast transitions, e.g. where no activation energy seems to limit the transition rate. An example is the clean surface W(100) which reconstructs from its bulk-like phase to a c(2x2) when the temperature is reduced from above room temperature to low temperatures. Even a very fast quench of the order of 100 K/s is not sufficient to decouple time and temperature dependences. This is demonstrated in Fig.1 where the background subtracted profiles of the 1/2 1/2 superstructure spot are displayed for the 1x1 → c(2x2) transition. It shows clearly that the transition develops during the temperature change, and that much higher cooling rates must be realized in order to observe the pure kinetics.

The temperature quench forces the system into a state which is thermodynamically unstable or metastable, in any case, far from equilibrium. The kinetics of the system's movement to its final state of equilibrium is of interest. It is triggered by thermal fluctuations by which atoms take random atomic

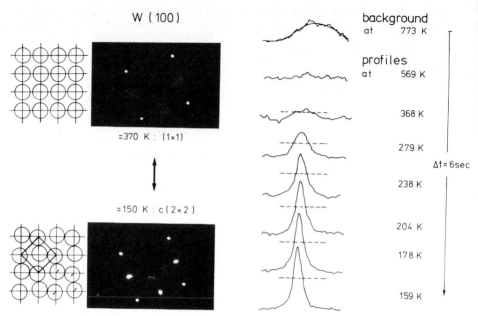

Fig.1: Development of the transition W(100)1x1 → c(2x2) during a temperature quench of about 100 K/s. On the left-hand side the surface atomic models and LEED patterns of the non-reconstructed (upper half) and the reconstructed (below) phases are displayed. On the right-hand side background subtracted profiles of the emerging 1/2 1/2 superstructure spot are given for various temperatures and times (part of the figure is taken from /10/).

arrangements which can lead to a nucleus of the new phase. If a critical size is exceeded, a growth process follows that results in a domain of the final phase. Different domains interact with each other when their domain walls meet and they may finally coalesce to form a larger domain. It is clear from this qualitative description that the transition kinetics can be influenced by a large variety of factors, e.g. substrate and overlayer symmetry, interaction between surface atoms, diffusion barriers, shape of domain walls, surface defects and others. Nevertheless, it apparently emerges from theoretical concepts that only a few features dominate the kinetics of transitions, which allows a classification of growth processes. So, as a quantitative result, theory gives the time dependence of the structure factor, which is believed to obey a scaling relation, according to which the average size L of growing domains should vary with time t as $L \sim t^x$. Fortunately, L can be obtained through diffraction experiments by measurement of the angular width or peak intensity of overlayer diffraction beams. Therefore, most experiments measure these quantities while others record the integral beam intensity, as this is a measure of the total number of scatterers in the ordered phase.

In spite of the restricted number of experiments dealing with the kinetics of surface phase transition, they fortunately cover a wide field, namely reversible and irreversible transitions, metals and semiconductors, adsorbate systems and clean surfaces. However, at the present stage of our understanding of the kinetics of 2D-phase transitions it is difficult to group the available experimental results according to theoretical classifications. On the one hand this is because the latter does not yet seem to be fully clear and complete, and on the other hand because many experiments do not show a single power law for the domain growth. Therefore, the structure of this paper follows more phenomenological criteria. In the next section, the kinetics of transitions in adsorbates is presented, where the substrate remains unaffected. In Sect.3 we concentrate on transitions of clean surfaces, i.e. on surface reconstructions. This is followed by the kinetics of adsorbate-induced reconstructions, where both adsorbate and substrate atoms relax into new equilibrium positions and so in this sense this is a combination of transitions described in the preceding sections. In Sect.5 we remain with this phenomenon of adsorbate-induced reconstruction but allow the interaction of the surface with a surrounding gas phase. Because of structure-dependent sticking coefficients, structural oscillations of the surface are observed whose time constants are influenced by the transition kinetics. Finally, in the last section we summarize our conclusions.

2. Ordering Kinetics in Adsorbates

It is well known that chemisorbed overlayers on a crystalline substrate frequently form well-ordered superstructures (for an overview see /19/). Different superstructures can develop with varying coverage and systems are known with up to nine commensurate structures (e.g. /20/). As pointed out above, the kinetics of adsorption interferes with the kinetics of transitions between different structures as a function of coverage when their characteristic time constants are comparable. Therefore, the available investigations trace the transition at fixed coverage as a function of time subsequent to a temperature quench. For an upquench this can lead from an ordered to a disordered state where the adatoms form, for example, a lattice gas. Of course, also the reverse process can be measured. If the initial adsorption is performed at low temperatures where the adatoms are immobile, the starting phase corresponds to a metastable disordered phase, possibly again a lattice gas. A temperature upquench then leads to an ordered equilibrium phase for which the kinetics can be recorded. The ordering mechanisms are believed to be different for the cases where the quench leads to a final equilibrium state with one single stable phase or with two coexisting phases. In the first case small domains are believed to be nucleated all over the surface, separated by their boundaries, and some of them grow at the expense of others. When quenched to a coexistence region, small islands of the ordered phase are nucleated and grow at the expense of the disordered area until equilibrium is established. The great majority of experiments correspond to a quench into a one-phase region of the phase diagram. The exponent x in the growth law $L \sim t^x$ is expected to be universal in the sense that it depends only on the degeneracy p of the ground state of the ordered phase. So for p = 2 the value x = 1/2 is expected and slower dependences result for larger p /21/. Fortunately, measurements for systems with p = 2 and larger (p = 8) are available and can be presented in Sect.2.1.

However, it is worth pointing out that the experimental situation is not without problems even if the coverage of the surface is conserved. This is because in diffraction experiments it is more convenient to measure the peak height I_p of a diffraction spot than to determine its half-width w, which is directly related to the mean domain size L of the developing ordered phase, $L \sim 1/w$. This is because the determination of w requires the measurement of the whole spot profile whilst I_p results from a single measurement at the spot center. Moreover, the measured profile is a convolution of the wanted physical profile with the instrument response function. If a Gaussian shape holds for both, the correction is simply $w = (w_{exp}^2 - w_{rf}^2)^{1/2}$, with w_{exp} and w_{rf} being the widths of the experimental profile and the response function, respectively, but in general a deconvolution must be performed.

If we assume the shape of the diffraction spot not to vary during the transition, the total spot intensity is $I \sim w^2 I_p$, which is proportional to N, the total number of adsorbed atoms contributing to the spot intensity. So, with $w^2 I_p \sim N$ the relation $I_p \sim NL^2$ results. For constant coverage $\Theta \sim N$ this means that $L \sim \sqrt{I_p} \sim t^x$ according to the assumed power law for L. However, this is only true if N is constant in the sense that all atoms occupy equivalent lattice sites and so contribute equally to the diffraction spot under measurement. So, for N required to be constant, the disordered adsorbate system must correspond to a lattice gas where the occupied sites are coherent. If this is not realized, N might change during the transition although the total coverage remains constant. In the extreme case of island growth by diffusion of atoms from non-coherent lattice sites, $L^2 \sim N$ results. Again, using $I_p \sim NL^2$ the relation $\eta = \sqrt{I_p} \sim N$ is obtained. As N represents the number of atoms in the ordered structure and is proportional to the area covered by it, η is frequently interpreted as the order parameter. Writing I_p in terms of L, the relation $L \sim \sqrt[4]{I_p}$ results, which is different from the case of conserved N. Of course, for real adsorbate systems neither of the two pure cases may be realized and the truth will be somewhere in between.

Presumably owing to this uncertainty there is no unique way of presenting results in the literature. Many authors demonstrate the dependence of $I_p(t)$, some of $\sqrt{I_p}(t)$ and others of the integrated intensity $I(t)$. Direct measurements of $L(t)$ through $w(t)$ are rare. In the following, examples are given for the transitions of oxygen adsorbed on W(110) and W(112) and of sulfur on Mo(110). Examples of intercalated graphite, which are related to adsorption systems, are also added.

2.1 Oxygen on W(110) and W(112)

Fortunately, measurements are available for a single adsorbate on two substrates which differ only in symmetry. In both cases, namely O/W(112) /22/ and O/W(110) /23, 24/ the adsorption is realized at low temperatures, resulting in an immobile lattice gas. By upquench of the temperature the development of superstructure spots is observed.

a) Oxygen on W(112)

The W(112) surface offers a lattice with a rectangular unit mesh on which oxygen adsorbs disorderedly at temperatures below 225 K. With increasing temperatures a p(2x1) ordered superstructure develops, as seen from the phase diagram given in Fig.2a, taken from /22/. The corresponding model of the surface is displayed in Fig.2b with the unit mesh of the clean surface (full line) and that of the (2x1) superstructure (broken line). It is evident that two antiphase domains of the overlayer are possible, increasing the degree of degeneracy to p = 2.

In the experiments described by Wang and Lu /22/ the oxygen was made to adsorb at 170 K with a final coverage of $\Theta = 0.5 \pm 0.05$. This initial state corresponds to the open circle in the phase diagram of Fig.2a. No superstructure spots were detected, indicating a high degree of disorder. The adsorbate system was then upquenched into the p(2x1) region of the phase diagram, whereby the coverage remained constant and different final temperatures were realized as indicated by the arrows in the phase diagram. The temperature quench took about 3-4 s and the measurement of superstructure spot intensities was started less than 1 s before the final temperature was reached. The corresponding growth kinetics of the 1/2 O spot for different final temperatures is given in Fig.3a for $\eta_N = \sqrt{I_p}$, which is believed to be the order parameter.

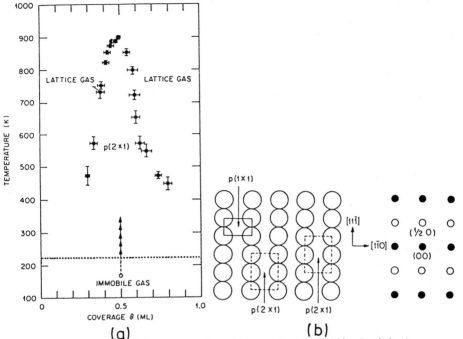

Fig.2: Phase diagram (a) of the adsorption system O/W(112). In (b) the surface unit cells of the clean and oxygen covered surface is given (full and broken lines, respectively) together with the schematic LEED pattern (filled circles: substrate beams, open circles: half-order beams) (as taken from /22/).

The spots are reported to be initially broad and to sharpen with time, but no quantitative measurements of the shapes or widths are available.

The time dependences given in Fig.3a obviously vary with changing temperature. The authors report that it is not possible to describe them quantitatively over the whole range of time measured. Therefore, they restrict themselves to the initial linear increase, which, according to $\eta_N \sim \sqrt{I_p} \sim N$, is interpreted as representing the linear growth of ordered domains as predicted by the LAC theory /21, 25/. So, implicitly the transition is assigned to the class of non-conserved N as discussed above, though the coverage Θ is constant. With $L \sim \sqrt[4]{I_p}$ the dependence $L \sim t^{1/2}$, i.e. $x = 1/2$, results. If the characteristic time of the initial linear increase of η_N is plotted over the reciprocal temperature, a linear dependence results, within the error of measurement (Fig.3b). This shows that the growth time is of Arrhenius type and an activation enthalpy of $\Delta H = 0.14 \pm 0.02$ eV is derived from Fig.3b.

Two features of the kinetics displayed in Fig.3a should be emphasized. The first and obvious one is that a fast initial increase is followed by a very sluggish time dependence for all temperatures. The second one is slightly hidden by the fact that η given in Fig.3a is normalised to its long but finite time value. So, it is not clear from the presentation that the long time values of η for different temperatures did not reach the equilibrium values realized by annealing, a fact which is only mentioned in a footnote in /22/.

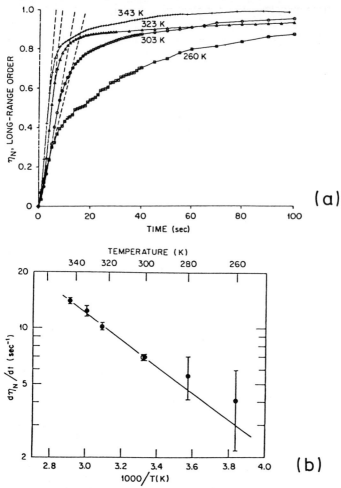

Fig.3: (a) Time dependence of $\eta_N = \sqrt{I_p}$ of the 1/2 0 beam of W(112)p(2x1)-0 for various temperatures and normalized to the respective long time value. The broken lines indicate the initial linear rates which are displayed in (b) in an Arrhenius plot (as taken from /22/).

b) Oxygen on W(110)

The system O/W(110) has been studied in detail /26-28/ and shows a more complex phase diagram than W(112). What is important in the present context is that oxygen is again disordered when adsorbed at low temperatures, for example, as described in the measurements by Wu et al. /23/ and Tringides et al. /24/.

For a coverage $\Theta = 0.5$ an upquench leads to a pure 2x1 phase, for $\Theta = 0.25$ to a 2x1 phase coexistent with a lattice gas. For the oxygen site a three-fold coordinated site on one of the two long-bridge sites was favoured /29/. This two-fold degeneracy is doubled by the possible realisation of two antiphase domains, and once more doubled by the existence of rotated domains, so p = 8

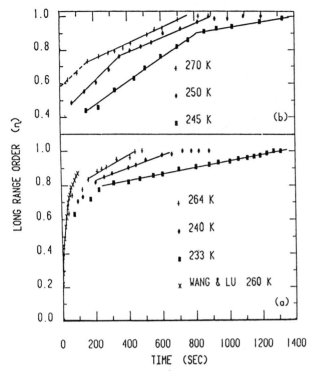

Fig.4: Development of $\eta = \sqrt{I_p}$ of the 1/2 1/2 beam of W(110)p(2x1)-O for different temperatures and coverage $\Theta = 0.5$ (a) and $\Theta = 0.25$ (b). In (a) measurements for W(112)p(2x1)-O by Wang and Lu /22/ are added (as taken from /23/).

results. However, the situation with respect to this point is not completely clear and the possibility of p = 4 is also assumed in the literature /23/.

Figure 4 displays the kinetics via $\eta(t) = \sqrt{I_p}(t)$ for both the quenches to the coexistence region with $\Theta = 0.25$ and to the one-phase region with $\Theta = 0.5$ including the comparison with the results of Wang and Lu for O/W(112). For the long-time behaviour the authors identify regions in which η changes linearly with time, as indicated by the straight lines and corresponding to $L \sim \sqrt[4]{I_p} \sim t^{1/2}$. The respective time constants are both temperature and coverage dependent and lower than for O/W(112). As can be seen from the figure, the segmentation into straight lines looks somehow arbitrary, and one could only speculate about its physical grounds. Moreover, the fast early time increase was not fitted in /23/, though interpreted by rapid ordering processes as "elimination of very unfavourable boundaries or the diffusion into lattice gas sites of atoms frozen during deposition into sites that are not lattice gas sites".

Recently the data shown in Fig.4 were reevaluated by Tringides et al. /24/ emphasizing the early time behaviour more by using a plot $\ln I$ vs $\ln t$, which results in an average slope of 0.66 as demonstrated in Fig. 5. So, with $I_p \sim t^{0.66}$ and now assuming N to be constant throughout the transition, $L \sim \sqrt{I_p} \sim t^{0.33}$ results. Fitting the time constant to a straight line in an Arrhenius plot, an activation enthalpy $\Delta H = 0.15 \pm 0.1$ eV results. This value is similar to that obtained for O/W(112) and the authors speculate that it might "reflect the difficulty with which boundaries are moved in the overlayer film".

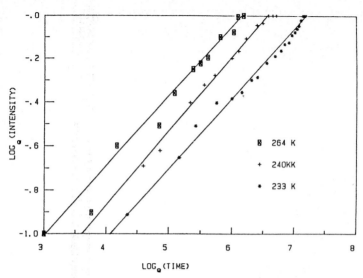

Fig.5: Reevaluation of the data presented in fig.4, here in a plot $\ell n I_p$ vs $\ell n t$ for different upquench temperatures. The average slope is 0.66, independent of temperature (as taken from /23/).

2.2 Sulfur on Mo(110)

With tungsten and molybdenum having the same lattice symmetry and only a 0.018 Å difference in the lattice constant, the substrate Mo(110) can be expected to have very similar properties compared to W(110). It is therefore interesting to compare the adsorption kinetics of S/Mo(110) as investigated by Witt and Bauer /30/ to that of O/W(110).

Figure 6 displays the phase diagram of S/Mo(110), taken from /30/. It appears that for low coverages p(2x2) islands coexist with a lattice gas, whilst in the region around $\Theta = 0.25$ a more or less one phase p(2x2) structure is found. For all values of Θ lattice gas disorder develops when a certain temperature $T_c(\Theta)$ is exceeded. Witt and Bauer /30/ report the kinetics of the reverse disorder-order transition when quenching from a temperature of 1040 K above the phase boundary to below it. Owing to experimental restrictions, initial and final temperatures were not varied and the temperature quench took more than 100 s, passing T_c only after about 40 s, a time which is taken as the starting time t_0 of the transition. (The authors proved their results given below to be only weakly dependent on the choice of t_0). Figure 7a gives the T(t) dependence and it appears that the temperature changes even after long times.

Unlike the other measurements, the kinetics of the transition were recorded not only for a single value of the sulfur coverage but for the set $\Theta = 0.12, 0.16, 0.19, 0.22, 0.24, 0.26$ and 0.32. The corresponding developments of the 1/2 1/2 superstructure LEED spot intensity I_p are presented in Fig.7b where t_0 acts as the time zero. Arrows above the time scale indicate the instants where the temperatures given were reached. At the long time end of the lower coverage curves, arrows lead to the intensities measured at 300 K in equilibrium.

210

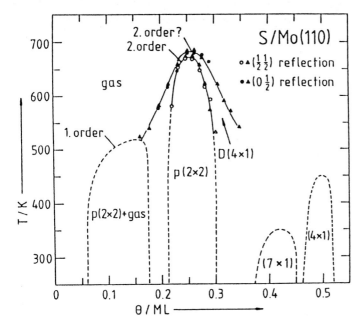

Fig.6: Phase diagram of S/Mo(110) (as taken from /30/).

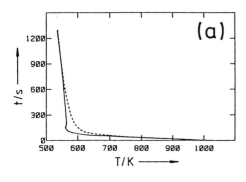

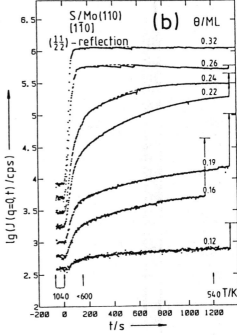

Fig.7: Temperature quench (a) of the adsorbate system S/Mo(110) with subsequent ordering of the p(2x2) structure displayed through $I_p(t)$ of the 1/2 1/2 superstructure spot for various coverages (b) (as taken from /30/).

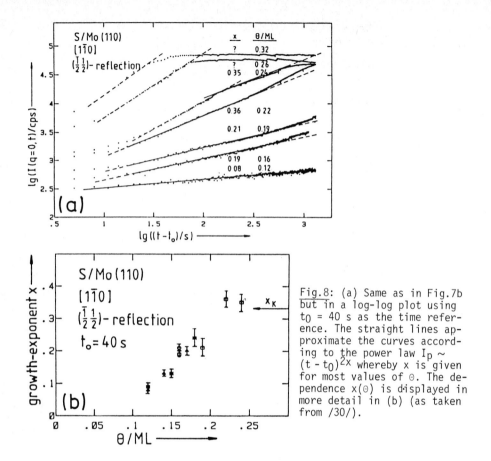

Fig.8: (a) Same as in Fig.7b but in a log-log plot using $t_0 = 40$ s as the time reference. The straight lines approximate the curves according to the power law $I_p \sim (t-t_0)^{2x}$ whereby x is given for most values of Θ. The dependence $x(\Theta)$ is displayed in more detail in (b) (as taken from /30/).

If $\ell n I_p$ is again plotted vs $\ell n(t-t_0)$ larger parts of the dependences can be fitted by straight lines $L \sim \sqrt{I_p} \sim (t-t_0)^x$, at least for the lower coverages (Fig.8a). It is evident that x depends strongly on Θ, as demonstrated more clearly by Fig.8b. For $\Theta = 0.25$ a value $x = 0.36 \pm 0.03$ results, i.e. close to $x = 1/3$. This is consistent with the result obtained for O/W(110) at $\Theta = 0.25$ where the same relation, $L \sim \sqrt{I_p}$, was used /24/.

2.3 Intercalated Graphite

Intercalated graphite presents a chance to observe 2D phase transitions similar to adsorption systems. Molecules intercalated between graphite layers are free to order, being thereby influenced by the graphite layer potential corresponding to an overlayer on a substrate. Usually intercalated graphite systems are distinguished according to their stage, stage-n meaning that there are n carbon layers between neighbouring intercalant layers.

As many such intercalant layers exist, the way is open to apply X-ray diffraction to trace their ordering kinetics. This was done by Homma and Clarke /31/ who studied graphite intercalated with $SbCl_5$ at stage-4 and stage-6.

212

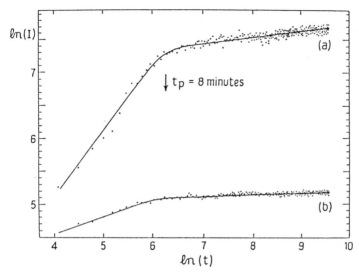

Fig.9: Development of the 10 beam peak intensity of the $(\sqrt{7} \times \sqrt{7})R \pm 19.11°$ structure formed by stage-6 $SbCl_5$ intercalated graphite after quenching to room temperature from 470 K (a) and 378 K (b). t_p indicates the time at which the initial fast growth is followed by a very sluggish development (as taken from /31/).

The molecules are reported to be disordered above 445 K. They order via an incommensurate phase near 338 K to a $(\sqrt{7} \times \sqrt{7})R \pm 19.11°$ superstructure with high degeneracy, p = 14, at ambient temperatures. The peak intensity of the 10 beam of this structure was recorded after a temperature quench from different elevated temperatures to room temperature. The quench, which was easily performed by blowing a cooled air stream over the sample, took less than 5 s. This is negligible compared to the characteristic times appearing in Fig.9, where the development of I_p with time is displayed for two prequench temperatures. Again, two different growing regimes show up, the initial one being fast, and the long-time behaviour being again very sluggish. The upper curve corresponds to a quench from 470 K to room temperature, i.e. starting from the disordered phase. The lower curve presents the development when starting at 378 K, i.e. at the incommensurate phase. Both the exponent x in $I_p \sim t^{2x}$ and the breakpoint t_p between the two growing regimes are different for the two curves. When starting from the disordered phase x = 0.52 ± 0.05 and $t_p \approx$ 8 min result, when starting from the incommensurate phase the values are x = 0.15 - 0.25 and $t_p \approx$ 5 min. In both cases the long-time behaviour is reported to be logarithmic for times up to 100 t_p. The crossover at t_p is interpreted as being caused by pinning of domain boundaries on defects introduced into the graphite matrix by the intercalation process. This is supported by the fact that t_p varies from sample to sample. It is emphasized that the power law for t < t_p and for quenching from the disordered stage is consistent with the LAC theory for antiphase domain coarsening. However, no interpretation for the low value of x is given for the quench from the incommensurate phase.

Another example of ordering kinetics is that of stage-7 $AlCl_3$-intercalated graphite, which shows an order-disorder transition at T_c = 170 K /32/, where the phase above T_c is liquid-like disordered and ordered below. The kinetics

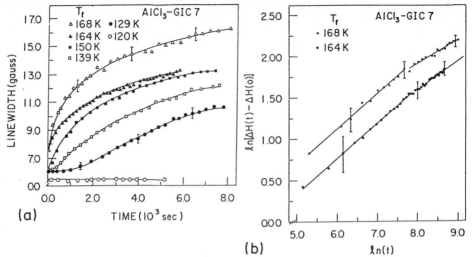

Fig.10: Time dependence of the ESR line width of AlCl₃ intercalated graphite when quenched from a disordered phase at T > 170 K to the temperature given (a). In (b) the same is presented in a log-log plot for two temperatures only and with subtraction of the zero-time half-width (as taken from /33/).

of the transition was investigated by Stein et al. /33/ using conduction-carrier spin resonance, one of the rare examples where no diffraction technique is used. The sensitivity of the method with respect to order arises from the assumption that the atomic shifts during the ordering process modify the charge density distribution on the carbon layers. This in turn is believed to shorten the mean free path of conduction carriers resulting in a broadening of the ESR line width, which is assumed to be directly proportional to the average domain size L. The corresponding results are shown in Fig.10a for a fast quench (400 K/min) from above T_c to the temperatures given. It appears that ordering is inhibited for low temperatures, indicating a freezing of microdomains. However, the two fast and slow growth ranges show up again, as is more clearly demonstrated in Fig.10b. For not too low temperatures the initial growth is $L \sim t^x$ with x = 0.42 ± 0.06 changing to x = 0.32 ± 0.06 for long times. The initial time value is interpreted to be consistent with results obtained by Monte Carlo calculations for large p. The slower long-time behaviour again is believed to be caused by pinning effects, when the average domain size becomes comparable to the average defect separation.

3. Reconstruction Kinetics of Clean Surfaces

If a solid is cut along a crystallographic plane in order to create the corresponding surface, chemical bonds are cut as well. This leaves the selvedge atoms generally in a state out of equilibrium and consequently they move towards new equilibrium positions. If the lateral symmetry of the first layer(s) is changed by this process the phenomen is called *reconstruction,* if only a collective vertical or horizontal shift of layer atoms is realized the term *relaxation* is used. Reconstruction can be *displacive,* i.e. atoms are only shifted away from their old equilibrium positions without changing the mean atomic density of the layer. However, reconstruction can also be accompanied by removal or addition of atoms, so that the mean atomic density is changed,

214

requiring some mass transport over quite large distances. Especially in semi-conductors with their covalent or at least partially covalent bonds, recon-struction is a common phenomenon (for a review see /35/). However, a con-siderable number of metal surfaces also show reconstruction, e.g. platinum, gold, iridium, molybdenum and tungsten (for a review see /36/). Fortunately, for both metals and semiconductors experiments are available following the kinetics of the reconstruction process.

3.1 Reconstruction of Si(111)

On cleaving the (111) surface of silicon at room temperature, a 2x1 super-structure develops. This is metastable and transforms to the well-known 7x7 structure on heating the sample to elevated temperatures. The geometric model of this structure was the subject of considerable efforts (for a review see /37, 38/) and it is only recently that different methods seem to agree with the so-called triangular-dimer stacking-fault model, i.e. a surface built up of dimers and vacancy defects produces the superstructure /e.g. 38/.

At temperatures elevated further, i.e. at about 1100 K, the surface under-goes a reversible order-disorder transition to a disordered 1x1 phase /39/. The transition is now believed to be of first order /37/. Investigations of its kinetics are available using LEED intensities /39/ as well as the new technique of LEERM (Low Energy Electron Reflection Microscopy) /40, 41/.

Bennett and Webb /39/ measured the time dependence of the peak intensity of a 7x7 superstructure spot developing on a temperature downquench from 1150 K. They determined the transition temperature to be 1140 K, which is slightly different to the value of 1100 ± 15 K reported more recently /40/. Figure 11a displays the time dependence on a logarithmic scale, where J_B is the background corrected value of I_p. It should be pointed out that the ver-tical axis gives the normalized difference to the final intensity, different from other presentations. Again, a fast initial change is followed by a much slower development. The initial time dependence is interpreted to be exponen-

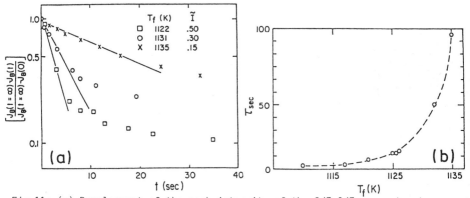

Fig.11: (a) Development of the peak intensity of the 3/7 3/7 superstructure spot of Si(111)7x7 upon a temperature quench from 1050 K to the temperatures given ($\tilde{I}$ denotes the value corrected by the Debye-Waller factor and normalized to the value of zero temperature). The initial time dependence is fitted by an exponential law (straight lines) from which the time constants displayed in (b) as function of the quench temperature are evaluated (as taken from /39/).

tial, the characteristic time of it is presented in Fig.11b as a function of temperature. A considerable slowing down appears when T_c = 1040 K is approached. However, if the transition is traced in the reverse direction by changing the sign of the temperature change, intensities respond very fast, i.e. within the instrumental time constant of 0.2 s. For the downquench the authors report $(J_B(\infty) - J_B(t))$ to follow a $t^{-1/2}$ law for long times, which they take as an indication of a diffusion-limited process. Because of the asymmetric time scale of the transition, the authors avoid using the term "critical slowing down", though symmetry is no necessary condition for this phenomenon.

Further insight into the kinetics of the transition is allowed by very recent measurements using the new LEERM technique. This was developed by Bauer and co-workers /42, 43/ and the principle is given in Fig.12a. The primary electron beam is focused into the back focal plane of a cathode lens, in which the originally high-energy electrons are decelerated to usual LEED energies in the hundred eV range. The specimen is illuminated by a parallel beam, and by back diffraction the LEED pattern is generated in the focal plane. Primary and diffracted beams are separated by a magnetic field and in the

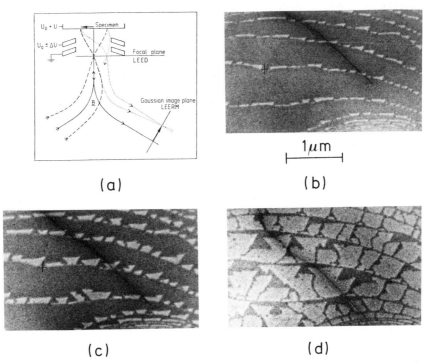

(a)

(b)

(c)

(d)

Fig.12: (a) Principle of the low energy electron reflection microscope (LEERM) with which the images in (b)-(d) were taken. The latter sequence corresponds to a temperature downquench of the Si(111) surface from the 1x1 disordered phase at T = 1112 K to ΔT = 6 K below the transition temperature of 1100 K. Bright areas correspond to developing reconstructed 7x7 phases (for more details see text). The black line across is caused by a crack in the microchannel plate (as taken from /41/).

Gaussian image plane the image of the surface can be observed. Using additional lenses it is up to the operator whether the LEED pattern or the LEERM image is displayed on a final screen. The resolution achieved so far is better than 20 nm, but 2-4 nm are believed to be possible /40-43/.

Telieps and Bauer used the method to observe the growth of 7x7 domains upon a downquench from $T_i = T_c + 12$ K to $T_f = T_c - \Delta T$ where ΔT was varied and T_c was determined to be 1100 ± 15 K. Only the 00 beam was used for imaging and the variation of its intensity with structure determined the contrast of the image. Figures 12b-d display micrographs after a quench of $\Delta T \approx$ 6 K which took about 6 s. The bright areas correspond to the 7x7 phase. It can be seen from Fig.12b, taken immediately after the quench, that nucleation of the 7x7 structure takes place along lines which can be shown to be steps separating terraces of 1x1 structure. The density of heterogeneous nuclei is reported to increase with ΔT, i.e. from $n = 4.5$ μm^{-1} ($\Delta T = 6$ K) to $n =$ 6.5 μm^{-1} ($\Delta T = 30$ K) to 25 μm^{-1} when quenched to room temperature. The heterogeneous nucleation is followed by growth of triangular-shaped domains until adjacent domains happen to interfere. Then new domains can be generated, mostly at the vertices of the old ones (Fig.12c). The coexistence of reconstructed and non-reconstructed areas is evident. Further growth leads to the image shown in Fig.12d where some of the 7x7 domains have grown at the expense of others. Between the domains, boundaries are visible which are broader than the steps, which show up as narrow dark lines. However, as also appears from Fig.12d, domains of the 1x1 structure which only slowly transform into the 7x7 phase are still present.

The speed of domain growth increases with ΔT to which the system was supercooled by quenching. For $\Delta T > 12$ K homogeneous nucleation was also observed. The growth rates of domains before interfering with each other could be roughly determined as function of ΔT and are given in Fig. 13. It appears that the domain growth velocity is proportional to the amount ΔT of supercooling. Consquently, the transition slows down when the quench ends near the critical temperature. This is very similar to the observation of Bennett and Webb, and, in fact, when using $v \sim 1/\tau$ the Figs.11b **and** 13 convert roughly into each other. Moreover, Telieps and Bauer confirm the observation made by Bennett and Webb, according to which the transition develops

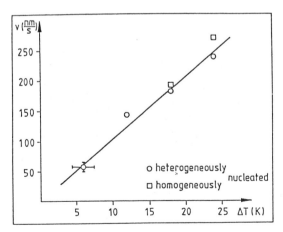

Fig.13: Velocity of domain growth of the 7x7 phase of Si(111) as function of supercooling ΔT (as taken from /41/).

much faster upon upquench than on downquench. In particular, no superheating is observed. Additionally, the authors report the absence of a significant hysteresis, i.e. temperature difference between up- and downquench transformations is less than 1.5 K.

3.2 Reconstruction of Pt and Ir (100) and (110) Surfaces

a) Ir(100) and Pt(100)

The (100) surfaces of both platinum and iridium are reconstructed in thermal equilibrium. Their structure was the subject of a large number of investigations by different methods. Today there is agreement that the selvedge consists of a quasi-hexagonally close-packed first layer which is arranged on the quadratic second and following layers. Of course a hexagonal arrangement cannot exactly fit on a quadratic one, and so considerable puckering is caused, as demonstrated by the surface model of Ir(100)1x5 displayed in Fig.14a, which results from LEED analysis /44-46/. The quasi-hexagonal structure coincides with the quadratic one after five atomic distances in the direction of one of the two orthogonal axes of the substrate unit mesh, whilst in the other direction atomic distances are unaffected. This creates a 1x5 superstructure which is realized in two symmetrically equivalent orthogonal domains, resulting in a LEED pattern as displayed in Fig.14b. It is evident from the surface model in Fig.14a that the reconstructed phase is highly degenerate: for example, the hatched atoms can be arranged in 5 equivalent but different bridge positions of substrate atoms, which together with the existence of orthogonal domains sets the degeneracy at $p = 10$. In the case of Pt(100) p is even larger, because for this surface there is an additional compression of the surface layer in the direction, in that in the case of Ir(100) the unit mesh length coincides with that of the substrate. Consequently, it is only after about 25 atomic distances in the substrate that the latter coincides with the overlayer, causing a 25x5 superstructure which, however, would be better called Pt(100)hex /47/ according to the real nature of the reconstruction. Still more complex, not only are two orthogonal domains realized but each of them is shown to be rotated by a small angle once the surface is heated above about 1100 K /47, 48/. This considerably increases the ground state degeneracy and creates a complicated LEED pattern, as demonstrated in Fig.14c. Its complexity hinders a structure determination by LEED but measurements using Rutherford backscattering /49/ and tunneling microscopy /50, 51/ lend strong support to the model. It is worth noting that the Au(100) surface shows a similar but not identical reconstruction.

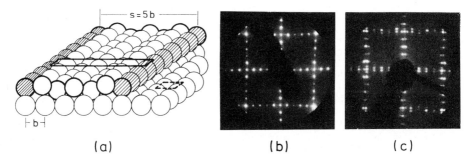

(a)   (b)   (c)

Fig.14: Model of the Ir(100)1x5 surface (a) and its LEED pattern (b) together with that of the reconstructed surface Pt(100)hex (c).

218

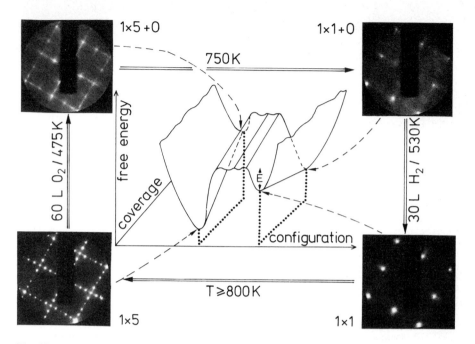

Fig.15: Dependence of the surface free energy (schematic) on surface configuration and coverage for Ir(100). LEED patterns are displayed for stable and metastable states.

The reconstructed phases of both Ir(100) and Pt(100) can be transformed back to their 1x1 bulk phases when appropriate gases are adsorbed. This corresponds to an adsorbate-induced phase transition, which will be treated in Sect.4. Here the original transition 1x1 → hex is presented, for which, however, the clean 1x1 has to be prepared first. The procedure is demonstrated in Fig.15 for Ir(100), from which it appears that the 1x5 structure corresponds to the absolute minimum of the free energy when the surface is clean. Subsequent to oxygen adsorption the surface is still in the reconstructed phase, but now in a metastable state. The barrier to the non-reconstructed (but oxygen-covered) phase is overcome by thermal activation, as indicated in the figure. After removal of oxygen by hydrogen the clean 1x1 phase results. The structure again switches from this metastable state to the reconstructed 1x5 phase by thermal activation, and this process was observed by the Erlangen LEED group using diffraction intensities /52/.

The initial 1x1 phase must not necessarily correspond to a flat and long-range ordered surface. Some disorder must exist because the substrate beams are slightly broadened compared to the reconstructed phase. However, the disorder must be weak enough not to modify the dynamics of scattering, because almost perfect agreement between experimental and calculated I(E) spectra could be obtained over a large energy range, assuming perfect order /53/. Disorder could therefore exist as 1x1 islands on the flat substrate which are large in diameter compared with the electron coherence length. Such islands were observed by means of scanning tunneling microscopy on Pt(100) with, however, CO adsorbed /54/.

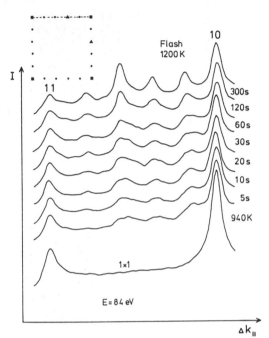

Fig.16: Development of LEED intensity profiles during the transition 1x1 →
1x5 of Ir(100) recorded at 84 eV after an upquench from about 100 K to 940 K,
which takes about 10 sec. The profiles are measured along the line in the
schematic diffraction pattern. The last profile results only after an additional flash to about 1200 K.

The kinetics of the transition is followed by a temperature upquench from
liquid air temperature, where the 1x1 phase is practically stable, to temperatures above 800 K. The time for the upquench is between 10 and 20 s. Due to
a fast computer-controlled video method /10, 55/ profiles through the changing
LEED pattern could be recorded with a resolution of 20 ms. Results are displayed in Fig.16 for a quench to 940 K and for various times after the quench.
Two main features show up. First of all it appears that the intensity level
between the integer-order spots increases. This is also visually observed by
the appearance of streaks in the pattern /52/. From the streaks superstructure
spots develop. Secondly, the superstructure spots do not appear at their
final positions right from the very beginning of the transition. They are
broad, showing some structure within themselves, and shift slowly to their
fifth-order positions, simultaneously changing shape and sharpening. Therefore it would not make much sense to measure peak intensities or beam halfwidths, as they are not well defined in the case of structured beam profiles.
The latter, as well as the appearance of streaks, indicate that the transition develops via an intermediate disordered state. Only one of the superstructure spots, the 6/5 0 and its equivalents, appears approximately at its
final position and sharpens during the transition. However, because of the
rather broad width of the response function in /52/ only poor information can
be drawn from the half-width dependence.

So, in the present case integral intensities are believed to be the appropriate quantities to be measured, as they reflect directly the number of

220

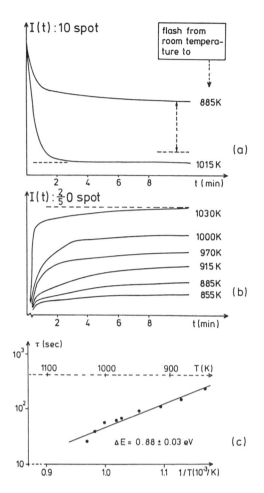

Fig.17: Development of integral beam intensities during the 1x1 → 1x5 transition, (a) for an integer order beam and (b) for a superstructure beam after up-quench from about 100 K to the temperature given. The initial increase of curves in (b) is linearly approximated resulting in a temperature-dependent time constant. This is plotted in (c) to give an activation energy of 0.88 ± 0.03 eV.

atoms ordered in the reconstructed phase. Figure 17 displays both the decrease of an integer-order spot intensity (a) and the increase of a super-structure spot intensity (b) for various quench temperatures. The broken horizontal lines give the intensity level detected at equilibrium. It is obvious that again there is a steep initial intensity change followed by very slow growth, which fails to reach the equilibrium intensities even after long times. From the initial steep increase of superstructure spot intensities a characteristic time is taken by linear approximation for each temperature. The resulting Arrhenius plot in Fig.17c gives an activation barrier of 0.88 ± 0.03 eV. This value is in agreement with results from work function measurements which, however, were performed at a linear temperature increase and so do not reflect the pure kinetics /56/.

Using LEED, similar experiments were performed for Pt(100) as for Ir(100), though in less detail /57/, with, however, similar results. Specifically, it was found that at first there is a development to a 1x5 superstructure as in the case of iridium. The splitting of superstructure spots (as shown in Fig.14c) appears only in the last stage of the transition, presumably

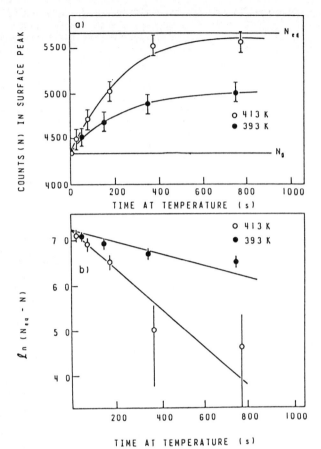

Fig.18: Kinetics of the transition 1x1 → hex through RBS (2.0 MeV $^4He^+$) for two temperatures in the linear (a) and logarithmic (b) plot, whereby the latter demonstrates the exponential growth at least for not too long times. N_0 and N_{eq} give the initial and final values of surface atoms density corresponding to the 1x1 and quasi-hexagonal phases, respectively. The initial 1x1 phase was not completely clean but covered with about 10^{14} cm^{-2} hydrogen (as taken from /49/).

owing to a final compression of quasi-hexagonal domains which have grown to a certain critical size /58/. The transition was also observed by Rutherford backscattering (RBS), however, starting with only an "almost clean" 1x1 surface /49/. The results are displayed in Fig.18a, where it again appears that equilibrium values are not approached in the case of low quench temperatures. Figure 18b plots the data to show that first-order kinetics holds, however, again failing for long times.

From the appearance of streaks along the unit mesh axis in the LEED pattern it can be concluded that the transition to the quasi-hexagonal order develops via a shifting of linear atomic rows. In an intermediate state they are disordered and then cling together in a quasi-hexagonal arrangement. The higher

222

density of this phase can be managed either by the building up of steps and/or
- as observed in the case of the CO-covered surface - by the existence of additional 1x1 islands on the flat substrate which disappear during reconstruction /54/ (see also Sect.4.2)

b) Pt(110)

The (110) surfaces of Au, Pt and Ir are known to show a 1x2 reconstruction
in thermal equilibrium. Various surface-sensitive techniques were used during
the last decade to determine the underlying structure and today the missing
row model is commonly accepted. It postulates the absence of every second
atomic row, thereby doubling the periodicity in the [001] direction. It is not
yet fully clear whether the 1x1 phase is ordered, creating, for example, steps
during the transition, or if lattice gas-like disorder holds, entailing a
transition of disorder-order type. The latter case was found to be true for
the high temperature reversible transition 1x1↔1x2 of Au(110), for which
critical exponents were derived /8/ but no investigations of the kinetics
exist. It is only recently that kinetics of the 1x1 → 1x2 transition have
been investigated for Pt(110) by the Erlangen group /59/, however, starting
from the low temperature metastable 1x1 phase and switching irresversibly
to the reconstructed phase. This procedure is again according to the principles demonstrated in Fig.15. The 1x1 phase can be prepared in a corresponding
way, being a bit more sophisticated for Pt(110) /59-61/. From the increased
background and beam widths of the 1x1 LEED pattern, disorder can be concluded,
in agreement with the failure of a LEED analysis based on perfect order /62/.

The kinetics of the transition was again traced by the measurement of beam
intensities developing on a temperature upquench, as shown in Fig.19a for the
1 3/2 superstructure spot. The time for the temperature quench was about
10 s, after which the curves begin to be displayed. Unfortunately, in this
way the initial intensity increase is partially cut away but it is clearly
visible. Evaluation of the corresponding slopes yields an activation barrier
of 0.41 ± 0.02 eV. It is evident that the initial steep increase is followed
by a sluggish time dependence, which again fails to approach the equilibrium
intensity even after long times. The decay of integer-order spots develops
correspondingly.

The curves in Fig.19a were fitted according to an Avrami type time dependence $I(t)/I(\infty) = 1-\exp(-kt^n)$, which was found by Avrami /63/ and Kolmogorov
/96/ assuming a linear growth rate but preventing nucleation in areas already
reconstructed. The integral spot intensity is taken to be proportional to the
number of atoms in the reconstructed phase. The dependence is equivalent to
an ansatz $\dot{x} = kt^{n-1}(1-x)n$ where $x = I(t)/I(\infty)$ and the term $(1-x)$ describes
impinging effects of boundaries. Figure 19b gives the result for one selected
temperature of 358 K; n is varied. It appears that the best fit results for
n = 1/4, which is also true for the other temperatures in Fig.19a. However,
no conclusion can be drawn from this value at the moment. Moreover, it must
be emphasized that Avrami predicted n = 3 for two-dimensional growth.

4. Kinetics of Adsorbate-Induced Reconstruction

Adsorbate-induced reconstruction can in a way be the reverse of clean surface
reconstruction. This was already indicated in the last section for the case
of CO/Pt(100) where the adsorbate restores the bulk-like structure on the surface. This is presented in more detail in Sect.4 2. However, adsorption can
cause surfaces to reconstruct also when the clean surfaces appear with their
bulk-like phase. This is observed for many examples of adsorbate systems.
However, investigations of the corresponding kinetics exist only for the
cases of oxygen and hydrogen on Ni(110) presented next.

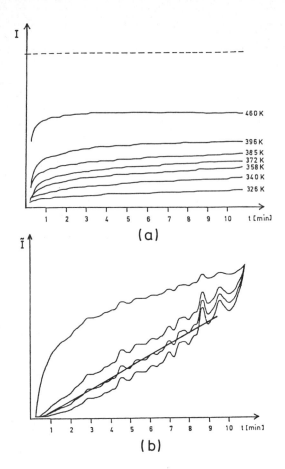

(a)

(b)

Fig.19. (a) Time dependence
of the 1 3/2 integral spot intensity of Pt(110)1x2 upon a
temperature upquench from
about 170 K to the temperatures given. The curves start
after the quench time of about
12 s. In (b) the dependence
for 358 K is plotted according
to an Avrami type dependence
$I(T)/I(\infty) = I-\exp(-kt^n)$ whereby
$I = \ln(1-I(t)/I(\infty))$ is used as
vertical axis. The straight
line corresponds to $n = 1/4$.

4.1 Oxygen and Hydrogen on Ni(110)

Oxygen is known to adsorb in a c(2x4) structure on the bulk-like Ni(110) surface when offered at temperatures below 250 K /64/. Upon annealing, however,
the adsorbate system reacts irreversibly by reconstruction of the nickel substrate. This is believed to be a first-order transition and to show a 2x1
missing row structure in its final state /65-68/. So, by an upquench from
below to above 250 K the kinetics of the transition can be observed, which was
done by Behm et al. /69/. It is believed that the removal of Ni atoms in producing the final missing row structure is the rate-limiting process, an assumption supported by the fact that oxygen orders already at low temperatures.
Its reordering during reconstruction can therefore be expected to be much
faster than that of Ni atoms. Figure 20a displays the development of the
1/2 0 superstructure beam intensity upon the upquench (shown as well). Again
there is a steep initial increase, however, overlapping with the temperature
rise. Once more, a very slow dependence follows which fails to reach the
equilibrium intensity even after 1 hour (not shown). The same data are re-
plotted in Fig.20b using a presentation $U = \ln|\ln (1-I(t)/I(\infty))|$ versus $\ln t$
according to Avrami's law discussed above. Independent of temperature, the ex-

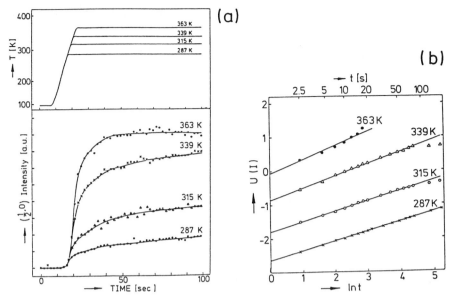

Fig.20: (a) Development of the 1/2 0 superstructure spot for the adsorbate-induced 2x1 reconstruction of O/Ni(110) upon the temperature quenches given above. (b) Same as in (a) using a plot $U = \ln|\ln(1-I(t)/I(\infty))|$ versus $\ln t$ (as taken from /69/).

ponent $n \approx 0.3$ results, which is approximately the same as that obtained for the case of Pt(110). From the ordinates of the straight lines in Fig.20b rate constants k can be taken, and by using an Arrhenius plot an activation barrier of 24 kJ/mol ($\hat{=}$ 0.25 eV) is deduced. This is interpreted as playing the role of the barrier involved when a Ni atom moves from one site to another, however, under simultaneous rearrangements of oxygen atoms which are believed to support this process.

Hydrogen is known to induce surface reconstruction on Ni(110) as well. Adsorbed below 180 K a series of ordered structures dependent on coverage is observed /70, 71/. With the coverage increasing beyond $\Theta = 1$ a reconstruction into a 1x2 phase is observed /72-75/. Competing with this, another irreversible reconstruction takes place when the adsorbate is annealed or hydrogen is adsorbed at temperatures higher than 180 K. Then streaks between integral order beams appear in the LEED pattern in the [01] direction, which are believed to be due to a reconstruction according to the suggested model given in Fig.21a /73, 75/. The streaks are caused by partially displaced rows which show order in the [110] direction but are disordered in the [001] direction. The reconstruction develops once a critical coverage is exceeded (5% at 300 K) and then shows itself to be independent of coverage, removable only through hydrogen desorption.

The kinetics of the transition was observed only recently by Penka et al. /76/. Figure 21b displays a typical temperature quench and the following development of integral streak intensities for various final temperatures and a constant coverage $\Theta = 0.15$. The data are reported to fit an Avrami-type law, however, only for t < 10 s, independent of coverage and temperature with $n \approx 1$, i.e. simply a first-order growth law. From the Arrhenius plot

225

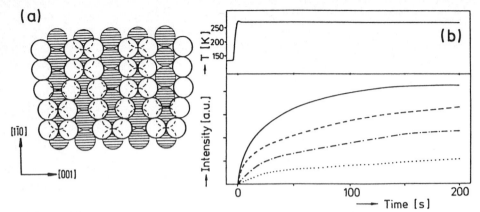

Fig.21: (a) Suggested model of Ni(110) after H-induced "streak phase" reconstruction, with H-atoms not shown (as taken from /75/). (b) Development of streak intensities during the transition for Θ_H = 0.15 subsequent to temperature quenches like the one shown above (....239 K, -.-.- 253 K, ---- 260 K, —— 270 K) (as taken from /76/).

of the rate constant k for two values of coverage, the same activation barrier of 25 kJ/mole (0.26 eV) is obtained. The preexponential factor of k is clearly coverage dependent, reflecting the necessity of hydrogen to induce the reconstruction. The constant value of the barrier height, however, is consistent with the assumption that the movement of Ni atoms is the time-limiting process. This is in agreement with the interpretation above for O/Ni(110) and in fact the measured activation barriers nearly coincide.

4.2 Carbon Monoxide on Pt(100)

As mentioned already in Sect.3.2a the quasi-hexagonal reconstruction of Pt(100) can be removed by adsorption of various gases such as CO or NO. Detailed investigations of the transition are due to Behm et al. /77/ using LEED, work function and thermal desorption measurements as well as to Norton et al. /49, 78/ using in addition Rutherford backscattering and nuclear microanalysis. Though no data could be obtained reflecting the pure kinetics, i.e. the development of the 1x1 phase with the CO coverage fixed, valuable information about the structural pathway of the transition resulted. This was recently completed by images obtained by scanning tunneling microscopy (STM) /54/.

The quasi-hexagonal reconstruction starts to be removed when a critical CO coverage of about 0.05 is exceeded. This is due to migration and clustering of CO molecules, which so form areas of high local coverage upon which the back transformation to the 1x1 phase is triggered locally. The CO molecules are then trapped on the 1x1 area developed, because the heat of adsorption is higher on the 1x1 phase than on the reconstructed phase. They order forming a c(2x2) superstructure observable by LEED. The c(2x2) patches are relatively small, containing less than 20 molecules /79/ according to LEED beam profiles. A large number of such (antiphase) domains eventually cover the surface, indicating homogeneous nucleation. This is impressively demonstrated by Fig.22 where in (a) the STM image of the reconstructed clean surface is displayed. The corrugation caused by the misfit of the first and second layers shows up

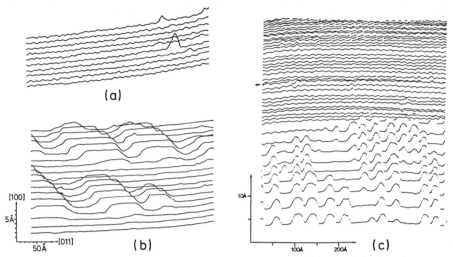

Fig.22: Scanning tunneling microscopy images of Pt(100) for (a) the clean reconstructed surface, (b) the surface as back transformed to the 1x1 phase through CO adsorption and (c) the surface during the transformation. An oscillatory structure on a profile demonstrates the first layer puckering due to reconstruction. CO molecules adsorbed on the 1x1 areas of (b) and (c) do not show up in STM (as taken from /54/).

clearly. Figure 22b presents the surface which has transformed fully back into the 1x1 structure, as demonstrated by the missing corrugation. The adsorbed CO is not visible in the image. It is clear that the lower atomic density of the 1x1 phase leads to the formation of 1x1 terraces of monatomic step height and 50-70 Å width on the 1x1 substrate. The image in Fig.22b was recorded at 460 K and a CO pressure of 5×10^{-6} mbar. At lower temperatures lower pressures are also sufficient, allowing Hösler, Ritter and Behm to follow the transition by time-resolved STM single line scans. This is displayed in Fig.22c where one scan takes about 5 s. The exposure at 375 K and a CO pressure of 2×10^{-8} mbar is started as indicated by the arrow. The corrugation, i.e. reconstruction, disappears within 3 scans after an exposure of about 0.7 L, and the unreconstructed phase is created with numerous 1x1 domains. Their size (20-30 Å) is considerable smaller than those created at the higher temperature (Fig.22b), so it appears that in the hex → 1x1 transition excess Pt atoms are transferred on top of the changing layer where they can form 1x1 islands by diffusion. The island size is strongly dependent on the temperature of the transition, and small islands developed at a moderate temperature can coalesce to larger ones only when the temperature is further increased.

When the reverse transition 1x1(CO) → hex is followed using STM, the atoms surprisingly do not order back to a flat reconstructed layer. Of course, excess atoms are needed to form the denser reconstructed phase. However, reconstruction is not nucleated simply at the positions where excess atoms would be available, i.e. at the 1x1 islands. It is reported that islands of the reconstructed phase are again formed, their positions being largely uncorrelated with that of the former 1x1 islands. Energy barriers for 1x1 island dissolution and coalescence of reconstructed domains must be overcome by considerable thermal activation. This is consistent with the findings for

227

the Ir(100)1x1 → 1x5 transition (Sect.3.2a) which was shown to be very similar to the transition of Pt. The LEED intensities showed quasi-saturation after an initial steep increase (Fig.17b) and approximation of the fully reconstructed phase could be realized only after additional higher thermal activation.

5. Kinetic Oscillations of Surface Structure

Section 3.2a described how the clean Pt(100) surface reconstructs irreversibly from its bulk-like 1x1 phase to a quasi-hexagonal arrangement of its first layer atoms. In the foregoing section it was demonstrated that the reconstruction can be removed by adsorption of gases, especially CO, when a critical coverage is exceeded. Consequently, by heating the adsorbate system e.g. in a constant ambient CO pressure, the coverage decreases owing to desorption. Falling below another critical value $\Theta_{CO} \approx 0.3$ /77, 78/ the 1x1 → hex transition is triggered for temperatures (> 400 K) high enough to overcome the activation barrier. Cooling down again, however, causes the reverse transition back to the 1x1 phase, only at lower temperatures. This hysteresis is caused by the fact that the heat of adsorption of CO is lower for the hex phase than for the 1x1 phase by about 25 - 40 kJ/mole, depending on coverage /77, 78/.

At constant temperature the CO coverage can be reduced by autocatalytic oxidation by oxygen dissociatively adsorbed when an additional oxygen partial pressure is provided. The adsorption kinetics of O_2 is known to depend strongly on surface structure, the sticking coefficient being of the order of 0.1 for Pt(100)1x1 /80, 81/ and about two orders of magnitude smaller for the reconstructed phase /49, 81-83/. This strong dependence of the adsorption properties of both carbon monoxide and oxygen was shown to lead to oscillations of the Pt(100) surface structure with time, combined with oscillations of the CO_2 production rate, when the partial pressures of CO and O_2 are properly adjusted /84-89/. Figure 23a demonstrates the structural oscillations via the time dependence of various LEED beams: The "hex" beam is the 3/5 1 beam of the reconstructed phase displaying the degree of reconstruction, the "c(2x2)" beam is the 1/2 1/2 beam of the c(2x2) CO superstructure developed on the non-reconstructed surface and so displaying the area of the 1x1 phase, and the 11 is a substrate beam which is of high intensity when the 1x1 phase is oxygen covered. The partial pressures for O_2 and CO were 4 x 10^{-4} Torr and 4 x 10^{-5} Torr, respectively. The data, taken by Ertl and co-workers /86/, were recorded using a high-speed video technique operated under computer control /10, 55/.

The oscillations can be qualitatively understood by the mechanisms displayed in Fig.24a starting with the CO covered 1x1 phase and the corresponding c(2x2) superstructure /79/. Because of the full coverage, dissociative oxygen adsorption can only take place at defects indicated by "X". Subsequently, neighbouring CO molecules are oxidized and CO_2 desorbs leaving a 1x1 area uncovered. Consequently, the c(2x2) beam will decrease and the 11 beam increase. Its high sticking coefficient allows oxygen to adsorb on the 1x1 patches. However, additionally arriving CO molecules and reaction keep the total coverage low enough for the transition 1x1 → hex to take place. Being activated this process is comparatively slow, so the "hex" beam in Fig.23a increases only gradually. Simultaneously the oxygen adsorption must decrease because of the low sticking coefficient on the reconstructed phase. Adsorbing CO is likely to diffuse to neighbouring oxygen-covered 1x1 areas, triggering the hexagonal transition there as well. However, when the reconstructed areas become sufficiently large to make diffusion impossible, the CO coverage can

228

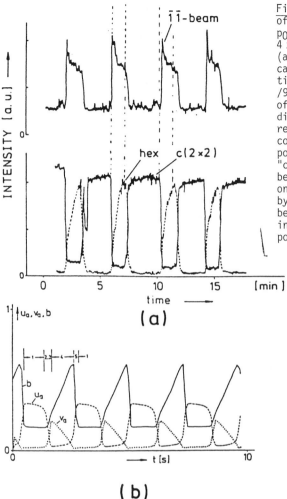

Fig.23: Structural oscillation of Pt(100) when exposed to $p_{O2} = 4 \times 10^{-4}$ Torr and $p_{CO} = 4 \times 10^{-5}$ Torr (a) experimental (as taken from /86/ and (b) calculated according to equations (1) - (4) (as taken from /91/). In (a) the oscillations of LEED beam intensities are displayed, whereby the 11 beam reflects the (partially) oxygen covered 1x1 surface and corresponds to curve v_a in (b), the "c(2x2)" beam is the 1/2 1/2 beam of the CO superstructure on the 1x1 phase as described by curve u_a in (b) and the "hex" beam is the 3/5 1 beam reflecting reconstructed areas corresponding to curve b in (b).

increase beyond its critical value again and removal of reconstruction and development of c(2x2)1x1 takes place. As no substantial barrier exists for this process it proceeds rapidly and closes the cycle.

It is clear from this description that the partial pressures of both oxygen and carbon monoxide, as well as the temperature, have to be properly adjusted to allow the oscillations. Figure 24b displays the existence region for oscillations at the fixed temperature of 480 K /90/. If the ratio p_{CO}/p_{O_2} is too low (region A) the CO adsorption is too slow to reduce the oxygen coverage at 1x1 patches and so a steady state coexistence of 1x1 and reconstructed areas is maintained. If, on the other hand, the p_{CO}/p_{O_2} ratio is too high (region B) a stable 1x1 structure is realized as oxygen adsorption is strongly inhibited.

Recently a mathematical model of 4 coupled differential equations was able to be established to describe the occurrence of oscillations more quantita-

229

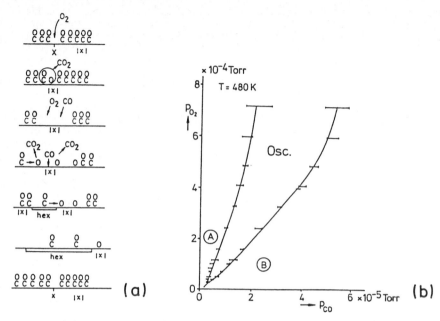

Fig.24: (a) Schematic of surface structures within one cycle of oscillations of Pt(100) (as taken from /79/) and (b) existence region of the oscillations at 480 K (as taken from /90/).

tively /91/. With h and q as the fraction of areas that are hexagonally (h) reconstructed or that have a quadratic (q) bulk-like structure, $\Theta h_{CO,O}^{h,q}$ being the local coverages of CO or O on h or q phases, and $CO^h = h\Theta_{CO}^h$, $CO^q = q\Theta_{CO}^q$, $O^h = h\Theta_O^h$, $O^q = q\Theta_O^q$ being the corresponding absolute coverages, the 4 equations /91/ may be written as

$$\dot{CO}^q = k_1 q p_{CO} - k_2 CO^q + k_3 q CO^h - k_4 CO^q O^q/q + k_5 \Delta \left(\frac{CO^q}{q}\right) , \qquad (1)$$

$$\dot{CO}^h = k_1 h p_{CO} - k_6 CO^h - k_3 q CO^h , \qquad (2)$$

$$\dot{O}^q = k_7 q p_{O_2} \left[\left(1 - 2\frac{CO^q}{q} - \frac{5}{3}\frac{O^q}{q}\right)^2 + \alpha \left(1 - \frac{5}{3}\frac{O^q}{q}\right)^2\right] - k_4 CO^q O^q/q , \qquad (3)$$

$$\dot{q} = \begin{cases} CO^q/\Theta_{CO,grow}^q & \text{if } \Theta_{CO}^q > \Theta_{CO,grow}^q \text{ and } CO^q > 0 \\ -k_8 q c & \text{if } c = \Theta_{CO}^q/\Theta_{CO,crit}^q + \Theta_O^q/\Theta_{O,crit}^q < 1 \\ 0 & \text{otherwise} \end{cases} \qquad (4)$$

In (1) and (2) the first two terms describe adsorption and desorption, and the third term accounts for the movement of CO from hexagonal areas to quadratic areas where it is trapped because of the higher adsorption energy. The remaining terms in (1) describe the second-order reaction CO + O → CO$_2$ and diffusion of CO. In (3) the first term again stands for adsorption, where the first part in parenthesis accounts for the dependence of the oxygen sticking coefficient on the O and CO coverage, being squared because of dissociative

230

adsorption. The second bracket allows for adsorption at defects, with a fitting parameter α in the range 0.1 - 0.5. Desorption of oxygen is neglected. Finally, (4) describes the phase transformation $1x1 \leftrightarrow$ hex ($q \leftrightarrow h$) where the upper part simulates the process hex $\rightarrow 1x1$ once a critical coverage $\Theta_{CO,grow}^q \approx$ 0.5 for growth is exceeded, and the middle part accounts for the reverse process when oxygen and CO are below critical coverages $\Theta_{CO,crit}^q \approx 0.3$ and $\Theta_{0,crit}^q \approx 0.4$, so that $c < 1$. The kinetic parameters can be taken from the literature /92/ as $k_1 = 2.94 \times 10^5$ ML Torr/s where ML stands for a monolayer, $k_2 = 9 \times 10^{-3}$/s at $\Theta = 0.3$ and 1.5/s at $\Theta = 0.5$, $k_4 = 10^3 - 10^5$/ML s, $k_5 = 10^{-7} - 10^{-5} cm^2$/s, $k_6 = 11$/s, $k_7 = 5.6 \times 10^5$ ML/Torr s and $k_8 = 2$/s. It should be emphasized that the only parameter related to the phase transition itself is k_8 for the $1x1 \rightarrow$ hex process. The reverse transition is believed to develop instantaneously once the critical CO coverage is reached.

In fact the numerical treatment of (1) - (4) with these parameters, but with k_5 and k_8 changed to $k_5 = 4 \times 10^{-4} cm^2$/s and $k_8 = 0.2$/s in order to save computer time, leads to the development of oscillations. The result is displayed in Fig.23b and compares very well with the experimental results given in Fig. 23a apart from the time scale, which is shortened due to the choice of k_5 and k_8.

Recently, spatial self-organization of the oscillations was also observed leading to a wave-like propagation of the alternating structures across the surface. The experiment was realized using a scanning low-energy electron diffraction technique /93/. The primary beam of 0.7 mm diameter was repeatedly deflected across the surface whereby an area of $4 \times 7 mm^2$ was scanned with 10 scans of 1 s each. Both the 1/2 1/2 spot intensity of the c(2x2) super-structure on the 1x1 phase and the 3/5 O spot intensity of the reconstructed phase were recorded as given in Fig.25. The wave-like propagation of structure is evident. It could be qualitatively described by (1) - (4) as well /91/, taking advantage of the observation that propagation always starts from the crystal edge. Therefore, an increased concentration of defects at the edge was introduced in the calculation. This locally increases the rate of oxygen adsorption, which in turn triggers the transition. Dividing the crystal up into compartments and giving α higher values at the edge, the wave-like propagation does in fact result from the calculation, where the spatial structure is mainly due to the coupling between CO diffusion and reactive removal of CO.

Though the oscillations described have been investigated for Pt(100) in detail, they should not be unique for this system. In fact, oscillations were recently detected also for the system CO on Pt(110) /94/, a surface which was shown in a preceding section to reconstruct. It should be emphasized, however, that reconstruction of the clean surface in principle is not necessary for the occurrence of the phenomenon, but adsorbate-induced reconstruction may be sufficient.

6. Discussion

As shown in this paper, nearly all measurements of the kinetics of surface phase changes study the transition upon a quench of an external thermodynamic parameter. In all cases this is the temperature because a temperature quench can be realized more easily than a quench of other variables. Nevertheless, in some cases the quench time could not be made short enough to avoid an interference of time and temperature dependences, at least during the start of the transition. In the case of the CO induced transition on Pt(100) the pure transition kinetics of the system is mixed with the adsorption/desorption

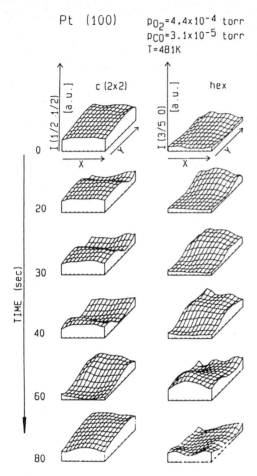

Pt (100) $p_{O_2}=4.4\times10^{-4}$ torr
            $p_{CO}=3.1\times10^{-5}$ torr
            $T=481K$

Fig.25: Structure map of a Pt(100) crystal in oscillation mode taken by scanning the surface with a primary beam of 0.7 mm diameter and recording LEED superstructure beam intensities of the hexagonally reconstructed (hex) and the unreconstructed by c(2x2) CO covered surface (as taken from /93/).

kinetics. Though this is also a very interesting case, which can even lead to the impressive temporal and spatial structural oscillations described, such investigations seem to be ahead of their time when the basics of the transition kinetics itself are under investigation and not yet fully understood.

For all the measurements presented the temperature quench leads to a point of the phase diagram away from the phase boundary, so to the author's knowledge there is no experimental material corresponding to nonlinear critical relaxation which occurs subsequent to a quench right to the phase boundary. The only investigation approaching this phenomenon is that of Si(111)1x1 ←→ 7x7 where the final state temperature was varied to approach the critical temperature. As a result, a pronounced slowing down was observed. However, this did not appear in the reverse direction of the transition.

The great majority of experiments correspond to quenches leading to a final state inside the phase diagram. Immediately after the quench the system is in a metastable or unstable state from which it relaxes to its final equilibrium

232

state. Therefore it seems to be unimportant whether the initial state from which the quench is started is stable or metastable. In any case, a state far from equilibrium results upon quenching. This is worth mentioning because some of the transitions presented, e.g. the reconstruction of clean metal surfaces, are in a metastable state before the transition is triggered.

The preferred method for investigating the kinetics of surface phase transitions was shown to be LEED, though other techniques are also applied. This is because LEED is strongly surface sensitive and beam profiles are believed to reflect directly the structure factor calculated in the theory, the width of which is proportional to the inverse of the average domain size L. In most cases theory and experiment are restricted to the time dependence $L(t)$, where t is the time after the quench. It is attempted to fit the experiment with a power law $L \sim t^x$, where x is a characteristic exponent resulting from scaling hypothesis. It is believed to be determined by the universality class of the transition under investigation.

By this, the experimental procedure to measure the kinetics seems to be very clear: Take the beam profile as a function of time, extract $L(t)$ and compare it with a power law to determine x. However, the measurement of beam profiles is tedious and time consuming and so time resolution could be lost. Therefore, in nearly all experiments, either the beam's peak intensity I_p or its integral intensity I is measured, which is possible with sufficient speed. From I no information on L can be extracted, but it is believed to be true for I_p. Here the trouble begins, and critism must start because this is true only for simple cases. One such case is realized when all atoms in the surface contribute to the diffraction spot under consideration and the ordering process develops through hopping of atoms from one of the corresponding preferred lattice gas sites to another. In this case $L \sim \sqrt{I_p}$ results as shown in Sect.2 and applied for the systems O/W(110) /24/, S/Mo(110) /30/ and SbCl$_5$ intercalated graphite /31/. However, an ordered domain can grow also from atoms which occupy non-coherent sites prior to joining the domain. Then the number of atoms is not conserved, though the coverage is constant, and $L \sim \sqrt[4]{I_p}$ results, as assumed for O/W(112) /22/ and the late time behaviour of O/W(110) /23/. It is not always clear which of these cases is realized in a certain experiment. Moreover, reality can correspond to any mixture of the extremes described.

Nevertheless, if we believe for the moment that L was correctly determined from I_p in the measurements presented here, it must be stated that it is only for not too long times that a power law was demonstrated. The exponents vary between 0.33 and 0.5 but no clear dependence on the degeneracy is evident, as shown in the table. Unfortunately, for early times the measured dependences can interfere with the temperature-time dependence when the quench is not fast enough. For longer times the time dependence becomes sluggish and in all cases the equilibrium values are not approached even after long times. Instead, some quasi-saturation is observed, the level of which increases with the quench temperature. The phenomenon may be caused by the pinning of domain boundaries at surface defects, which can only be removed by heating to elevated temperatures.

It is necessary to discuss the role of the peak intensity I_p in some more detail. It was mentioned that its connection to L is not a priori clear. However, the situation is still worse from several points of view. First of all the angular resolution of any LEED detector is limited. Therefore, instead of I_p some central part of the beam is measured. If this is small compared to the beam width, I_p is well approximated. In the other case, however, more or less the integrated intensity is measured and the information about L is

233

lost. In many publications, however, information concerning the angular width with which I_p is measured is missing. Yet the measurement of I_p is problematic for more fundamental reasons, too. It is well known that I_p is affected by thermal diffuse scattering, a fact which is known to complicate the determination of critical parameters by LEED considerably /30/. In the case of temperature quenches, i.e. for temperatures being subsequently constant, one could believe that this is of no importance. However, one has to remember that the Debye-Waller factor depends strongly on the surface Debye temperature Θ_s. This can change considerably during the transition, as it depends among other things on the mutual interaction of surface atoms. So, for the phase transition of clean W(100) it could be shown that Θ_s changes from 210 K to about 400 K /95/. Consequently, even for constant temperatures thermal diffuse scattering may change during the transition and so modify I_p. Moreover, apart from this kinematic effect also dynamic effects could influence I_p. It is well known that LEED is dominated by multiple scattering, which in turn depends strongly on the arrangement of atoms. Therefore, the rearrangement of atoms during a phase transition must be expected to modify I_p dynamically, too.

The measurement of I_p becomes fully meaningless when the shape of a diffraction beam changes during the transition, as observed in the case of Ir(100) 1x1 → 1x5 /52/ (see Fig.16) or when the beam is not well defined, as in the streak reconstruction of H/Ni(110) /76/. Then, only the integrated intensity seems to be a reasonable quantity reflecting the number of atoms in the phase under investigation. Unfortunately, at the moment there is no scheme provided by theory to extract information from $I(t)$, apart from Avrami's fromula. The application of the latter, however, gives results (see also Table 1) which are far from the values predicted by Avrami for the 2D case and cannot be interpreted at the moment.

In conclusion, it must be said that, apart from clear cases, the confidence which can be attributed to I_p measurements and subsequent evaluation of L is restricted. On the other hand, information from integrated intensities is limited, too. So, the only general way out seems to be the measurement of spot profiles. Their shape is only poorly influenced by thermal diffuse and multiple scattering, which in good approximation only modify the intensity scale. However, this requires fast measurement methods in order to guarantee the necessary time resolution. As demonstrated in Fig.1 this is possible at least in principle when for example computer-controlled and high-sensitivity

Table 1: Results of peak or integral intensity evaluation (I_p or I in second column) for different systems of degeneracy p, where m, x and n correspond to the exponents in $L \sim I_p^{1/m}$, $L \sim t^x$ and in Avrami's law, respectively (for Ir(100) data in /52/ were reevaluated to give n).

Surface system	P	I/I_p	m	x	n	Ref.
O/W(112)	2	I_p	4	0.5		/22/
O/W(110)	(4)	I_p	4	0.5		/23/
	8	I_p	2	0.33		/24/
S/Mo(110)	3	I_p	2	0.36		/30/
$SbCl_5$-GIC6	14	I_p	2	0.5		/31/
Ir(100)	10	I			0.25	/52/
Pt(110)	2	I			0.25	/59/
O/Ni(110)	2(?)	I			0.3	/69/
H/Ni(110)	?	I			1	/76/

techniques are applied. This must be combined with detectors of sufficiently narrow experimental response function to save the necessary resolution. Of course, imaging techniques such as STM and LEERM could develop into very powerful methods, as demonstrated. They do not suffer from reciprocal space presentation but deliver a direct impression of the surface with near atomic spatial resolution. If they could be speeded up to follow a transition with sufficient time resolution the kinetics could even be the subject of a movie.

References

1. S. K. Sinha: Ordering in Two Dimensions (North-Holland, Amsterdam 1980)
2. E. Bauer: In Phase Transitions in Surface Films, ed. by J. G. Dash and J. Ruvalds (Plenum, New York 1980) p. 267
3. Symposium on Statistical Mechanics of Adsorption, Surf. Sci. 125, 1-325 (1983)
4. D. P. Woodruff, G. C. Wang and T. M. Lu: In The Chemical Physics of Solid Surfaces and Heterogeneous Catalysis, Vol. 2 (Elsevier, Amsterdam 1983) p. 259
5. P. Bak: In Chemistry and Physics of Solid Surfaces IV, Springer Series in Chemical Physics 35 (Springer, Berlin 1984) p. 317
6. L. D. Roelofs, A. R. Kortan, T. L. Einstein and R. L. Park: Phys. Rev. Lett. 46, 1465 (1981)
7. T. L. Einstein: In Chemistry and Physics of Solid Surfaces IV, Springer Series in Chemical Physics, Vol. 20, ed. by R. Vanselow and R. Howe (Springer, Berlin 1982) p. 251
8. J. C. Campuzano, G. Jennings and R. F. Willis: Surf. Sci. 162 484 (1985)
9. L. J. Clarke: Surface Crystallography (John Wiley & Sons Ltd. 1985)
10. K. Heinz and K. Müller: In Structural Studies of Surfaces, Springer Tracts in Modern Physics, Vol. 91, ed. by G. Höhler (Springer, Berlin 1982) p. 1
11. M. G. Lagally, J. A. Martin: Rev. Sci. Instr. 54, 1273 (1983)
12. J. B. Pendry: Low Energy Electron Diffraction (Academic Press, London 1974)
13. M. A. Van Hove and S. Y. Tong: Surface Crystallography by LEED, Springer Ser. Chem. Phys., Vol. 2 (Springer, Berlin 1979)
14. K. Heinz: Appl. Phys. A41, 1 (1986)
15. W. Moritz and M. G. Lagally: Phys. Rev. Lett. 56, 865 (1986)
16. M. Henzler: In Dynamical Phenomena at Surfaces, Interfaces and Superlattices, Springer Ser. Surf. Sci. 3, ed. by F. Nizzoli, K.-H. Rieder and R. F. Willis (Springer, Berlin 1985) p. 14
17. M. G. Lagally: In Chemistry and Physics of Solid Surfaces IV, ed. by R. Vanselow and R. Howe, Springer Ser. Chem. Phys. 20 (Springer, Berlin 1982) Chap. 12
    M. G. Lagally: Appl. Surf. Sci. 13, 260 (1982)
18. I. K. Robinson: In The Structure of Surfaces, Springer Ser. Surf. Sci. 2, ed. by M. A. Van Hove and S. Y. Tong (Springer, Berlin 1985) p. 60
19. G. A. Somorjai and M. A. Van Hove: Adsorbed Monolayers on Solid Surfaces, Structure and Bonding 38 (Springer, Berlin 1979)
20. L. Peralta, Y. Berthier and J. Oudar: Surf. Sci. 55, 199 (1976)
21. I. M. Lifshitz: Sov. Phys. JETP 15, 939 (1962)
22. G.-C. Wang and T.-M. Lu: Phys. Rev. Lett. 50, 2014 (1983)
23. P. K. Wu, J. H. Perepezko, J. T. McKinney and M. G. Lagally: Phys. Rev. Lett. 51, 1577 (1983)
24. N. Tringides, P. K. Wu, W. Moritz and M. G. Lagally: Ber. Bunsenges. Phys. Chem. 90, 277 (1986)
25. S. M. Allen and J. W. Cahn: Acta Metall. 27, 1085 (1979)
26. T. Engel, H. Niehus and E. Bauer: Surf. Sci. 52, 237 (1975)

27. G.-C. Wang, T.-M. Lu and M. G. Lagally: J. Chem. Phys. 69, 479 (1978) and Phys. Rev. Lett. 39, 411 (1977)
28. M. G. Lagally, T.-M. Lu and G.-C. Wang: In Ordering in Two Dimensions, ed. by S. K. Sinha (North-Holland, Amsterdam 1980) p. 113
29. M. A. Van Hove and S. Y. Tong: Phys. Rev. Lett. 35, 1902 (1975)
30. W. Witt and E. Bauer: Ber. Bunsenges. Phys. Chem. 90, 248 (1986)
31. H. Homma and R. Clarke: Phys. Rev. Lett. 52, 629 (1984)
32. R. M. Stein, L. Walmsley, G. M. Gaulberto and C. Rettori: Phys. Rev. B 32, 4774 (1985)
33. R. M. Stein, L. Walmsley, S. Rolla and C. Rettori: Phys. Rev. B33, 6524 (1986)
34. P. S. Sahni, D. J. Srolovitz, G. S. Grest, M. P. Anderson and S. A. Safran: Phys. Rev. B28, 2705 (1983)
35. C. B. Duke: CRC Crit. Rev. Solid State Mater. Sci. 8, 69 (1978)
36. P. J. Estrup: In Chemistry and Physics of Solid Surfaces V, Springer Ser. in Chem. Phys. 35, ed. by R. Vanselow and R. Howe, (Springer, Berlin 1984) p. 205
37. W. Witt and E. Bauer: Surf. Sci., to be published
38. E. G. McRae: In The Structure of Surfaces, Springer Ser. in Surf. Sci. 2, ed. by M. A. Van Hove and S. Y. Tong (Springer, Berlin 1985) p. 278
39. P. A. Bennett and M. B. Webb: Surf. Sci. 104, 74 (1981)
40. W. Telieps and E. Bauer: Surf. Sci. 162, 163 (1985)
41. W. Telieps and E. Bauer: Ber. Bunsenges. Phys. Chem. 90, 197 (1986)
42. E. Bauer: Ultramicroscopy 17, 51 (1985)
43. W. Telieps and E. Bauer: Ultramicroscopy 17, 57 (1985)
44. E. Lang, K. Müller, K. Heinz, M. A. Van Hove, R. J. Koestner and G. A. Somorjai: Surf. Sci. 127, 347 (1983)
45. W. Moritz: Habilitationsschrift, München (1983)
46. N. Bickel and K. Heinz: Surf. Sci. 163, 435 (1985)
47. P. Heilmann, K. Heinz and K. Müller: Surf. Sci. 83, 487 (1979)
48. M. A. Van Hove, R. J. Koestner, P. C. Stair, J. P. Biberian, L. L. Kesmodel, I. Bartos and G. A. Somorjai: Surf. Sci. 103, 189 (1981)
49. P. R. Norton, J. A. Davies, D. K. Creber, C. W. Sitter and T. E. Jackman: Surf. Sci. 108, 205 (1981)
50. R. J. Behm, W. Hösler, E. Ritter and G. Binnig: J. Vac. Sci. Technol. A4, 1330 (1986)
51. R. J. Behm, W. Hösler, E. Ritter and G. Binnig: Phys. Rev. Lett. 56, 228 (1986)
52. K. Heinz, G. Schmidt, L. Hammer and K. Müller: Phys. Rev. B32, 6214 (1985)
53. G. Besold, K. Heinz, E. Lang and K. Müller: J. Vac. Sci. Technol. A1, 1473 (1983)
54. W. Hösler, E. Ritter and R. J. Behm: Ber. Bunsenges. Phys. Chem. 90, 205 (1986)
55. P. Heilmann, E. Lang, K. Heinz and K. Müller: In Determination of Surface Structure by LEED, ed. by P. M. Marcus and F. Jona (Plenum, New York 1984) p. 463
56. J. Küppers and H. Michel: Appl. Surf. Sci. 3, 179 (1979)
57. K. Heinz, E. Lang, K. Strauss and K. Müller: Appl. Surf. Sci. 11/12, 611 (1982)
58. K. Heinz, E. Lang, K. Strauss and K. Müller: Surf. Sci. 120, L401 (1982)
59. A. Barthel: Diploma work (Erlangen, 1986)
60. C. M. Comrie and R. M. Lambert: JCS Faraday Trans. 72, 1659 (1976)
61. S. Ferrer and H. P. Bonzel: Surf. Sci. 119, 234 (1982)
62. U. Teschler: Diploma work (Erlangen, 1986)
63. M. Avrami: J. Chem. Phys. 7, 1103 (1939); 8, 212 (1940); 9, 177 (1941)
64. R. J. Behm, G. Ertl, V. Penka and R. Schwankner: J. Vac. Sci. Technol. A3, 1595 (1985)

65. J. A. Van der Berg, L. K. Verheij and D. G. Armour: Surf. Sci. 84, 408 (1979)
66. R. G. Smeenck, R. M. Tromp and F. W. Saris: Surf. Sci. 107, 429 (1981)
67. T. Engel, K. H. Rieder and I. P. Batra: Surf. Sci. 148, 321 (1984)
68. N. Niehus and G. Comsa: Surf. Sci. 151, L171 (1985)
69. R. J. Behm, G. Ertl and J. Wintterlin: Ber. Bunsenges. Phys. Chem. 90, 294 (1986)
70. V. Penka, K. Christmann and G. Ertl: Surf. Sci. 136, 307 (1984)
71. T. Engel and K. H. Rieder: Surf. Sci. 109, 140 (1981)
72. J. H. Onuferko and D. P. Woodruff: Surf. Sci. 91, 400 (1980)
73. K. Christmann, F. Chebab, V. Parka and G. Ertl: Surf. Sci. 152/153, 356 (1985)
74. R. J. Behm, K. Christmann, G. Ertl, V. Penka and R. Schwankner: In The Structure of Surfaces, Springer Ser. in Surf. Sci. 2, ed. by M. A. Van Hove and S. Y. Tong (Springer, Berlin 1985) p. 257
75. K. Christmann: Ber. Bunsenges. Phys. Chem. 90, 307 (1986)
76. V. Penka, R. J. Behm and G. Ertl: J. Vac. Sci. Technol. A4, 1411 (1986)
77. R. J. Behm, P. A. Thiel, P. R. Norton and G. Ertl: J. Chem. Phys. 78, 7437 (1983); 78, 7448 (1983)
78. T. E. Jackman, K. Griffiths, J. A. Davies and P. R. Norton: J. Chem. Phys. 79, 3529 (1983)
79. G. Ertl: Ber. Bunsenges. Phys. Chem. 90, 284 (1986)
80. G. Pirug, G. Broden and H. P. Bonzel: Chem. Phys. Lett. 73, 306 (1980)
81. M. A. Barteau, E. I. Ko and R. J. Madix: Surf. Sci. 102, 99 (1981)
82. G. Kneringer and F. P. Netzer: Surf. Sci. 49, 125 (1975)
83. P. R. Norton, K. Griffiths, T. E. Jackman, J. A. Davies and P. E. Bindner: Surf. Sci. 138, 125 (1984)
84. P. Hugo: Ber. Bunsenges. Phys. Chem. 74, 121 (1970), H. Beusch, P. Fieguth and E. Wicke: Chem. Ing. Techn. 44, 445 (1972)
85. G. Ertl, P. R. Norton and J. Rüstig: Phys. Rev. Lett. 49, 177 (1982)
86. M. P. Cox, G. Ertl, R. Imbihl and J. Rüstig: Surf. Sci. 134, L517 (1983)
87. P. R. Norton, P. E. Bindner, K.Griffiths, T. E. Jackman, J. A. Davies and J. Rüstig: J. Chem. Phys. 80, 3859 (1984)
88. G. Ertl: Surf. Sci. 152/153, 328 (1985)
89. R. Imbihl, M. P. Cox and G. Ertl: J. Chem. Phys. in press
90. M. Eiswirth, R. Schwankner and G. Ertl: Z. Phys. Chem. Neue Folge 144, 59 (1985)
91. R. Imbihl, M. P. Cox, G. Ertl, H. Müller and W. Brenig: J. Chem. Phys. 83, 1578 (1985)
92. The detailed references are given in /91/
93. M. P. Cox, G. Ertl and R. Imbihl: Phys. Rev. Lett. 54, 1725 (1985)
94. M. Eiswirth and G. Ertl: Surf. Sci. in press
95. P. Heilmann, K. Heinz and K. Müller: Surf. Sci. 89, 84 (1979)
96. A. N. Kolmogorov, Bull. Acad. USSR, Phys. Ser. 3, 355 (1938)

Kinetics of Adatom Ordering on Surfaces

J.D. Gunton

Physics Department and Center for Advanced Computational Science,
Temple University, Philadelphia, PA 19122, USA

1. INTRODUCTION

Chemisorbed and physisorbed systems provide in principle a rich testing
ground for the study of pattern formation in systems undergoing phase transitions. This topic belongs to the general area of the kinetics of first
order phase transitions, which involves the evolution of systems initially
far from equilibrium to equilibrium, ordered states. The basic physics involves an understanding of interface instabilities in particular and the
kinetics of topological defects in general. The field as such has a long
history, involving experimental studies of systems such as liquid-gas, binary fluids, multicomponent alloys, intercalated compounds, and superfluids
and superconductors. Only rather recently have detailed studies begun of
similar phenomena in surface science. The article by K. Heinz in this volume provides a comprehensive summary of the current status of experimental
work in this field. It also provides a thorough analysis of the difficulties involved in typical diffraction experiments. The general field has
profited by computer simulation studies, which in surface science was initiated by a study of phase separation in a lattice gas model of O/W(110)
by SAHNI and GUNTON [1]. The difficulties involved in such Monte Carlo
studies have been critically reviewed recently by MILCHEV et al [2].

On the theoretical side, initial work dealt with nucleation and growth
studies relevant to the decay of metastable states. Subsequently theoretical work treated the evolution of unstable states, which in contrast to
the nucleation of localized droplets involves a long wavelength instability. Our theoretical understanding of this latter topic is still rather
limited, although progress has been made recently based on models for the
motion of interfaces which separate the various possible phases of the system. An alternative approach involves a combination of renormalization
group ideas as implemented by Monte Carlo techniques. It is fair to say
that at the moment many theoretical issues remain unresolved. In one case,
there is currently a controversy concerning the validity of the classical
Lifshitz-Slyozov theory for systems such as binary alloys. MAZENKO and
VALLS [3] have argued that the Lifshitz-Slyozov theory (which describes
the late stage growth of droplets in phase separating systems) is incor-
rect, and that the characteristic length scale (e.g. average droplet size)
$L(t)$ satisfies $L(t) = A \ln t$ for late times (where t denotes the time following the quench to an unstable state), rather than the Lifshitz-Slyozov
prediction, $L(t) = B t^{1/3}$. Reviews and discussions of various of these
theoretical issues are given by GUNTON et al [4], BINDER [5], KAWASAKI [6],
VOORHEES [7], and MAZENKO and VALLS [3,8].

In this article I will concentrate primarily on Monte Carlo studies of
models relevant to the ordering of adatoms on surfaces, including both
physisorption and chemisorption. Chemisorbed systems are probably better
suited for most experimental studies, as the time scale for ordering of

physisorbed systems is generally quite short. Among the topics I will discuss are dynamical scaling and the role of parameters such as ground state degeneracy, conservation laws, impurities, and domain walls in determining growth laws for L(t) and (possibly) dynamical universality classes.

2. DYNAMICAL SCALING AND UNIVERSALITY CLASSES

In a typical experiment (or computer simulation) the system is quenched from a disordered state to an unstable point below a critical point (or line of critical points). The experimental procedures are reviewed by Heinz. Due to the instability the adatoms on the surface begin to rearrange their positions on the substrate in order to achieve their equilibrium, ordered structure appropriate for the final quench temperature T_f and coverage θ. This rearrangement leads to the formation of local regions (or islands) of order. As a consequence, if one studies the scattering intensity due to radiation incident upon this system, one observes the development of Bragg peaks associated with these local superstructures. Let $\vec{k} = \vec{q} - \vec{Q}_i$, where $\vec{Q}_i$ denotes one of the Bragg wavenumbers and $\vec{q}$ is the wavenumber of the scattered radiation. Then as the adatoms evolve to their equilibrium ordered structure, the scattering intensity $S(\vec{k},t)$ approaches its equilibrium value $S(\vec{k}) = \lim_{t \to \infty} S(\vec{k},t)$ which for small k consists of two parts:

$$S(\vec{k}) = \psi^2 \, \delta(\vec{k}) + \chi(\vec{k}), \qquad (2.1)$$

where ψ denotes the order parameter. The first term describes the well-known Bragg scattering due to the long-range order associated with ψ, while the second term is the scattering due to the short-range order. Its form usually can be described by the well-known Ornstein-Zernike function, $\chi/(1 + k^2\xi^2)$, where χ is the order parameter susceptibility and ξ is the correlation length.

The question with which we are concerned, however, is the behavior of $S(\vec{k},t)$ for finite t following the quench. In this case, the first term in (2.1) is no longer a Dirac delta function, since due to the finite size (and distribution of island sizes), the δ-function is smeared out over a finite range of k. The inverse of the width of this smeared out Bragg peak can be used as one definition of a characteristic, time-dependent length scale of the system. One of the more interesting results to emerge from experimental, theoretical and computer studies of $S(\vec{k},t)$ for a variety of different systems is that of dynamical scaling, i.e.

$$S(\vec{k},t) = L(t)^d F(kL(t)), \quad t > t_o . \qquad (2.2)$$

In (2.2) F(x) is called a scaling function and in principle can depend on the direction of $\vec{k}$, as well as on the quench temperature and coverage. Here d denotes the dimensionality, which in surface science is d = 2, while t_o denotes an initial "transient" time. Presumably for $t < t_o$ the process of domain formation is occurring, with the sizes of the domains (islands) being small and comparable to the equilibrium correlation length (which is of the order of a lattice constant unless the quench temperature T_f is close to the phase transition temperature T_c). The value of T_0 is system-dependent and ranges from very short times (e.g. possibly picoseconds) in models of physisorbed systems, to thousands of minutes in certain binary alloys. It is my opinion that in much of the Monte Carlo and experimental literature in which scaling is reported, there are probably small departures from scaling present at early times which could be seen if one looked carefully at the data. This would be due to attempting to fit the data to a scaling form in too early a time regime. A good example of this is provided in a recent Monte Carlo study of the Q-state Potts model (KUMAR et al) [9], where one can see a weak time dependence in $F(x)$, which eventually disappears as one reaches the scaling regime.

One is also interested in the time dependence of $L(t)$, which is often assumed to be a power law, i.e.

$$L(t) = A(t - t_1)^n \quad , \quad t > t_1 \quad , \tag{2.3}$$

where t_1 denotes the earliest time at which the power law fit is used and n is the dynamical exponent. It is also possible that $L(t)$ is a logarithmic function of time in certain cases.

Important questions, then, are what parameters are relevant in determining n and $F(x)$. In critical phenomena one would ask, what are the relevant variables for determining a universality class, so that by analogy one might think that there are dynamical universality classes in this nonequilibrium problem. It is clear from many studies that whether or not the order parameter is conserved can affect the value of n. As well, the degeneracy of the ground state might be a relevant parameter. Also, it is possible that the uniaxiality of the ground state might be a relevant parameter, as mentioned in the article on O/Pd(110) by Ala-Nissila and Gunton in this volume. It seems clear that we are far from an understanding of such dynamical universality classes (assuming that they do indeed exist in these nonequilibrium problems).

3. LATTICE GAS MODELS

Before discussing specific results of model calculations, we present a short summary of the kinetic lattice gas models which have been used in the bulk of the Monte Carlo simulations to date. The reader will note that these models are rather simplified representations of the kinetics of adatoms on substrates. Nevertheless, they have provided a useful starting point for developing at least a qualitative understanding of some aspects of the kinetics of adatom ordering.

The usual lattice gas Hamiltonian for a gas of adatoms adsorbed at various sites of a two-dimensional Bravais lattice can be written as

$$H = -\{\Sigma\phi(\vec{r}_i) \, c_i + \sum_{i,j} \phi(\vec{r}_{ij}) \, c_i \, c_j + \Sigma\phi(\vec{r}_{ijk}) \, c_i \, c_j \, c_k + \ldots \} \quad , \tag{3.1}$$

where $\phi(\vec{r}_i)$ is usually taken to be independent of $\vec{r}_i$ and is the binding energy of an adatom to the substrate. (If one models a situation in which the adatom coverage is not conserved, as in adsorption-desorption dynamics for physisorbed systems, $\phi(\vec{r}_i) = \phi = \mu + \varepsilon$, where μ is the chemical potential). The interaction energies $\phi(\vec{r}_{ij})$, $\phi(\vec{r}_{ijk})$,...describe pairwise, triplet and higher order interactions. In principle these should be obtained from a first principles quantum mechanical calculation. In practice, however, given the difficulty of doing such a calculation, these parameters are rather determined by comparing the experimental phase diagram for the system with theoretical predictions obtained from (3.1). Occasionally (in the absence of experimental results) the parameters are chosen rather arbitrarily, with the constraint that they yield the correct ground states for the system of interest. The occupation variable $c_i = 0$ if the lattice site i is empty and $c_i = 1$ if it is occupied by an adatom. If there are N sites, the coverage $\theta = N_a/N$, where N_a is the number of adatoms. Monte Carlo simulations consider coverages of a monolayer or less.

One of the interesting aspects of surface science is that Hamiltonians such as (3.1) display a rich variety of two-dimensional phase transitions. An excellent review of this subject, within the context of the Landau-Lifshitz theory, has been given by SCHICK [10].

We must now specify the dynamical model corresponding to a given Hamiltonian such as (3.1). This choice of dynamical model depends on whether we are trying to study a physisorbed system, in which the coverage is not conserved, or a chemisorbed system, in which the coverage is conserved during the experiment. In both cases one assumes that the dynamics is governed by a master equation for the time-dependent probability distribution function describing the instantaneous configuration of adatoms. The different models correspond to different choices for the transition rates in the master equation. In the physisorbed case, the physisorbed layer is initially in a disordered thermal equilibrium state, in contact with its vapor. After the quench, the subsequent development of order involves both the evaporation of adatoms into the vapor and the condensation of atoms from the vapor onto the substrate. In this case one uses Glauber dynamics, in which the transition rate per unit time, $W(c_i \to c_i')$ describes an adatom at site i evaporating to leave a vacancy, or the inverse process, depending on whether $c_i = 1$ or $c_i = 0$. The only restriction on the form of W is that it satisfies detailed balance. Thus there are in principle many kinetic Glauber models which one can choose to model the adsorption-desorption dynamics. One standard choice is

$$W (c_i \to c_i') = \frac{\alpha}{2} [1 - \tanh (\delta H/2kT] , \qquad (3.2)$$

where α^{-1} is a system-dependent parameter which sets the unit of time and δH is the change in the energy corresponding to the process $c_i \to c_i'$.

If, on the other hand, one is modelling a system in which the coverage is constant during the experiment, as in chemisorption, then one uses the so-called kinetic Kawasaki model. This is a very simple model of surface diffusion in which an adatom at site i ($c_i = 1$) hops to a nearest neighbor site j which is empty ($c_j = 0$). The transition probability for unit time for this to occur is $W(c_i c_j \to c_i' c_j')$, where $c_i' = 0$ and $c_j' = 1$. Obviously such a model conserves the total coverage. As in the Glauber model, there are many different kinetic Kawasaki models, one of which involves a transition rate similar to (3.2).

Note that in both these models of kinetics, only the change in energy between the initial and final configuration of adatoms determines the transition rate.

4. A MODEL FOR SYSTEMS LIKE O/W(112)

In this section we discuss the one model which is well understood, both in terms of its growth law and scaling function. As might be expected, it is perhaps the simplest model which one could define in this general area. For convenience we rewrite (3.1) in an Ising spin representation (well known in critical phenomena), by making the transformation $S_i = 2c_i - 1 = \pm 1$. The model we will discuss has a Hamiltonian with only nearest-neighbor pairwise interactions,

$$H = -J \sum_{<ij>} S_i S_j , \qquad (4.1)$$

where the sum over runs over nearest neighbor pairs on a square lattice. In magnetism, this model is known as the ferromagnetic or antiferromagnetic Ising model, for $J \gtrless 0$ respectively. The Hamiltonian has a p = 2 degenerate ground state, which for the ferromagnet consists of either all spins up (all sites empty) or all spins down (all sites occupied).

Domain growth in this model has been studied extensively for the case of attractive interactions and Glauber dynamics. This would be a simple model of adsorption-desorption dynamics for a physisorbed system. Monte Carlo results show that the average size of the domains grows like $L(t) \sim t^{1/2}$, i.e. the growth exponent n = 1/2. This is in agreement with a variety of theoretical predictions for this model (see for example [4]). The simplest way to understand this result based on theory is due to ALLEN and CAHN [11]. Consider a sufficiently long time following a quench such that one has regions of opposite order (i.e. regions of adatoms and empty sites) separated by gently curved interfaces. They showed that under these circumstances the normal component of the velocity of the interface at a given point on the interface is proportional to the mean curvature at that point, i.e.

$$v = D K , \qquad (4.2)$$

where D is an effective diffusion constant. The growth law can be understood by using the following simplified argument. Assume $K \propto L^{-1}$, where L is the average domain size and further assume that v = dL/dt. Then one immediately finds from (4.2) that $L(t) \sim t^{1/2}$. This simple argument shows that in the case of a two-fold degenerate ground state with a nonconserved order parameter, the domain growth is curvature driven.

In addition, Monte Carlo studies showed that dynamical scaling for S(k,t), (2.2), is satisfied. Subsequently it has been shown [9] that this scaling function and the interface-interface correlation function are in excellent agreement with theories [12], [13]. The simplest way to explain these theories is to use the arguments of reference [13]. These authors rewrote (4.2) as an explicit nonlinear equation of motion for the interfaces. They then linearized this equation, which yielded a simple diffusion equation for the motion of the interfaces. They further assumed that the distribution of these interfaces was given by a Gaussian. This then leads to explicit expressions for the interface-interface correlation function and structure function S(k,t) which are in agreement with the Monte Carlo data. It should also be noted that real space renormalization group work has also yielded the growth exponent n = 1/2 [14], [15], and a reasonably good approximation to F(x) [14].

There is another dynamical model which has been studied using (4.1). In this model one considers repulsive interactions (J <0) and diffusion (Kawasaki) dynamics. Monte Carlo simulation of this model has shown that not only is n = 1/2, but within the accuracy of the study, the scaling function is essentially the same as for the Glauber model described above [16]. Thus both dynamical models seem to belong to the same dynamical universality class. Since the diffusion dynamics for (4.1) with (J < 0) is a reasonable first approximation to O/W(112) it would seem that the exponent n ≈ 1/2 obtained for this system by WANG and LU [17] (see Heinz's article for details) is consistent with theory. The lattice gas Hamiltonian which has been proposed for O/W(112), however, has a second neighbor pair interaction in the direction of one of the two basis vectors for the unit cell. Thus one might expect anisotropic growth, with a scaling function which might depend on the direction of $\vec{k}$. A Monte Carlo study of domain growth for this model is currently underway [18]. It would be most useful to have an experimental measurement of $S(\vec{k},t)$ (rather than just the peak height) to compare with Monte Carlo and theoretical results.

242

5. DOMAIN GROWTH IN A LATTICE GAS MODEL WITH A SYMMETRIC, FOUR-FOLD DEGEN-ERATE GROUND STATE

SADIQ and BINDER [19] have made a very detailed study of a model on a square lattice described by (3.1) with repulsive (negative) nearest (ϕ_{nn}) and next-nearest neighbor interactions (ϕ_{nnn}) taken to be of equal strength. The other interactions in (3.1) are taken to be zero, except for the term $\phi(\vec{r}_i) = \phi = \mu + \epsilon$. For a coverage $\theta = 1/2$ this model undergoes a second-order transition to a (2x1) phase at a temperature of $T_c \simeq 2.1$ (measured in units of $|\phi_{nn}|/4k$). This model is of interest because the 2x1 ground state has a degeneracy $p = 4$, corresponding to the different ways one can arrange the adatoms in alternating columns or rows. As a consequence, there are many different types of walls between the ordered regime that develop during growth. Thus, it is of interest to determine whether this model with a two-component order parameter (m=2) and p=4 satisfies the same domain growth law as the one discussed in the previous section, with a single component order parameter (m=1) and degeneracy p=2.

The answer is that for Glauber dynamics, in which neither the order parameter nor the coverage θ is conserved, the growth law is $L(t) \sim t^{1/2}$, i.e. the same as for the model with n=1 discussed in the previous section. This is not a trivial result, however, since there are four types of walls which are energetically degenerate at T = 0 which remain present during the growth. Thus the situation is more complicated than for the case m=1, p=2. Nevertheless, the presence of these walls does not affect the growth law exponent n = 1/2. It should be noted however, that although the domain growth exponent is the same for these two models with Glauber dynamics, both their static and dynamic critical phenomena are different. (The model with m=1 is in the Ising universality class, whereas the model discussed by Sadiq and Binder with m=2 belongs to the universality class of the xy-model with cubic anisotropy, whose exponents are nonuniversal.)

The more interesting result is that for the case of conserved coverage (Kawasaki diffusion dynamics) appropriate for chemisorbed monolayers, the domain growth exponent is no longer n = 1/2. Thus although the Glauber and Kawasaki models are similar in that the order parameter is nonconserved in both cases, the conservation of adatoms is a relevant variable for determining the dynamic universality class in this case. Sadiq and Binder find that at high enough temperatures $L(t) \sim t^{1/3}$, i.e. n = 1/3. At lower temperatures they cannot determine the asymptotic growth law, but find that their data can be fit with an effective exponent, $n_{eff}(T)$, which varies with temperature. They attributed this as a crossover effect from the exponent which characterized their results or a quench to zero temperature, n=o. (That is, for a quench to T = 0, the system eventually became trapped in a metastable state and stopped evolving, thus never reaching equilibrium). They suggested that the asymptotic growth law for all T > 0 would be characterized by n = 1/3, but that this behavior occurs at much later times as one lowers T. Sadiq and Binder also suggested that the physical origin of the n = 1/3 behavior was the existence of excess (or deficit) coverage present in two types of walls present in the domain growth. They argued that for domains to grow (on average), adatoms had to diffuse from regions with excess coverage to regions with deficit coverage, over distances of the order of L(t). Such long-range diffusion is known to give rise to n = 1/3, as explained by the well-known theory of LIFSHITZ and SLYOZOV [20] for phase-separating binary alloys.

Two further points should be made in this regard. The first is that somewhat similar results were obtained by VINALS and GUNTON [21] for a lat-

243

tice gas model of H chemisorbed on Fe(110) surfaces. This model has both a 2x1 phase as well as a 3x1 phase. Their Monte Carlo results for Kawasaki diffusion dynamics gave n = 1/2 for the 2x1 phase. For the 3x1 phase, however, they were only able to determine an effective exponent. Thus even though walls with excess density were present during domain growth in the 3x1 phase, they were unable to determine the asymptotic growth law. In particular, they were unable to verify the Sadiq-Binder conjecture [19] that n = 1/3 for this case.

The second point to be made is that an alternative scenario has been proposed for the domain growth law for systems which get permanently trapped in metastable states at T = 0. Namely, MAZENKO and VALLS [14] have proposed that at least for the case m=1 with Kawasaki dynamics (as defined in Section 4) the asymptotic growth law is logarithmic in time. Thus it is possible that for the Sadiq-Binder model, e.g., the true growth law at finite temperature is L(t) ~ ln t, and that the Monte Carlo studies have not reached this late stage growth regime. Clearly the role of the conservation of coverage in the kinetics of domain growth of such chemisorbed systems remains rather poorly understood.

6. EFFECTS OF A WETTING TRANSITION ON DOMAIN GROWTH IN A TWO-DIMENSIONAL ANNNI MODEL

In this section we discuss a model [22-24] which has no current realization in surface science. Nevertheless it displays two interesting properties which will undoubtedly be found in chemisorbed systems, namely a uniaxial ground state and a wetting transition. This is the two-dimensional anisotropic next nearest neighbor Ising (ANNNI) model, which has been studied in detail. The Hamiltonian for this model in spin representation is given by an analogue of (3.1)

$$H = -\sum \left(J_1 s_{ij} s_{i+1j} - J_2 s_{ij} s_{i+2j} + J_0 s_{ij} s_{ij+1} \right) \quad , \quad S_{i,j} = \pm 1, \qquad (6.1)$$

where $J_i > 0$, i = 0,1,2, and the summation goes over all sites of a square lattice with Ising spins $s_{ij} = \pm 1$. The model could also be thought of as a lattice gas model using the variables c_i introduced in Section 3, with c = 1/2(1 + S). The indices i and j correspond to the x and y-directions, respectively. In Fig. 1 we depict the phase diagram of the model with the standard parametrization $\alpha = J_2/J_0$ and $J_1 = (1-\alpha)J_0$. The dashed line indicates the approximate position of the wetting line within the (2,2) antiphase. This phase is a uniaxial (4x1) phase with a ground state degeneracy p=4. It consists of an alternating sequence of two ferromagnetic layers of up and down spins in the x-direction.

Consequently, the order parameter Ψ has two components Ψ_1 and Ψ_2, as in the model discussed in the previous section. The wetting transition occurs when a soft superheavy-light (A|D) wall decays to three heavy-light (A|B) walls [23]. Symbolically, A|D →A|B|C|D, i.e. phases B and C wet A and D (see also Fig. 5(b)) below. The properties of this wetting transition have been studied in detail [24,25]. At T = 0 the wetting occurs at $\alpha = 1/2$.

In this model there is a growth exponent given by n ≈ 0.5 for low-temperature quenches everywhere in the (2,2) phase of the model for the case of Glauber dynamics. However, there is an abrupt change in the anisotropy of growth as one crosses the wetting line [22,23]. Namely, in the wet region the domains grow more rapidly in the y-direction than along the x-direction,

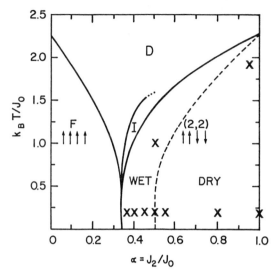

Figure 1   Phase diagram of the ANNNI model showing the disordered (D) phase, the commensurate (2,2) and ferromagnetic (F) phases and the incommensurate (I) phase. The dashed line indicates the wetting line. Crosses display the values of α and $k_B T_f/J_0$ for which quenches from the disordered phase were performed.

whereas just the opposite occurs in the dry region. Both the universal exponent n = 1/2 and this rather sudden change in the nonuniversal features of the domain growth can be explained by invoking a growth mechanism based on the existence of vertex-antivertex pairs in the system, which become unfavorable due to the wetting transition, as is discussed below. To my knowledge, this is the first time that a wetting transition has been shown to play a significant role in the kinetics of domain growth. Since systems with uniaxial (px1) phases are often found in physisorbed and chemisorbed systems, such as for example in O/Pd(110) [26-28] one would expect that the phenomena observed in the ANNNI model might be observed in some two-dimensional surface systems as well [28]. One also expects that the results obtained for this model have some relevance in clarifying the nature of the dynamical universality classes. Namely, it has also been found [23] that the ANNNI model with conserved (Kawasaki) dynamics has a growth exponent n≈ 0.5, in contrast to the earlier work of Sadiq and Binder discussed in Section 5. As we noted there, these authors suggested n = 1/3 for a model with a symmetric p=4 phase. It is probable that this difference in exponents is a manifestation of different growth mechanisms in the two models.

The Monte Carlo simulations for the domain growth of the ANNNI model were performed by quenching from a high temperature ($k_B T/J_0 = \infty$) disordered phase to a low temperature $k_B T_f/J_0 = 0.2$ within the (2,2) antiphase region for several values of α. These quench points are indicated in Fig. 1. Standard Glauber "spin flip" dynamics was used to model adsorption-desorption events in the system. Two independent measures of a characteristic length L(t) were used. First, the square root of the peak of the structure function, $\sqrt{S(0,t)}$, defines an average length. Secondly, to estimate the anis-

245

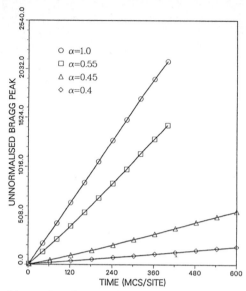

Figure 2   Time development of the Bragg peak S(0,t) as a function of the anisotropy parameter α. All results are for 128x128 systems at $k_B T_f / J_0 = 0.2$. (MCS: Monte Carlo step)

otropy of the growth one can use the inverses of the squares of the perimeter densities to define effective domain areas $A_x(t)$ and $A_y(t)$ in the x and y-directions, respectively. The results are summarized in Fig. 2 [23].

A dynamical exponent $n \simeq 0.5$ is obtained, independent of the value of the anisotropy parameter α (to a good degree of accuracy) both in the dry and wet regions. However, there is a strong dependence of both the average growth rate and the anisotropy of the growth on the parameter α. Namely as is evident from Fig. 2, in the dry region the growth in the x-direction is faster than in the y-direction. This anisotropy changes monotonically with decreasing α. In addition, there is a more rapid change across the wetting line. In the wet region the y-direction always grows faster. These features are also evident in Figs. 3(a)-(b) and 4(a)-(b), in which we show typical domain configurations for $\alpha = 1$ in the dry region and $\alpha = 0.5$ in the wet region [23]. The growth mode "rotates" by an amount of $\pi/2$ due to the wetting transition.

The domain configurations of Figs. 3 and 4 suggest that there are two main ingredients in the growth of the ANNNI model. First, a reduction of curvature characteristic of the Allen-Cahn mechanism [11] is clearly visible. However, there are restrictions on the configurations of the domains due to the high degeneracy of the ground state. These are apparent in vertices which are present for all values of α and at all times studied in our system. Recently, Kawasaki [27] has derived explicit equations of motion for p-state clock models which take into account the coupling between interfaces and vertices. By averaging over anisotropies he has shown that both for the two- and three-dimensional cases a $t^{1/2}$ growth law follows, since the problem can be reduced to dissipative dynamics of opposite "Coulomb charges". The existence of these vertices has been demonstrated for the clock model with several values of p [28]. The presence of analogous vertices is evident from Figs. 3 and 4

246

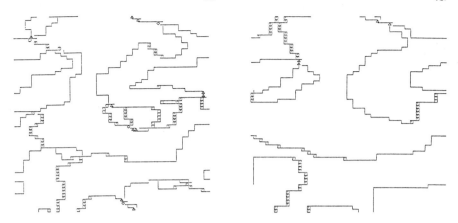

Figure 3   Time evolution of domains for a 64x64 system at the decoupling
point $\alpha = 1$, with $k_BT_f/J_o = 0.2$:   (a)   20 MCS/site and (b)
40 MCS/site.   There is an exact degeneracy of the heavy-light
walls (squares).   Other walls and excitations are also shown.

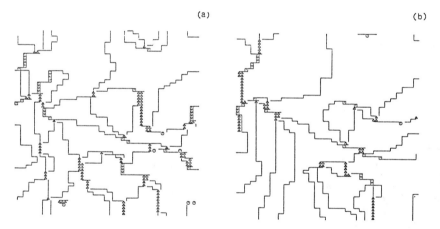

Figure 4   Typical configurations in the wet region of the phase diagram
at $\alpha = 0.5$, $k_BT_f/J_o = 0.2$ for a 64x64 system: (a)   30 MCS/
site and (b) 60 MCS/site.   Some of the dry walls persist in
the vertex junctions but there is a change in the direction of
the anisotropy of the growth.

for the ANNNI model as well (see also Fig. 5b),as was pointed out in an
earlier study [22].   Moreover, the topology of growth in the four-state
clock and ANNNI models is very similar for $\alpha = 1$.   Assume that the dynamics of the ANNNI model can be approximated by the equations derived by
Kawasaki.   These can be written as

$$\partial\phi/\partial t = L(\nabla^2 - \hat{n}\hat{n}: \nabla\nabla)\phi$$

247

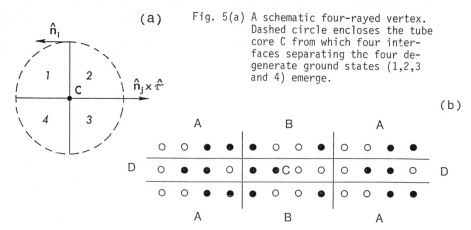

(a)

$\hat{n}_1$

1 | 2
C
4 | 3

$\hat{n}_j \times \hat{\tau}$

Fig. 5(a) A schematic four-rayed vertex.
Dashed circle encloses the tube
core C from which four interfaces separating the four degenerate ground states (1,2,3
and 4) emerge.

(b)

A   B   A

○ ○ ● ● │ ● ○ ○ ● │ ○ ○ ● ●

D   ○ ● ● ○ │ ● ●C ○   ○ │ ○ ● ● ○   D

○ ○ ● ● │ ● ○ ○ ● │ ○ ○ ● ●

A   B   A

Figure 5(b) A dominant vertex-antivertex configuration in the ANNNI
model shown here in a quadrupole configuration. There are
two vertex-antivertex pairs consisting of a heavy (A│B),
light (C│D), soft superheavy (D│C) and soft superlight
(C│D) walls. The energies of the vertex interfaces are
$0.5J_0$ and $0.5\alpha J_0$ in the y and x-directions, respectively.
After wetting these vertices become unfavourable due to
the instability of the soft superheavy-light walls.

$$L^{-1}\ \underline{\varepsilon}(\tau,t)\ \cdot\ \partial\vec{q}(\tau,t)/\partial t\ =\ \sum_{j=1}^{4}\sigma_j\ (\hat{n}_j \hat{x}_\tau),$$

(6.2)

for four-rayed vertices of the type shown in Fig. 5(a).

The first equation is written for a phase variable which is singular at
the interfaces, with $\hat{n}\equiv\nabla\phi/|\nabla\phi|$, where L is a kinetic coefficient. The
second equation describes explicitly the motion of four-rayed vertices,
with q denoting the position of a <u>defect line</u> where the four interfaces
meet, and τ is the unit vector parametrizing the line. Again, L is the
kinetic coefficient, σ_j is the surface tension and $\underline{\varepsilon}$ denotes the mass density tensor associated with the defect line. The right-hand side of (6.2)
can be considered to be the net surface tension force acting on the defect
tube, while the left-hand side is the friction force due to the motion of
the defect tube. Equation (6.2) neglects both thermal noise and the
force arising from the misfit parameter δ in the <u>chiral</u> clock model which
was included in Kawasaki's original derivation. This latter term would
also be present in the ANNNI model for $\alpha < 1$, since $\delta\sim(1-\alpha)$. It is easy
to see intuitively from (6.2) how the anisotropy between the x- and y-
directions arises in the ANNNI model. Namely, even at $\alpha = 1$ where
there is an energetic degeneracy between all the relevant domain walls,
the microscopic structure of the walls is very different in the x- and
y-directions. Thus, the coupled equations of motion are expected to be
different in the x and y-directions. In addition to this, for $\alpha<1$ the
degeneracy is broken and there is a uniaxial chirality which is proportional to $(1-\alpha)$ along the x-axis, which changes the anisotropy.

A more quantitative, albeit heuristic, version of this argument can be made [23], based on (6.2), which yields the $t^{1/2}$ growth law. As well, the theory yields an estimate of the anisotropy which is in reasonable agreement with the Monte Carlo results.

In order to understand the rather sudden change both in the anisotropy and the average growth rate as the wetting transition takes place, let us next examine the vertex mechanism in the ANNNI model in detail. In Fig. 5(b) we show two typical vertex-antivertex combinations of heavy-light and soft superheavy-light walls.

Due to the wetting transition this becomes unfavorable because of the instability of the latter walls. Thus there is a jump in the driving force of the vertex motion. However, some of the unstable walls still do exist far from equilibrium and seem to be concentrated near the vertex junctions. This may be an indication that the system will nevertheless try to use the same vertex mechanism even in the wet region. It may also be possible that a pure curvature effect plays a more important role in the wet region then in the dry region.

Finally, we briefly discuss the soliton theory of the ANNNI model. It can be shown that for the static case the relevant approximate solution which can be obtained for the order parameter $\psi = \psi e^{i\phi}$ is a phase soliton [29] of the form $\phi_{\pm} (x) = - \pi/4 + \arc[\tan(e^{\pm x/\zeta})]$, where ζ is a "correlation length". If this is applied to the dynamical problem, it can be shown that a form of a dissipative Sine-Gordon equation is obtained. The ordinary two-dimensional Sine-Gordon equation is known to have vertex-type solutions [30]. A further study of the existence of similar solutions for the dissipative case would clearly clarify the nature of the dynamical problem for the ANNNI and related models.

We conclude this section by noting that both the model discussed by Sadiq and Binder, as well as the one discussed here, should illustrate some of the quite interesting phenomena that one should observe in domain growth studies in monolayers adsorbed on surfaces. The possibility of studying nonequilibrium phenomena in systems with such a variety of topological defects makes these systems particularly important to study.

7. MOLECULAR BEAM EPITAXIAL GROWTH ON THE Si(100) SURFACE

We now briefly describe a rather different class of problems involving the kinetics of phase transitions. Namely, the effects of surface reconstruction on crystal growth. We limit our discussion to a recent molecular dynamics simulation [31] of molecular beam epitaxial growth of the Si(100) surface, as an example of the type of phenomena we have in mind. References concerning the theoretical and experimental studies of the Si(100) surface are given in [31].

The question which we address here is the role of this surface phase transition (from the bulk structure to the 2x1 reconstructed surface) on crystal growth. This is a particularly interesting theoretical question, given that there is experimental evidence that there is an epitaxial temperature T_e. Namely, for temperatures above T_e Si(100) has been found to grow epitaxially, but for temperatures less than T_e it grows in an amorphous state.

In an attempt to gain some insight about this subject, a molecular dynamics study of the crystal growth under MBE conditions, was carried out for a model of Si(100). The model potential is due to STILLINGER and WEBER [32] and involves both a two-body and three-body term. The three-body term includes a factor which is a function of the angle between two of the three particles subtended by the third particle. The ideal tetrahedral angle appropriate for the silicon crystal structure is favored by this term. The molecular dynamics simulation of Newton's equations of motion for this potential was carried out at two temperatures, $T_L = T_m/8$ and $T_H = 2T_m/3$, where T_m is the bulk melting temperature. The first notable result obtained is that prior to the deposition of new silicon (from the beam) the 2x1 surface reconstruction of the truncated bulk silicon was observed. The other interesting results are that at T_L growth of an amorphous overlayer and the persistence of surface reconstruction was found. At T_H, on the other hand, the growth is characterized by the formation of more ordered epilayers and the disappearance of the 2x1 surface reconstruction. These results are in qualitative agreement with the experimental studies of the MBE growth of the Si(100) surface by GOSSMAN and FELDMAN [33].

Although this molecular dynamics study needs to be extended to include more realistic potentials and more powerful computer facilities than the conventional supercomputer used in this study, it seems clear that in the future such simulations should provide a powerful tool for investigating this important class of surface science problems.

8. CONCLUSIONS

We end by pointing out some of the experimental difficulties involved in studying the kinetics of domain growth. As already pointed out by BINDER [34], there is always a finite upper limit on the domain size, L_{max}, which can be observed in experiments. (The same is true in the computer simulation studies.) Since characterizing the kinetics of domain growth requires $L(t) \ll L_{max}$ in order for (2.3) to be correct, this upper bound on $L(t)$ imposes severe limitations on the time regime in which one can study domain growth. There are a variety of effects which determine L_{max}. For example, point defects such as adsorbed impurity atoms play the role of a random ordering field. This field has been shown to destroy true long-range order in two dimensions (see [34] for some relevant references). As a consequence, even if one has a surface which is free of steps, there is a maximum size, L_{max}, which is determined by the ordering energy and the impurity concentration [34]. As well, interfaces between different ordered regions are pinned by the impurities. As a consequence the domain growth is predicted to satisfy a logarithmic growth law [35,36]. In addition to these effects, one always has to worry about substrate inhomogeneities, such as steps, which provide an upper limit on the maximum domain size. In this case physisorbed systems on exfoliated graphite provide a better testing ground for experiments, since L_{max} can be quite large in such cases. (However, this is balanced by the fact that in physisorbed systems, the typical time scale for equilibration is rather short.)

In conclusion we note that computer simulation and theoretical studies of rather simplified models of physisorbed and chemisorbed systems have led to some recent progress in understanding the kinetics of interface motion, etc., in such systems. Much remains to be done, however, since the field is really in its infancy. More realistic models must be developed. More accurate computer simulation studies over longer time intervals than currently available have to be carried out. More progress has to be made in our theoretical understanding of these far-from-equilibrium phenomena. Most important, however, at this point in time, is the need for experimental studies in this most challenging field of research.

Acknowledgements

The author wishes to acknowledge the support of ONR Grant #N00014-83-K-0382. As well, he wishes to thank Dr. T. Ala-Nissila, Dr. D. Chowdhury, Professor E. T. Gawlinski and Professor M. Grant for many stimulating discussions on various aspects of this work.

REFERENCES

1. P. S. Sahni and J. D. Gunton, Phys. Rev. Lett. 47, 1754 (1981)
2. A. Milchev, K. Binder and D. W. Heermann, to be published in Zeit. Phys. B (1986)
3. G. F. Mazenko and O. T. Valls, Phys. Rev. 33, 1823 (1986)
4. J. D. Gunton, M. San Miguel and P. S. Sahni, Vol. 8, p. 267 (1983), edited by C. Domb and J. L. Lebowitz (New York, Academic Press)
5. K. Binder, Condensed Matter Research Using Neutrons, p. 1 (1985) edited by S. W. Lovesey and R. Scherm (New York, Plenum)
6. K. Kawasaki
7. P. W. Voorhees, J. Stat. Phys. 38, 231 (1985)
8. G. F. Mazenko, O. T. Valls and F. C. Zhang, Phys. Rev. B32, 5807 (1985)
9. S. Kumar, J. D. Gunton and K. Kaski, Temple University preprint (1986)
10. M. Schick, in Progress in Surface Science, Vol. 11, 245 (1981)
11. S. M. Allen and J. W. Cahn, Acta Metall. 27, 1085 (1979)
12. K. Kawasaki, M. C. Yalabik and J. D. Gunton, Phys. Rev. A17, 455 (1978)
13. T. Ohta, D. Jasnow and K. Kawasaki, Phys. Rev. Lett. 49, 1223 (1982)
14. G. F. Mazenko and O. T. Valls, Phys. Rev. B27, 6811 (1983); Phys. Rev. B30, 6732 (1984)
15. J. Vinals, M. Grant, M. San Miguel, J. D. Gunton and E. T. Gawlinski, Phys. Rev. Lett. 54, 1264 (1985)
16. K. Kaski, M. C. Yalabik, J. D. Gunton and P. Sahni, Phys. Rev. B28, 5263 (1983)
17. G. C. Wang and T. M. Lu, Phys. Rev. Lett. 50, 2014 (1983)
18. K. Diff, T. Ala-Nissila and J. D. Gunton, unpublished
19. A. Sadiq and K. Binder, J. Stat. Phys. 35, 617 (1984); Phys. Rev. Lett. 51, 674 (1983)
20. I. M. Lifshitz and V. V. Slyozov, J. Phys. Chem. Solids 19, 35 (1961)
21. J. Vinals and J. D. Gunton, Surface Science 157, 473 (1985)
22. T. Ala-Nissila and J. D. Gunton, Temple University preprint (1986)
23. K. Kaski, T. Ala-Nissila and J. D. Gunton, Phys. Rev. B31, 310 (1985); T. Ala-Nissila, J. D. Gunton and K. Kaski, Phys. Rev. B33, 7583 (1986); T. Ala-Nissila, J. D. Gunton and K. Kaski, Temple University preprint (1986)
24. T. Ala-Nissila, J. Amar and J. D. Gunton, J. Phys. A19, L41 (1986)
25. P. Rujan, G. Uimin and W. Selke, Z. Phys. B Reprint (1986)
26. P. Rujan, G. V. Uimin and W. Selke, Phys. Rev. B32, 7453 (1985)
27. G. Ertl and J. Kuppers, Surface Sci. 21, 61 (1970); P. Rujan, W. Selke and G. Uimin, Z. Phys. B53, 221 (1983)
28. T. Ala-Nissila and J. D. Gunton, this volume
29. P. Bak and J. von Boehm, Phys. Rev. B21, 5297 (1980)
30. O Hudak, Phys. Lett. 89A, 245 (1982); S. Takeno, Prog. Theor. Phys. 68, 992 (1982); A. B. Borisov, A. P. Tankeyev, A. G. Shagalov and G. V. Bezmaternih, Phys. Lett. 111A, 15 (1985)
31. E. T. Gawlinski and J. D. Gunton, Temple University preprint (1986)
32. F. Stillinger and T. Weber, Phys. Rev. B31, 5262 (1985)

33. H. J. Gossmann and L. C. Feldman, Phys. Rev. B32, 6 (1985)
34. K. Binder, in Berichte der Bunsen - Gesellschaft für Physikalische Chemie 90, 257 (1986)
35. J. Villain, Phys. Rev. Lett. 52, 1543 (1984)
36. G. Grinstein and J. F. Fernandez, Phys. Rev. B29, 389 (1984)

Kinetics of Domain Growth in a Model of O/Pd(110)

T. Ala-Nissila and J.D. Gunton

Physics Department and Center for Advanced Computational Science,
Temple University, Philadelphia, PA 19122, USA

1. INTRODUCTION

When a two-dimensional surface system is quenched from a high-temperature disordered phase to a temperature below its phase transition point a dynamical process of ordering takes place. This problem of domain growth is of fundamental importance in nonequilibrium statistical mechanics and has practical applications in many fields, including surface science [1]. In this paper we present preliminary results of a study of domain growth for a model of O/Pd(110), which exhibits a uniaxial (3x1) phase [2]. This system is modeled by the anisotropic or axial next nearest neighbor Ising (ANNNI) model in a field [3-6]. The Hamiltonian of this model in a lattice-gas representation is

$$H=-\sum_{<ij>} [-\phi_1 n_{ij} n_{i+1j} -\phi_2 n_{ij} n_{i+2j} +\phi_0 n_{ij} n_{ij+1} +(\mu+\varepsilon) n_{ij}] , \qquad (1)$$

where $\phi_i > 0$, $i=0,1,2$, μ is the chemical potential, ε is the binding energy of oxygen on Pd(110) and the summation goes over all sites of a square lattice. The indices i and j correspond to the x and y directions, respectively. The variables n_{ij} are zero or one corresponding to empty or occupied lattice sites. In Fig. 1(a) we display the phase diagram of this model with coverage θ vs. reduced temperature $k_B T/J_1$, where $J_1=\phi_1/4$. This phase diagram is based on a parametrization of the ANNNI Hamiltonian with $\phi_0=\phi_1$ and $\kappa=\phi_2/\phi_1=0.3$, and reproduces fairly well the observed (3x1) and (2x1) phases around the coverage values of $\theta=1/3$ and $\theta=1/2$, respectively. We note, however, that recent more accurate measurements suggest that this phase diagram may be more complicated, in particular near $\theta=1/2$ and for higher coverages [7,8].

The (3x1) phase near a coverage $\theta=1/3$ consists of an alternating sequence of rows of two empty lattice sites followed by a single row of occupied sites in the x-direction (see Fig. 1(b)). The degeneracy of this phase is p=3, and the relevant two-component order parameter can be written as

$$\psi_\alpha=(1/NM)^{1/2} \sum_{n,m} s_{nm} \exp(i(\vec{Q}_\alpha \cdot \vec{r}_{nm})), \quad \alpha=1,2, \qquad (2)$$

253

where we have used the Ising spin representation, with $s_{ij}=2n_{ij}-1$. Here $\vec{Q}_1=2\pi/a(1/3,0)$ and $\vec{Q}_2=2\pi/a(-1/3,0)$ are the two positions of the Bragg peaks and N and M are the number of sites in the x and y directions, respectively. The anisotropic (unnormalized) structure factor $S(\vec{k},t)$ is defined accordingly as

$$S_\alpha(\vec{k},t)= <|\sum_{n,m} s_{nm}\exp(i((\vec{Q}_\alpha+\vec{k})\cdot\vec{r}_{nm}))|^2>, \quad \alpha=1,2, \tag{3}$$

where $\vec{k}$ is the deviation from the Bragg positions. $S_1(\vec{k},t)$ and $S_2(\vec{k},t)$ are equivalent and here we compute S_1 and limit ourselves to the case $k_x,k_y>0$.

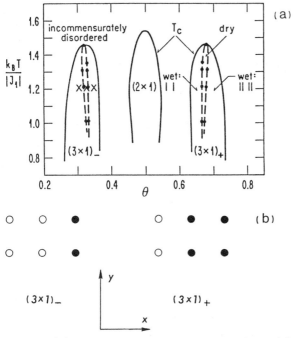

Figure 1(a) Schematic phase diagram of the model of O/Pd(110) with the parametrisation discussed in the text [5]. The quench points with coverages θ=0.30, 0.325 and 0.35 are indicated with crosses within the (3x1)_ phase. Finer details of the phase diagram, including the floating incommensurate phases present in the model, are not shown here.

Figure 1(b) Schematic picture of the (3x1)_ and (3x1)+ phases. Open and filled circles are empty and occupied lattice sites, respectively. For simplicity it is assumed in the text that the underlying lattice is a square lattice.

2. RESULTS

We have studied the kinetics of the domain growth of the model of O/Pd(110) by quenching from a high temperature $(k_B T/J_1 = \infty)$ disordered phase to a temperature $k_B T_f/J_1 = 1.2$ below the order-disorder transition line of the (3x1)_ phase. We have used standard Kawasaki (nearest neighbor) particle exchange dynamics which corresponds to random diffusion events and the conservation of the initial coverage in the lattice gas version of the model. The order parameter is not conserved in our model, however. Three different coverages were used to study the effect of the wetting transitions on the domain growth, namely $\theta = 0.30$, $\theta = 0.325$ and $\theta = 0.35$. The first and last values of θ correspond to the two different wet regions in the phase diagram while $\theta = 0.325$ is located within the narrow dry region (see Fig. 1(a)). The preliminary results we present here are based on an analysis of 289 quenches on 120x120 square lattices.

2.1 Dynamical Behaviour of the Domain Walls and Excitations

In Fig. 2 we present all the domain walls and excitations which we have followed in detail during the domain growth. Since the system is far from equilibrium most of these rather low energy excitations are expected to be present and have nontrivial effects on the growth. The two wetting transitions in the system occur due to the competition between the (3x1) phase and

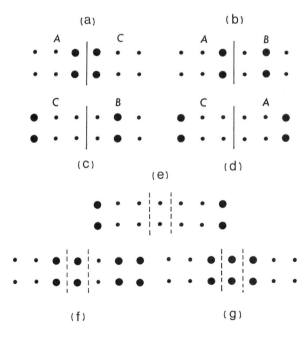

Figure 2 All the different domain walls and excitations which have been followed during the growth of the (3x1) phase: (a) a superheavy wall; (b) a heavy wall; (c) a light wall; (d) a superlight wall; (e) a vacancy defect; (f) an adatom defect and (g) a triple adatom excitation

the simple antiphase (+-) and (2,2) antiphase (++--) configurations [5,6].
In the wet "| |" region of the phase diagram a light or C|B wall becomes un-
stable and is wet by the B phase. Thus two heavy A|B walls appear (A|B and
B|C walls are equivalent). In the other wet "|| ||" region for smaller
coverages heavy A|B walls are unstable and two light walls arise.

In Figures 3(a)-(d) we display the absolute and relative amounts of domain walls and excitations shown in Fig. 2 during growth for θ=0.30 and

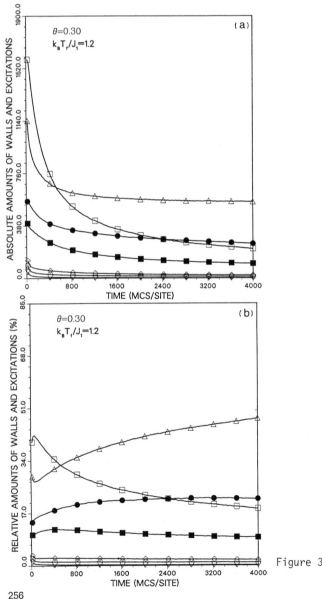

Figure 3. Caption see
opposite page

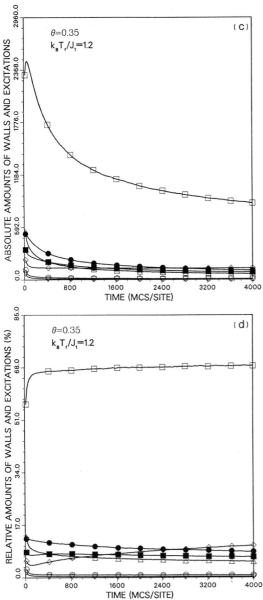

Figure 3   The statistics of domain walls and excitations for 120x120 systems at $k_BT_f/J_1=1.2$ with two different coverages: (a) absolute
and (b) relative statistics for the coverage $\theta=0.30$; (c) absolute
and (d) relative amounts for $\theta=0.35$.   The symbols corresponding
to distinct walls and excitations are as follows: superheavy
walls (open circles), heavy walls (open squares), light walls
(filled circles), superlight walls (filled squares), vacancy defects (triangles), adatom defects (diamonds) and triple adatom
excitations (arrowheads)

257

θ=0.35. As expected, most of the domain walls and excitations persist at all times. In the case of the highest coverage θ=0.35, the heavy walls are by far the most dominant. This is consistent with the wet nature of this portion of the diagram, in which the heavy walls are favoured. In the narrow dry region (not shown here) the results are very similar except that the number of heavy walls is reduced and other walls and excitations appear in larger amounts. Finally, in the other wet region with θ=0.30 the number of heavy walls is strongly reduced. As expected, the light walls are now abundant at late times. However, in addition to this superlight walls also appear together with vacancy defects, which actually dominate at later times. This latter mechanism seems to be dominant in order that a low coverage be stabilized.

2.2 Growth Law

In order to analyze the dynamical growth law, we have studied the behaviour of two independent definitions for the length scale $\bar{R}(t)$. The peak of the anisotropic structure factor has the dimensions of the square of a length, so one possible measure of a length is [9]

$$\bar{R}^2(t)=S(0,t)/\psi_T^2 , \tag{4}$$

where ψ_T denotes the equilibrium value of the order parameter. This definition measures the long wavelength correlations of the system, since $S(0,t)=NM<|\psi|^2>$. In addition to this we have computed the squares of the normalized inverse perimeter densities $A_i(t)$ (i=x,y), which define the corresponding effective domain areas in the x and y directions. These quantities are sensitive to short-range correlations. To test whether a simple power law of the form $\bar{R}(t)\sim t^n$ is valid, we have fitted our data to the expression

$$y(t)-y(t_0)=D_i(t-t_0)^{2n_i} , \quad i=x,y, \tag{5}$$

where $y(t)\equiv\bar{R}^2(t)$. Note that a "log-log" fitting procedure may be misleading here since it tends to overweight the initial part of the data. In Table 1 we show the results for $S(0,t)$ and $A_i(t)$ for θ=0.30 and θ=0.325. The results are most consistent with a dynamical exponent n≈0.5 throughout the phase diagram. Also, there is not much difference in the average growth

Table 1. Results of numerical least-squares fits of the Monte Carlo data
to (5) for coverages θ=0.30 and 0.325. The first two rows are
fits to S(0,t) while the other results are fits to the effective
domain areas $A_x(t)$ and $A_y(t)$. The anisotropy factor D_y/D_x is
also estimated.

Coverage Θ	Δt/MCS	D_x D_y	$2n_x$ $2n_y$	D_y/D_x	±Δ
0.30	20-600	2.63	0.99	-	0.06
0.325	20-600	2.63	1.01	-	0.06
0.30	20-1000	0.15 0.62	1.06 0.96	4.2	0.06
0.325	20-1000	0.06 0.57	1.10 0.86	9.3	0.06

rates in different parts of the phase diagram. The growth is also strongly
anisotropic, as indicated by the ratio of the growth amplitudes D_y/D_x from (5).
This ratio is larger than one which entails a faster growth in the y-direction. This situation is very similar to the results obtained for a (4x1)
phase in an earlier study where the effect of a wetting transition was studied [10,11]. A faster growth in the y-direction is also expected to reflect the existence of "slablike" domain configurations and a strong anisotropy in the structure function.

2.3 Generalised Anisotropic Scaling

The anisotropic domain growth manifests itself not only through the prefactors D_x and D_y but also in the shape of the structure function $S(\vec{k},t)$. In
Figs. 4(a)-(d) we show the time development of the structure functions a-
long the k_x and k_y axes in the different regions of the phase diagram for
Θ=0.30 and Θ=0.35. As the coverage increases the structure function becomes broader in the x-direction, and actually shifts its maximum to an incommensurate value in the finite systems studied. This is an indication of
strongly slablike configurations where the domain walls in the x-direction
exist in large amounts. This anisotropy can be made quantitative by de-
fining a generalized second moment in the (k_x,k_y) plane [12].

$$k_2(\theta,t) = \sum_{k=0}^{k_c} k^2(\theta)S(k(\theta),t)/ \sum_{k=0}^{k_c} S(k(\theta),t),$$ (6)

259

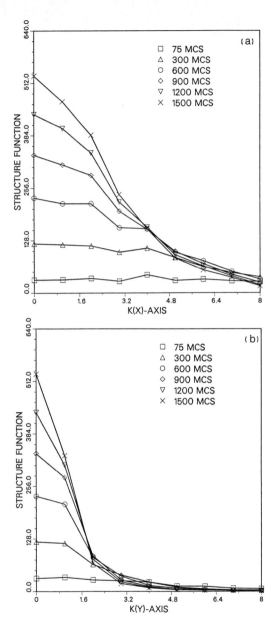

Figure 4. Caption see opposite page

where Θ is the angle between $k(\Theta)$ (a line in the k-space) and the k_x direction and k_c is a cutoff parameter. Thus $\Theta = 0$ and $\Theta = \pi/2$ correspond to second moments in the x $[k_2^{(x)} \equiv k_2(0)]$ and y $[k_2^{(y)} \equiv k_2(\pi/2)]$ directions, respectively. We have tested the dynamical scaling form with several dif-

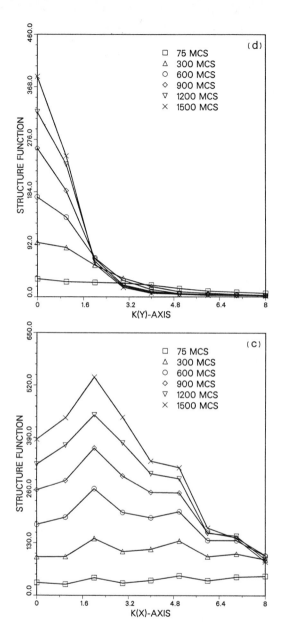

Figure 4 Time development of the anisotropic structure function $S(\vec{k},t)$ for two different coverages along k_x and k_y axes: (a)-(b) are for $\theta=0.30$, and (c)-(d) for $\theta=0.35$

ferent definitions of the length scale $\tilde{R}(t)$. Using the second moments we have calculated the scaling functions

$$F(x) = k_2(\theta,t)S(k(\theta),t), \quad x=k(\theta)/\sqrt{k_2(\theta,t)} \qquad (7)$$

for the x and y directions, and the scaling function

$$\tilde{F}(\tilde{x}) = S(k(\theta),t)/S(0,t) \; , \; \tilde{x} = k(\theta)\sqrt{S(0,t)} \qquad (8)$$

along the k_x and k_y axes. In Figs. 5(a) and (b) we present two typical scaling functions calculated for the coverage $\theta=0.30$. The generalised scaling holds to a good degree of accuracy. However, it is obvious that with increasing coverage it becomes increasingly difficult to reach the asymptotic scaling regime for finite systems. We note that this may also have relevance to some experimental studies. Namely, due to the strong anisotropy, the peak of the Bragg reflection may appear at an incommensurate value even if the phase is commensurate, if complete equilibrium is not reached. Thus it may not be easy to detect the true incommensurate phase with simple diffraction measurements in a system with uniaxial phases.

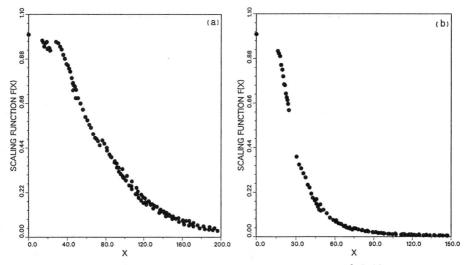

Figure 5   Some typical scaling functions for the coverage $\theta=0.30$:
(a) scaling along k_x axis with the generalised second moment $k_2^{(x)}(t)$ and (b) scaling along k_y axis with the Bragg peak $S(0,t)$.

3. SUMMARY

Our results show how a relatively simple model of a chemisorbed system, namely a model of the (3x1) phase of O/Pd(110), displays a richness of topological features which affect the domain growth in a nontrivial way. Despite these complications we find a dynamical exponent which seems to hold everywhere in the phase diagram, namely $n \approx 0.5$. This result is very similar to a recent study of a (4x1) (or (2,2)) uniaxial phase [10,11]. However, in this case the exponent 0.5 was only reached in the limit of very long times due to the roughening effect of a large number of lattice defects present in the system. Additionally, the growth in the wet region was found to slow down so much compared to the dry region that a quantitative estimation of the dynamical exponent was unsuccessful. In our case the wetting transitions do not cause such a dramatic change in the growth rates, presumably due to the relatively small excess density present in the relevant walls. Thus it seems that some universal growth mechanism is present which gives $n \approx 0.5$ for all coverages within the (3x1) phase. Recently, it has been suggested [11] that a generalised form of a theory originally developed by Kawasaki [13,14] applies to the results obtained in the (2,2) antiphase. In this theory a growth exponent $n = 1/2$ is obtained due to the coupled motion of interfaces and vertices present in the system. A more detailed study of the dynamics of the (3x1) phase with both diffusion and adsorption-desorption dynamics will be published elsewhere, together with some theoretical considerations.

We want to thank Professor M. Grunze for useful discussions. This work was supported by ONR Grant #N00014-83-K-0382.

1. J. D. Gunton, M. San Miguel, and P. S. Sahni: In Phase Transitions and Critical Phenomena, edited by C. Domb and J. L. Lebowitz, Vol. 8 (Academic, London, 1983)
2. G. Ertl and J. Kuppers: Surface Sci. 21, 61 (1970)
3. P. Rujan, W. Selke and G. Uimin: Z. Phys. B53, 221 (1983)
4. P. Rujan and G. V. Uimin: J. Phys. A17, L61 (1984)
5. P. Rujan, G. V. Uimin and W. Selke: Phys. Rev. B32, 7453 (1985)
6. P. Rujan, G. V. Uimin and W. Selke: Z. Phys. B Preprint (1986)
7. M. Wolf, A. Goschnick, J. Loboda-Cackovic, M. Grunze, W. N. Unertl and J. H. Block: Preprint (1986)
8. J. Goschnick, M. Wolf, M. Grunze, W. N. Unertl, J. H. Block and J. Loboda-Cackovic: Preprint (1986)
9. A. Sadiq and K. Binder: J. Stat. Phys. 35, 517 (1984)
10. T. Ala-Nissila, J. D. Gunton and K. Kaski: Phys. Rev. B33, 11 (1986)
11. T. Ala-Nissila: Ph.D. Thesis, Temple University (1986) (Unpublished)
12. K. Kaski, T. Ala-Nissila and J. D. Gunton: Phys. Rev. B31, 310 (1985)
13. K. Kawasaki: Ann. Phys. (N. Y.) 154, 319 (1984)
14. K. Kawasaki: Phys. Rev. A31, 3880 (1985)

The Influence of Hydrogen on the Infrared Spectrum of N_2 Chemisorbed on the Ni(110) Surface

M.E. Brubaker, I.J. Malik, and M. Trenary

Department of Chemistry, University of Illinois at Chicago,
Chicago, IL 60680, USA

I. Introduction

The N_2/Ni(110) system has been extensively studied with a wide variety of experimental techniques. Grunze, et al., [1] have written a comprehensive review of studies of N_2 on Ni(110) through 1985. We have recently studied the N-N stretching vibration of N_2 on the clean Ni(110) surface with the technique of Fourier transform infrared reflection absorption spectroscopy (FT-IRAS) [2]. At a saturation coverage of 0.72 monolayer (ML) the IR spectrum consists of an intense band at 2194 cm^{-1}, a weak shoulder at 2204 cm^{-1} and a still weaker peak at 2220 cm^{-1}. Although the high coverage N_2 spectrum was very reproducible, the low intensity of the additional high-frequency bands suggests that they may not be characteristic of the ideal high coverage N_2 overlayer but are due rather to some imperfection. A major concern in any surface study is the role of background contamination from the residual gas in the vacuum system. As hydrogen is the major constituent of the residual gas in most stainless steel ultrahigh vacuum systems, including ours, we conducted the experiments reported here to specifically determine the influence of hydrogen on our N_2/Ni(110) IR spectra.

Numerous experimental studies have revealed the complex nature of hydrogen adsorption on the Ni(110) surface [3-5]. Molecular beam scattering studies show that H_2 has a dissociative sticking probability of nearly unity on Ni(110) [3]. LEED and He scattering experiments reveal that for H coverages between 1.0 ML and 1.5 ML the Ni(110) surface undergoes a reconstruction in which the rows of Ni atoms along the [1Ī0] direction are paired in the perpendicular [001] direction [4,5]. This reconstruction yields a very sharp (1x2) LEED pattern. At submonolayer coverages and temperatures below 130 K adsorbed H forms lattice gas phases in which no reconstruction of the substrate takes place.

II. Experimental

We have given a complete and detailed description of our apparatus and of our sample preparation and cleaning procedures elsewhere [2]. We therefore give only a brief description here. The apparatus consists of a stainless steel ultrahigh vacuum chamber combined with a commercial FTIR. The chamber is equipped with an Auger electron spectrometer, a quadrupole mass spectrometer and low-energy electron diffraction (LEED) optics. The base pressure of the chamber is typically $4x10^{-11}$ Torr as measured by a Bayard-Alpert ion gauge calibrated for N_2. The accuracy of the gauge for N_2 was verified through the dependence of coverage on exposure for the Ni(110) surface using published values for the sticking coefficient [1]. The mass spectrum of the chamber residual gas showed the m/e = 2 peak to be a factor of five greater than the next highest peak, the m/e = 28 peak of CO. The fact that H_2 was the dominant component of the residual gas was verified by noting how long it took to achieve a given hydrogen coverage from the residual gas background. The hydrogen coverage was measured from H_2 thermal desorption peak areas calibrated to a saturation coverage of 1.5 ML. All IR spectra reported here were obtained at 2 cm^{-1} resolution using triangle apodization. A photovoltaic indium antimonide detector was used. The crystal was exposed to both N_2 and H_2 by back-filling the chamber.

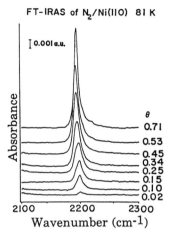

FT-IRAS of N_2/Ni(110) 81 K

Fig. 1. FT-IRAS spectra of N_2/Ni(110) at 81 K as a function of coverage. Each spectrum is a ratio of 1024 (8.5 min) sample scans to 1024 background scans.

Fig.2. The N-N stretching band of 0.1 ML N_2 before and after exposing the sample to H_2 at 115 K.

III. Results

Figure 1 shows the IR band of the N-N stretch as a function of coverage at 81 K on the clean Ni(110) surface. We have discussed this result in detail elsewhere [2]. At the highest coverage the spectrum shows an intense peak at 2194 cm^{-1}. Our results are in basic agreement with previous vibrational studies of N_2/Ni(110) with respect to the frequency of the main peak [6]. In addition to the main band at 2194 cm^{-1}, the high resolution and sensitivity available with FT-IRAS reveals a weak shoulder at 2204 cm^{-1} and a small peak at 2220 cm^{-1}. No extra features appear on the low frequency side of the main band. If we assume that the nitrogen molecules responsible for the 2220 cm^{-1} band have the same absorption coefficient as the N_2 in the 0.02 ML spectrum of fig. 1, then the coverage of N_2 molecules contributing to the 2220 cm^{-1} IR peak would be only a few hundredths of a monolayer. It is not unreasonable to suspect contamination or defects to be present at such low levels. Furthermore, at the high coverage where the 2220 cm^{-1} peak is observed there should be appreciable coupling among the N-N vibrations. Coupling should lead to an enhancement of the intensity of the high-frequency bands. Thus the concentration of N_2 molecules contributing to the higher frequency peaks may be much smaller than would be implied by assuming a concentration simply proportional to

IR peak areas. Dipole-dipole coupling has been shown to enhance the intensity of the high-frequency peaks of adsorbed CO relative to the lower frequency bands [7].

To determine the influence of hydrogen on the saturation coverage N_2 overlayer at 81 K, we recorded N_2 spectra before and after exposing the sample to several Langmuirs of H_2. We observed no change in the infrared spectrum. Leaving the N_2 saturated surface at 81 K in the residual gas ambient for a period as long as 7 hours caused the IR peak area to decrease by 30% and the peak center to shift from 2194 cm^{-1} to 2191 cm^{-1}. The relative heights of the 2204 cm^{-1} and 2220 cm^{-1} peaks, however, were not affected. These observations indicate that hydrogen does not adsorb on the N_2 saturated surface at 81 K. Similarly, if the surface is saturated with hydrogen at 81 K, in which case we observed the reconstructed (1x2) LEED pattern, a high exposure to N_2 yields a N_2 coverage of no more than .01 ML. The small amount of N_2 which does adsorb has a broad IR band at 2230 cm^{-1}.

At low N_2 coverages, hydrogen will adsorb on the surface. The spectra in fig. 2 are of 0.1 ML N_2 at 115 K before and after exposing the sample to 1 L H_2. After the H_2 exposure the H induced reconstructed (1x2) LEED pattern was observed. From the integrated IR peak area, we estimate that the N_2 coverage for the spectra in fig. 2 decreased by a factor of two upon H_2 exposure. The effect of the hydrogen adsorption is to shift the N_2 IR band from 2194 cm^{-1} to 2207 cm^{-1} and to decrease the FWHM from 14 cm^{-1} to 9 cm^{-1}. In light of the fact that the surface has undergone a major reconstruction, the change in the N_2 spectrum is rather modest.

The spectra in fig. 3 show the behavior of the N_2 band over a period of time long enough to ensure substantial contamination from the background gas. The surface was initially exposed to 0.1 L N_2 which yields a N_2 coverage of about 0.1 ML [2]. After 2.5 hours the band has reached its final frequency of 2205 cm^{-1} and FWHM of 7.0 cm^{-1}. Although LEED observations were not made for this particular experiment, we have observed the development of the H reconstructed (1x2) pattern from background contamination after comparable periods of time. The N_2 spectrum of fig. 4 shows the result of exposing the sample at 81 K to a gas phase mixture of N_2 and H_2.

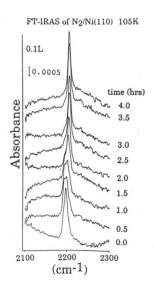

FT-IRAS of N_2/Ni(110) 105K

0.1L

[0.0005

time (hrs)
4.0
3.5
3.0
2.5
2.0
1.5
1.0
0.5
0.0

2100   2200   2300
(cm^{-1})

Fig. 3. The dependence of the N_2 IR band on time showing the influence of hydrogen contamination from the residual gas background. Each spectrum was obtained in 31 seconds (64 scans).

266

This yields a split peak with components at 2193 cm^{-1} and 2201 cm^{-1}. Unfortunately we do not know the coverage of hydrogen or nitrogen for this spectrum nor do we know the ratio of N_2 to H_2 partial pressures of the gas phase mixture. The integrated area of the band is 90% of the integrated area of the 0.72 ML band of fig. 1. The results of figs. 1-3 suggest that the 2193 cm^{-1} component corresponds to N_2 on clean areas of the surface with the 2201 cm^{-1} component associated with adsorbed hydrogen.

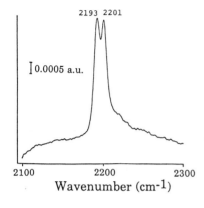

Fig. 4. N_2 spectrum obtained from gas phase mixture of N_2 and H_2.

IV. Discussion and Conclusion

Our main objective in these experiments was to determine if the small 2204 cm^{-1} and 2220 cm^{-1} components on the high-frequency side of the 2194 cm^{-1} band for the saturated N_2 overlayer on a nominally clean Ni(110) surface are due to hydrogen contamination from the residual gas background. These results indicate that the 2204 cm^{-1} component, but not the 2220 cm^{-1}, band, could be due to small amounts of coadsorbed hydrogen. Other observations however indicate that it is not correct to simply associate the 2204 cm^{-1} peak with hydrogen contamination on the surface. Recording a hydrogen thermal desorption spectrum after a N_2 IR spectrum provides a quantitative measure of the relative amount of hydrogen on the surface. By varying the amount of time after the sample had been cleaned to the time it was dosed with N_2 we could vary the small amount of H_2 contamination. Two saturation coverage N_2 IR spectra showed essentially the same intensity for the 2204 cm^{-1} shoulder even though the H coverage, as determined by H_2 thermal desorption, differed by a factor of two. The 2220 cm^{-1} peak was also insensitive to the H_2 coverage.

At this time, the most likely explanation for the 2204 cm^{-1} peak is that it corresponds to N_2 adsorption at a site present either at a defect on the clean surface or made available by H induced reconstruction of the surface. It is known that low hydrogen coverages on Ni(110) can cause localized reconstruction at temperatures above 130 K [5]. Since our experiments were all conducted below 130 K, we would have to assume that the presence of adsorbed N_2 lowers the temperature for the local reconstruction. More definitive conclusions require additional experimental work.

The role of contamination and of defects is central to our discussion of the high-frequency components of our N_2 IR band. We should emphasize, however, that this concern stems from the fact that the high sensitivity and resolution of FT-IRAS makes the technique sensitive to very subtle effects. Certainly, the widely used technique of electron energy loss spectroscopy (EELS) would be completely insensitive to the effects reported here. The EELS spectra of $N_2/Ni(110)$ of Horn, et al., [6] do reveal a peak at 580 cm^{-1} which is almost certainly due to an Ni-H stretch from background hydrogen contamination, although the authors identified the band as due to dissociated N_2. Our base pressure of 4×10^{-11} Torr and fast data acquisition times make us less susceptible to contamination than usual in a UHV surface experiment. Our use of Auger spectroscopy to determine surface cleanliness is standard practice, as was the method of preparing our crystal. We routinely observed LEED patterns of the clean well-annealed surface in which the spots were very sharp and the background was quite low. It is only the IR results which suggest non-ideal surface conditions. It may well be that as the technique of FT-IRAS becomes more widely used, surface scientists will find it necessary to develop better methods of obtaining clean well-ordered single crystal surfaces.

Acknowledgements

Acknowledgement is made to the donors of the Petroleum Research Fund, administered by the American Chemical Society, for partial support of this research. This work was also partially supported by grants from Research Corporation and the National Science Foundation (CHE-8603891).

References

1. M. Grunze, W. N. Unertl and M. Golze, Progr. Surf. Sci., In Press.
2. M. E. Brubaker and M. Trenary, J. Chem. Phys., 85, 6100 (1986).
3. H. J. Robota, W. Vielhaber, M. C. Lin, J. Segner and G. Ertl, Surf. Sci. 155, 101 (1985).
4. N. J. DiNardo and E. W. Plummer, Surf. Sci. 150, 89 (1985), T. Engel and K. H. Reider, Surf. Sci. 109, 140 (1981), V. Penka, K. Christmann and G. Ertl, Surf. Sci. 136, 307 (1984).
5. K. Christmann, F. Chelab, V. Penka and G. Ertl, Surf. Sci. 152, 356 (1985).
6. M. Grunze, R. K. Driscoll, G. N. Burland, J. C. L. Cornish and J. Pritchard, Surf. Sci. 89, 381 (1979), B. J. Bandy, N. D. S. Canning, P. Hollins and J. Pritchard, J. Chem. Soc., Chem. Commun. 58 (1982), K. Horn, J. DiNardo, W. Eberhart, H.-J. Freund, E. W. Plummer, Surf. Sci. 118. 465 (1982).
7. P. Hollins and J. Pritchard, Surf. Sci. 89, 486 (1979), R. Ryberg, Surf. Sci. 114, 627 (1982).

The Reaction of Carbon Monoxide with Chemisorbed Oxygen on Pd(110): An Example of a Structure-Sensitive Reaction

J. Goschnick[1], *J. Loboda-Cackovic*[1], *J.H. Block*[1], *and M. Grunze*[2]

[1]Fritz-Haber Institut der Max-Planck Gesellschaft,
 Faradayweg 4-6, D-1000 Berlin 33, Germany
[2]Department of Physics and Laboratory for Surface Science and Technology,
 University of Maine, Orono, ME 04469, USA

1. Introduction

The oxidation of carbon monoxide on palladium surfaces comprises one of the most extensively studied model systems in heterogeneous catalysis. The reaction has been investigated on the low index crystal surfaces by a variety of experimental techniques, and it was found that the temperature dependence of the steady state reaction on Pd {111}, {110}, {210}, {100} and polycrystalline palladium shows almost no difference [1]. Because of this apparent insensitivity with respect to surface structure, it was concluded that the reaction is "structure-insensitive" and it is therefore widely regarded as the prime example of a heterogeneous catalytic reaction not dependent on surface topography.

The results by ENGEL and ERTL on Pd {111} [2] show that the reaction proceeds along the domain boundaries of chemisorbed oxygen and carbonmonoxide. In this communication we will show that the reaction mechanism between chemisorbed CO and O on Pd {110} is unique, since the domain boundary reaction proceeds in a unidirectional fashion along the [110] direction of the crystal substrate at low substrate temperatures (T < 420 K), and only at higher temperatures (T > 420 K) a two-dimensional domain boundary reaction is observed. Further, the activation energy is different on Pd {110} than on Pd {111} and only due to a compensative effect between the activation energy and the pre-exponential in the reaction rate expression a similar overall reaction rate results.

In a previous report [3], we described our LEED data for the clean Pd {110} surface which shows, at T ~ 250 K, a phase transition into a disordered overlayer, as also reported by FRANCIS and RICHARDSON using He-atom scattering experiments [4]. Oxygen adsorption onto the Pd {110} surface leads, via a complex "1x3" structure, to a commensurate c(2x4) overlayer which saturates at $\Theta_0 = 0.5 \pm 0.03$ [5]. The experiments discussed in this communication were conducted on a c(2x4) oxygen overlayer which was reacted off at different temperatures. Formation of a sharp "1x3" structure is an activated process and in the time intervals of the measurements and the temperature range discussed in this report, only weak diffraction features for a "1x3" phase were observed with decreasing coverage during the reaction. We believe that the intrinsic disorder of the clean Pd {110} surface is not relevant to our present discussion.

2. Experimental Procedure

The experiments were conducted in a stainless steel UHV system with a base pressure of $p < 8 \times 10^{-11}$ mbar as described previously [3-5]. Briefly, it contained a four grid LEED optics and two quadrupole mass spectrometers, one of which was separated from the chamber by a differentially pumped aperture system (Fig. 1A). The LEED data were recorded with a video LEED system which allowed us to monitor the diffraction pattern during the reaction. The crystal was cleaned by Ar-ion sputtering and reactive oxygen cleaning cycles as described in reference [5].

For the experiments described in this paper the clean Pd {110} surface was exposed to 5 L O_2 at 6 x 10^{-8} mbar and T = 482 K until a saturated c(2x4) oxygen overlayer structure ($\Theta_0 = 0.5 \pm 0.03$) was established. Then a CO-pressure of $p_{CO} = 5 \times 10^{-8}$ mbar was introduced by a pressure jump into the system and the LEED pattern together with the CO_2 signal measured with mass spectrometer 2 (see Fig. 1) was followed during the reaction, or the crystal was rotated within 1 mm in front of the

269

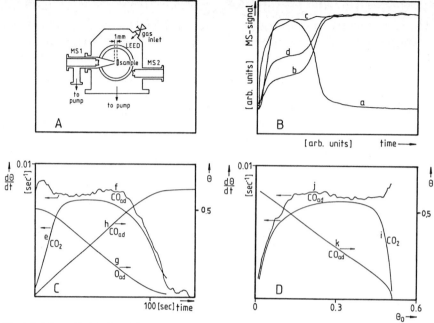

Fig. 1 **A**. Schematic drawing of the experimental geometry; **B**. typical experimental results: (a) flux of CO_2 (j_{CO_2}) measured with mass spectrometer 1; (b) backscattered CO-flux from the sample measured with mass spectrometer 1; (c) isotropic pressure of CO in the chamber measured with mass spectrometer 2; (d) total CO flux from the sample (see text); **C**. fluxes and coverages versus reaction time: (e) j_{CO2}; (f) net adsorption rate of CO; (g) oxygen coverage; (h) CO coverage; **D**. fluxes and CO coverage versus oxygen coverage: (i) j_{CO2}; (j) net CO-adsorption rate; (k) CO coverage.

apertured mass spectrometer 1. In this position, both the reaction and the uptake of CO by the sample were measured versus reaction time by the signals at m/e = 44 (trace a, Fig. 1B) and m/e = 28 (trace b, Fig. 1B), respectively. During these measurements mass spectrometer 2 was used to record the isotropic pressure change (trace c, Fig. 1B) in order to obtain the impingement rate onto the sample. Trace d gives the total flux of CO from the surface, including CO contained in CO_2, corrected for the true impingement rate and the m/e = 28 fragment of CO_2 produced in the mass spectrometer.

According to the stoichiometry of the reaction

$$CO_{ad} + O_{ad} \rightarrow CO_{2(g)},$$ (1)

the flux of CO_2 (j_{CO_2}) derived from the mass spectrometer signal I_{CO_2} is related to the decrease of oxygen coverage

$$-\dot{\theta}_0 = - \frac{d\theta_0}{dt} = I_{CO_2} / \Phi_{CO_2} = j_{CO_2},$$ (2)

where Φ_{CO_2} accounts for a sensitivity correction and the coverage calibration.

270

The intensity of the mass spectrometer signal at $m/e = 28$, I_{28}, has to be corrected by the fragmentation ratio y^{28} of CO_2 to derive the intensity I_{CO} due to CO.

$$I_{CO} = I_{28} - (y^{28} \cdot I_{CO_2}) . \qquad (3)$$

The increase in CO coverage on the crystal $\dot{\Theta}_{CO} = \dfrac{d\Theta_{CO}}{dt}$ is given by the uptake rate j^e_{CO}, which is the difference of the incident flux of CO (j^i_{CO}) and the backscattered flux (j^b_{CO}), reduced by the reaction flux j_{CO_2} :

$$\dot{\Theta}_{CO} = j^e_{CO} - j_{CO_2} . \qquad (4)$$

In terms of the mass spectrometer signals, this gives

$$\dot{\Theta}_{CO} = (I^i_{CO}(t) - I_{CO}(t)) / \Psi_{CO} - \left(\frac{I_{CO_2}(t)}{\Psi_{CO_2}} \right) , \quad \text{where} \qquad (5)$$

$$I^i_{CO}(t) = \frac{J_{CO}(t)}{J_{CO}{}^\infty} I_{CO}{}^\infty . \qquad (6)$$

The superscript ∞ refers to the final signal intensities at $(t \rightarrow \infty)$, J_{CO} is the signal of mass spectrometer 2, and Ψ_{CO} is equivalent to Ψ_{CO_2}. With these relationships, the CO-coverage $\Theta_{CO}(t)$ and oxygen coverage $\Theta_O(t)$ can be calculated as

$$\Theta_{CO}(t) = \int_0^t \dot{\Theta}_{CO} \, dt , \qquad (7)$$

$$\Theta_O(t) = \Theta_O^0 - \int_0^t \dot{\Theta}_O \, dt , \qquad (8)$$

with Θ_O^0 the initial oxygen coverage.

In Fig. 1C the respective quantities are plotted. Curve f is the net adsorption rate of CO, e is the CO_2 reaction rate, g and h represent the O and CO coverage, respectively. From Fig. 1C it is straightforward then to plot the respective quantities as a function of oxygen coverage (Fig. 1D) or CO-coverage. Calibration of coverages was done by comparison to known CO coverages as described in ref. [5]. The errors made in the evaluation of the data are less than 5% for the reaction rate and the oxygen coverage, while for the absolute values of the CO flux and coverages calculated from differences (see eq. 5), the errors are estimated to be about 10% [6]. For the data shown here, corrections were also made for the angular dependence of adsorption and desorption fluxes as discussed elsewhere [6].

3. Results

A. Mass Spectrometric Reaction Studies.

In Fig. 2A we show an overview of the reaction rate as a function of oxygen coverage for the temperature range studied. On first inspection, we can discriminate between three different regions, i.e. the initial steep rise in reaction rate at $0.4 < \Theta_O < 0.5$, a region of nearly constant reaction rate with

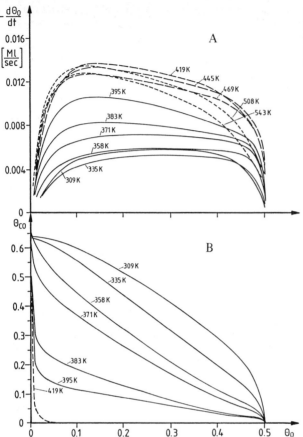

Fig. 2 **A.** CO_2 reaction rate as a function of oxygen coverage and temperature; **B.** CO-coverage as a function of oxygen coverage during reaction.

changing oxygen coverage ($0.1 < \Theta_0 < 0.4$) and a final steep decrease in rate at $\Theta_0 < 0.1$. An increase in reaction rate with temperature is found up to T = 419 K, at higher temperatures the reaction rate decreases and shows a stronger dependence on oxygen coverage. This general pattern of behaviour parallels the known steady state reaction rate as a function of temperature [1], where also a steep increase in rate between 350–450 K is followed by a continuous decrease in rate with substrate temperature. The reason for this change in steady state reaction rate is the depletion of CO through desorption at higher temperatures which reduces the reaction probability.

In this communication, however, we want to restrict the discussion mainly to the temperature range below T = 419 K, where a considerable CO coverage is established under reaction conditions and the desorption of CO need not to be considered in the evaluation of the rate parameters. As was confirmed in separate experiments, desorption of CO occurs at no measurable rate during the reaction measurements at T < 419 K and the coverage of Θ_{CO} is large enough to be directly measurable from the experiments described in Fig. 1. The CO coverage is given by the difference between the adsorption flux onto the surface (where adsorption refers to all molecules sticking and remaining or reacting on the surface) and, at temperatures where desorption is negligible, the reaction flux of CO_2 which depletes the CO-coverage (eq. 4 and 7). The final CO coverage at $\Theta_0 = 0$ is given by the equilibrium coverage at the respective temperature and we confirmed this by comparison with CO equilibrium adsorption data on clean Pd{110}[9]. Further, below 419 K a c(2x4) oxygen LEED pattern persists at $\Theta_0 > 0.1$ because the transition into the complex "1x3" structure is kinetically hindered and therefore does not become observable during the time interval of our measurements. This eliminates the necessity to consider the phase transition in the interpretation of our reaction measurements.

272

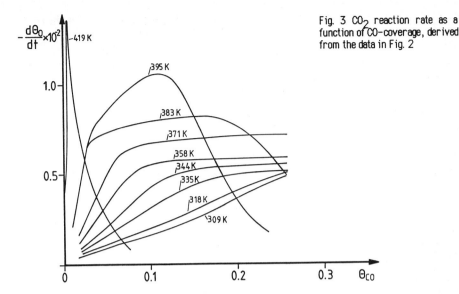

Fig. 3 CO_2 reaction rate as a
function of CO-coverage, derived
from the data in Fig. 2

The initial strong increase of the CO coverage in the region $\Theta_0 = 0.5$–0.45 is associated with the establishment of stationary conditions. The initial incident CO flux is primarily adsorbed, producing a rapid rise of the CO coverage. In this phase the turnover number is proportional to the CO coverage as Fig. 3 demonstrates, according to a 1st order reaction with respect to CO. But as a consequence of the rising CO coverage, the reaction rate increases and causes in turn a decreasing adsorption rate, since more of the (constant) incoming CO flux is consumed by the reaction. The initial reaction phase ends, when the process is approaching the stationary state.

In order to explain the intermediate coverage range in Figs. 2 and 3, we need to distinguish between the local coverage of the adsorbates within their domains which determine the reaction rate, and the total coverages which are evaluated by the procedures described in section 2. If we assume that the adsorbed CO spreads homogeneously in the surface regions depleted of oxygen by the reaction, the sum of the relative coverages of the reactants is equal to one. The total coverages Θ_{CO} and Θ_0 are then related to their local coverage Θ_{CO}^{loc} and Θ_0^{loc} by

$$(\Theta_{CO} / \Theta_{CO}^{loc}) + (\Theta_{CO} / \Theta_{CO}^{loc}) = 1$$

or

$$\Theta_{CO} = \Theta_{CO}^{loc} - (\Theta_{CO}^{loc} / \Theta_0^{loc}) \cdot \Theta_0 . \qquad (9)$$

A nearly constant reaction rate is found in the intermediate coverage range at the lower temperatures. This indicates, together with the persistence of the oxygen c(2x4) LEED pattern, that the local coverage of CO remains approximately constant whereas the total CO coverage increases. Since desorption can be neglected below $T \sim 420$ K, the total CO coverage and therefore the slope of Θ_{CO} vs. Θ_0 is determined by the balance between net adsorption and reaction rate. With temperature (Fig. 2A) the reaction rate increases and the slope of the Θ_{CO} vs. Θ_0 curves decreases accordingly. Higher reaction rates will also lead to a lower stationary local CO coverage, i.e. the local coverage depends on temperature.

Deviations from a linear increase in Θ_{CO}/Θ_0 in the intermediate coverage range are produced by the coverage dependence of the activation energy of the surface reaction. The activation energy increases with oxygen depletion at $T > 350$ K, leading to a corresponding increase of the stationary local CO

273

coverage. This tendency is cancelled below T ≈ 350 K, because in this temperature regime an effective activation energy of nearly zero was found for all oxygen coverages $\Theta_0 < 0.4$.

The final phase is characterized by an approximate linear dependence of the reaction rate on oxygen coverage as shown in Fig. 2A. The final steep rise of the CO coverage in Fig. 2B, especially for the high temperature traces, is due to the strong decline of the rate at the end of the reaction. This enables an increasing part of the incident CO flux to be adsorbed on the surface leading to a saturated CO layer.

In conclusion, the coverage dependence of the CO_2 formation contains three characteristic regions. In the initial stage (0.5-0.4), the reaction is 1st order with respect to the CO coverage. Further progress of the reaction leads to the establishment of stationary conditions with approximately constant local coverages of both reactants and a constant reaction rate. During the intermediate range ($\Theta_0 = 0.4-0.1$) the stationary situation is caused by the local balance of reactive consumption and net adsorption rate of CO in the oxygen-free patches. In this reaction regime the CO_2 is formed with a nearly constant rate. Model calculations carried out by us [6], show that such a behaviour is *inconsistent* with a mechanism where the reaction occurs in a statistically mixed adlayer or in a two-dimensional fashion on the boundaries of CO and O domains. An independence of reaction rate on O (and CO) coverage is expected for a mechanism, where the reaction proceeds preferably in a one dimensional fashion only at the ends of O or CO chain-like domains, thus maintaining a nearly constant number of reaction sites after an induction period. Any two dimensional reaction propagation would give a dependence on CO and O coverage, since the length of the domain circumference depends on its size [8]. In the final stage the reaction rate approximately evanesces following 1st order behavior with respect to the oxygen coverage. This can be interpreted by the disappearance of whole oxygen domains. As a consequence of the declining reaction rate, an increasing part of the incoming CO flux leads to adsorption of CO until finally the CO saturation coverage is reached.

B. LEED Data

As described in reference [5], upon adsorption of O_2 on Pd {110} first a pseudo "1x3" structure is formed which changes, at $\Theta \sim 0.27$, into a c(2x4) structure. This sequence of LEED structures is observed when the crystal temperature is held between 400 to 540 K during oxygen exposure.

The experiments described below were conducted in the same fashion as the reaction experiments in front of the mass spectrometer 1. During reaction the LEED pattern was monitored to obtain the intensity and width of the (0, -1) substrate reflex, the (1/2, -3/4) and (0, -1/2) beams of the c(2x4) structure, the (0, -1/3) beams of the oxygen-induced "1x3" phase, and the background intensity. Correlation with the CO_2 reaction rate as a function of time and subsequent integration gave the dependence of the LEED data as a function of oxygen coverage, as plotted in Fig. 4. A detailed description of the experimental parameters and data evaluation procedure is given in [6].

Figure 4A. shows the change in intensity for the (0,-1/2) beam of the c(4x2) structure and the appearance and disappearance of the (0,-1/3) intensity of the "1x3" phase as a function of oxygen coverage. We choose to normalize the intensity of the (0,-1/2) beam of the unreacted c(2x4) phase to $I^M_L \equiv 1$ for each reaction temperature and to plot the square root of the normalized intensity $\sqrt{I^M_L / I^M_L}$ as a function of oxygen coverage. For a constant number of oxygen domains this quantity should be proportional to the number of scatterers in the domains. The (0, -1/3) intensities were also normalized to the initial (0, -1/2) intensity. For clarity, Fig. 4A. shows only data for four selected temperatures. At the lowest temperatures (T = 335 K) and up to ~400 K no intensity, and at 400 K < T < 419 K only weak intensity, is found in the (0, -1/3) position. Only at T > 420 K a clear transition from the c(2x4) into the "1x3" oxygen phase is observed during the reaction. As shown in Fig. 2 for reaction temperatures below T = 400 K, both oxygen and CO are coexisting in the adsorbate phase during reaction. This is apparent also in our LEED data, where at T < 370 K a CO induced weak c(4x2) pattern is observed to coexist with the c(2x4) oxygen phase. This leads to additional diffraction intensity into the (0, -1/2) position and is the reason why the data for T = 335 K (shown in Fig. 4A) do not approach zero for $\Theta_0 \to 0$, whereas the oxygen c(2x4) intensity at (1/2, -3/4) (not shown here) does approach the baseline at $\Theta_0 \sim 0.1$ as for the higher temperatures. Between 420 and 370 K, a very diffuse c(2x2) CO overlayer structure was observed after removal of the oxygen adsorbate at the end of the reaction.

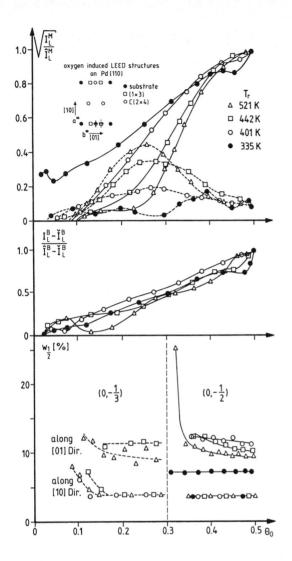

Fig. 4 **A.** Normalized intensities of the (0, −1/2) beam of the c(2x4) structure (solid line) and the (0, −1/3) beam of the "1x3" structure (dashed line) as a function of oxygen coverage at different reaction temperatures; **B.** normalized baseline intensity for the (0, −1/2) beam as a function of oxygen coverage; **C.** full width at half maximum of the (0, −1/3) and the 0, −1/2) beam along the [01] and [10] direction as a function of temperature and oxygen coverage.

The coexistence of a c(2x4) oxygen phase and a c(4x2) CO phase at T < 370 K support a model where the reaction takes place on the domain boundaries. We suggest that also at higher temperatures such a mechanism prevails, although due to the smaller CO coverage (compare Fig. 2B) CO induced LEED patterns are not observed.

A closer inspection of the intensity versus θ_0 plots in Fig. 4A for T > 400 K shows a slight initial decrease in intensity followed by an almost linear decrease between $0.2 < \theta_0 < 0.4$. The regime of linear decrease is steeper the more prominent the appearance of the "1x3" phase is during reaction. The gradual suppression of the "1x3" phase with decreasing reaction temperature is believed to be caused mainly by an activation barrier in the structural rearrangement [6], but an additional stabilization of the c(2x4) phase by coadsorbed CO at the lowest temperature (T < 420 K) cannot be ruled out at present. Support for an activated transition is obtained from the observation that the intensity in the (0, −1/3) position continues to increase even beyond the temperatures where the stationary CO coverage is practically zero.

Figure 4B shows the noncoherent intensity derived from the baseline of the $(0, -1/2)$ profile along the [10] axis, normalized at each temperature to the difference between initial and final value of the baseline intensity. This is the background intensity of the beam profiles shown in Fig. 4C. We note that this baseline intensity shows a linear decrease with decreasing oxygen coverage, nearly independent of temperature, except for the initial stronger decrease at T = 335 K.

This continuous and, from the beginning, linear decrease in baseline intensity versus the initial slow and then steeper decrease in the coherent diffraction intensity indicates that the reactivity of oxygen adsorbed in the ordered domains must be initially smaller than the reactivity of oxygen adsorbed in regions without long-range order. This initial higher reactivity of disordered oxygen atoms as seen in the LEED data parallels the initial steep increase and proportionality of the reaction rate to CO coverage shown in Fig. 2. Once the disordered oxygen has reacted off, the linear decrease in the coherent diffraction intensity reveals that the reaction involves $c(2x4)$ oxygen domains. The decrease in coherent diffraction intensity can be caused either by a decrease in domain sizes, or by a statistical removal of oxygen atoms out of the ordered domains. In the latter case, however, we should also observe an increase in incoherent diffraction intensity, which is obviously not the case. Thus, our observations support a domain boundary reaction after the initial induction period.

Information on the average domain size in the [10] direction parallel to the densely packed Pd-atom rows on the {110} surface and in the orthogonal [01] direction is contained in the widths of the diffraction beams. The transfer width of our instrument was experimentally determined to be ~100 Å in the [10] direction and ~150 Å in the orthogonal [01] direction for the given experimental setup.

Figure 4C. shows the FWHM (W_k) in percent of the reciprocal lattice vectors a^* and b^* (see insert Fig. 4A.) as a function of oxygen coverage and temperature during reaction. Along [10], the halfwidth remains constant for $\Theta_0 > 0.15$ at a value of 3.3% of a^*, which is the experimental resolution of the instrument. This means that the average size of the oxygen domain in the [10] direction exceeds ~100 Å. Only at $\Theta_0 \leq 0.15$ the average long-range order parallel to the rows of the Pd {110} surface decrease below ~100 Å as is evident in the increase in W_k. Along [01], the halfwidth is distinctly greater and not at the limit of the experimental resolution, indicating initially smaller domain sizes in the direction perpendicular to the [110] rows. That the initial average domain size at T > 400 K decreases with increasing reaction temperature is explained by the higher average temperature of the oxygen overlayer between the termination of oxygen exposure and introduction of CO. We further note, that both for the $c(2x4)$ and the "1x3" phase the increase in W_k at depletion of the respective phase is more strongly apparent at the higher reaction temperatures.

The constant halfwidth for $\Theta > 0.15$ in the [10] direction together with the continuous decrease in incoherent scattering intensity is consistent with our model of a domain boundary reaction where the reaction propagates preferentially in the [10] direction. The long-range order in the [10] direction is expected to exceed the instrumental resolution over a wide coverage range, when the reaction proceeds faster in this direction than in the orthogonal [01] direction, and the probability of forming new reaction centers within the ordered domains is small. The final increase in halfwidths sets in at a somewhat higher coverage ($\Theta_0 \sim 0.15$) than the decrease in reaction rate ($\Theta_0 \sim 0.1$) (Fig. 2A). The final reduction in reaction rate is therefore explained by the complete disappearance of chain-like domains. This, of course, is preceded by a decrease in their length, which becomes observable in the LEED experiments at a somewhat higher oxygen coverage than in the mass spectrometer studies.

The increase in W_k in the [01] direction which is more pronounced the higher the reaction temperature indicates that with increasing temperature a reaction propagation perpendicular to the [110] rows sets in. A contribution by the "1x3" structure to the broadening, however, cannot be excluded. For those temperatures where a reaction in the [01] direction is evident in the LEED data also a stronger dependence of the CO_2 formation rate on oxygen coverage (Fig. 2A) is observed.

At the lowest temperature displayed in Fig. 4C (T = 335 K), the halfwidth of the $(0, -1/2)$ beam is initially smaller and remains constant as compared to the higher temperatures. The initially lower halfwidth is explained [6] by an ordering of oxygen domains into layer domains due to coadsorbed CO. We were unable to observe a compression into a new LEED structure as for the low temperature coadsorption phases for CO and O on Pd {111} [7]. The constant halfwidth as a function of oxygen

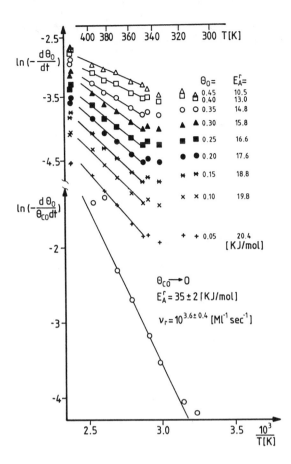

Fig. 5 Arrhenius plots for the temperature dependence of the reaction rate. The reaction rate isosters for different oxygen coverages are displayed in the upper part of Fig. 5. The temperature dependence of the rate constant in the initial reaction regime is presented by the open circles below.

The figure plots:

Upper part: $\ln\left(-\dfrac{d\theta_0}{dt}\right)$ vs $T[K]$ (400 380 360 340 320 300)

$\theta_0=$	$E_A^r=$
0.45	10.5
0.40	13.0
0.35	14.8
0.30	15.8
0.25	16.6
0.20	17.6
0.15	18.8
0.10	19.8
0.05	20.4
	[KJ/mol]

Lower part: $\ln\left(-\dfrac{d\theta_0}{\theta_{co}dt}\right)$

$\theta_{co}\longrightarrow 0$

$E_A^r = 35 \pm 2 \,[KJ/mol]$

$\nu_r = 10^{3.6\pm 0.4}\,[Ml^{-1}sec^{-1}]$

x-axis: $\dfrac{10^3}{T[K]}$ (2.5, 3.0, 3.5)

coverage in both the [01] and [10] direction indicates that reaction propagation perpendicular to the rows is substantially smaller than at the higher temperatures. In summary, the LEED data support a model for a domain boundary reaction which preferentially proceed along the [10] direction of the crystal substrate at reaction temperatures below 400 K.

C. The Reaction Rate Equations

The data shown above reveal the complex behavior of the reaction mechanism, which depends not only on O and CO coverage, but also on the reaction temperature. In Fig. 5, we show the Arrhenius plots for the temperature dependence of the reaction rates for T < 420 K and the activation energies calculated from the slopes of the reaction rate at constant oxygen coverage. As discussed in section 3A (eq. 9), the total CO-coverage in the intermediate reaction regime is nearly linearly related to the oxygen coverage. Therefore, the coverage dependence in this reaction regime can be reduced to depend on oxygen coverage only and the rate can be expressed as

$$- d\theta_0/dt = \nu_r \cdot f(\theta_0) \cdot \exp\left(- E^r(\theta_0)/RT\right)$$

However, the local CO coverage , which determines the reaction rate, depends on temperature as discussed in section 3A. Thus, the activation energy derived from the rate expression above is only equal to the activation energy of the elemental reaction step, when the order of the reaction with respect to θ^{loc} is zero. A detailed analysis of the data [6] confirm the approximate independence of reaction probability on local coverage. Also, an independence of the reaction probability on local CO coverage is plausible in our model where the reaction proceeds in a one-dimensional fashion at the end of linear oxygen and CO domains. As long as the collision frequency of CO (given by Θ^{loc} and the surface diffusion rate) onto the ends of oxygen chains is higher than the reaction probability, the local CO coverage does not effect the turnover number. Therefore, the apparent activation energies derived from Fig. 5 can be considered in a first approximation to represent the activation energy for the elemental reaction step on the surface. The dependence of E^{app} on oxygen coverage is given by $E^{app} = 22-20 \cdot \Theta_0$ [kJ/mol] and the preexponential is related in a compensative fashion to E^{app} by $v_r \cdot f(\Theta_0) = 4.7 \cdot 10^{-3} \exp[E^{app}/(R \cdot 347)]$ where $T = 347$ K is the isokinetic temperature. The decrease in activation energy with increasing oxygen coverage is probably related to the change in adsorption energy of the reaction gases through lateral interactions within and between the two domains. A decrease in activation energy with oxygen coverage has also been noted for the reaction on Pd {111} [2].

In the initial reaction regime at $\Theta_0 > 0.45$, the reaction is proportional to the CO coverage as shown in Fig. 3. We recall that the LEED data suggested a preferential reaction of disordered oxygen which implies that the initial reaction rate will not depend on total oxygen coverage. A plot of $d\Theta_0/[dt\Theta_{CO}]$ versus reciprocal temperatures gives an activation energy of $E^r = 35 \pm 2$ kJ/mol and $v_r \cdot f(\Theta_0) = 7.5 \times 10^3$ sec^{-1} for this reaction regime. In this case, the experimental activation energy should be equal to the elemental activation energy of reaction, since the local coverages of CO and O are not determined by the reaction itself. For the reaction of CO with a saturated oxygen overlayer at 360 K < T < 500 K on Pd{111} ENGEL and ERTL found an activation energy of $E^r = 59$ kJ/mol and $v_r \cdot f(\Theta_0) = 8.0 \times 10^7$ sec^{-1}[2]. This activation energy is about 50% lower than the one found at T > 500 K and was interpreted by the compression of oxygen domains by coadsorbed CO leading to a reduction of adsorption energy and consequently activation energy of reaction. ENGEL and ERTL also observed an induction period in their low temperature measurements, contrary to our results, which indicated that reaction involved oxygen within or at the boundaries of ordered domains. However, the net reaction rates at T ≤ 400 K on the two surface orientations are only slightly different and at T = 310 K identical. This can lead to the impression that the reaction is not dependent on surface structure.

As discussed elsewhere [6], the kinetic parameters and therefore the net rates become identical on Pd {110} and {111} at T > 480 K, where the adsorption/desorption equilibrium of CO has to be considered in the evaluation of the kinetic data. At these higher temperatures, the LEED data showed that the reaction between the dense O-domains and chemisorbed CO proceeds also in the [01] direction of the surface, i.e. it is described by a two-dimensional domain boundary reaction as on the Pd{111} surface [2]. Under these conditions, the reaction is structure insensitive. Only under reaction conditions where desorption is negligible and where both CO and O domains coexist, the effect of the Pd {110} surface topography is reflected in the rate parameters and in the LEED data.

We note that the reaction conditions established in our experiments are different than under steady state conditions, where both CO and O are present in the gas phase. Under such circumstances, the reaction as a function of increasing substrate temperature only sets in when surface sites become available for oxygen adsorption through CO desorption (T > 350 K) and the maximum of reaction rate is obtained (depending on the pressures) at ~ 500 K [1] where both reactants are present in substantially lower concentrations than under our experimental conditions.

4. Conclusions

In summary, the reaction of chemisorbed CO and O on Pd{110} is strongly effected by the surface topography if the reaction conditions allow the coexistence of both reactants. For the initial phase of CO_2 formation out of a c(2x4) oxygen overlayer we found an activation energy of $E^r = 35 \pm 2$ kJ/mol. Subsequently the reaction becomes stationary and the local coverages are kept approximately constant, whereas the overall coverages of the two reactants are changed through the variation in domain size. In this reaction regime the activation energy depends on oxygen coverage as $E^r = 22-20 \cdot \Theta_0$ [kJ/mol], if a dependence of the rate on the local coverage is neglected because of the unidirectional character of the reaction. Such a unidirectional propagation of the reaction is supported by the LEED and mass spectroscopic results. The comparison to the corresponding data for the Pd {111} surface

[2] shows that the reaction parameters are different for the two orientations, but due to a compensation effect the overall rates are nearly equal. We suggest that this is the reason why in steady state experiments [1,9] no pronounced influence on the surface geometry is observed. A similar conclusion was derived for the CO-oxidation reaction on Pt surfaces by HOPSTER, IBACH and COMSA [10].

5. Acknowledgements

This work was supported by the Deutsche Forchungsgemeinschaft through Sonderforschungsbereich 6. We thank W. Feige for his assistance in building and maintaining the equipment.

6. References

1. G. Ertl and J. Koch, Proc. 5th Int. Congr. Catalysis, Palm Beach, 1972, 969; G. Ertl, in "Chemistry and Physics of Solid Surfaces, Vol. III, ed. R. Vanselow, W. England, CRC Press, Boca Raton, (1982).

2. T. Engel and G. Ertl, J. Chem. Phys. 69 (1978) 1267.

3. M. Wolf, J. Goschnick, J. Loboda-Cackovic, M. Grunze, W.N. Unertl and J.H. Block, Surf. Sci., in press.

4. S.M. Francis and N.V. Richardson, Phys. Rev. B 33 (1986) 662.

5. J. Goschnick, M. Wolf, M. Grunze, W.N. Unertl, J.H. Block and J. Loboda-Cackovic, Surf. Sci., (1986). Proceedings of ECOSS 8, Jülich (1986).

6. J. Goschnick, Dissertation, F.U. Berlin (1987).

7. H. Conrad, G. Ertl and J. Küppers, Surf. Sci. 76 (1978) 323.

8. D. Mukesh, W. Morton, C. N. Kenney and M.B. Catlip, Surf. Sci. 138 (1984) 237.

9. J. Koch, Ph.D. Thesis, T.U. Hannover, (1972).

10. H. Hopster, H. Ibach and G. Comsa, J. Catal., 46, (1977) 37.

Conformal Invariance, Multiple Scattering and the Au(110) (1 x 2) to (1 x 1) Phase Transition

P. Kleban[1], *R. Hentschke*[1], *and J.C. Campuzano*[2]

[1]Department of Physics and Astronomy and Laboratory for
  Surface Science and Technology, University of Maine, Orono, ME04469, USA
[2]Department of Physics, University of Illinois, Chicago, IL60680, USA

1. Introduction

Conformal invariance predicts the scattering lineshape and full dependence on scattering parameters for LEED (Low-Energy Electron Diffraction) intensities from finite regions at second-order phase transitions in 2-D. Existing data for the Au(110) (1x2) to (1x1) phase transition provide a partial test of the theory. Good agreement for the lineshape is obtained over a range of temperatures including the expected "bulk" (infinite plane) transition temperature T_c. Such comparisons constitute important tests of the conformal principle, since fundamental quantities in field theory are probed. The predicted intimate relation between finite size and intensity variations may provide an important tool in surface characterization.

In this work we confine ourselves to an outline of the main new results. Detailed treatment will be published elsewhere (KLEBAN, HENTSCHKE and CAMPUZANO (1); HENTSCHKE and KLEBAN (2)).

2. Theory of Scattering at Surface Critical Points

Consider a surface Bragg peak at parallel wave number q_0, which is induced by an ordered overlayer or (reversible) reconstruction of different spatial symmetry than the substrate. As the temperature of the system is raised, a phase transition may be approached, and the intensity will decrease. If the transition is second order, the intensity will decrease continuously, even on an ideal (perfect and infinite) substrate. At the transition temperature T_c, the intensity is no longer properly described as a Bragg peak, since there is no corresponding overlayer or reconstruction induced long-range order; i.e., the delta function part of the intensity will vanish. But diffuse scattering intensity will remain, due to the critical fluctuations. It is this intensity that we can calculate, including multiple scattering contributions, when one is at T_c, the transition temperature of the "bulk" system (infinite plane). The computations are based on conformal invariance (CARDY (3)), which is believed to hold generally at critical points, and has some particularly strong consequences in two dimensional systems. The theoretical techniques employed are intimately related to string theory in elementary particle physics. The results are based on (local) symmetry only, and thus depend (with one exception) on the universality class of the phase transition. This, in turn, is generally determinable by observable symmetry breaking (the relative symmetry of the ordered vs. high-temperature phases). Thus this powerful method may be employed without detailed knowledge of the forces underlying the phase transition.

Although there is no long-range order at T_c, long range correlations remain. These correlations determine the diffuse intensity for q near q_0. Now any real surface will be composed of finite regions, of size L (typically L ~ 100 Å for metal surfaces), which are undergoing the phase transition. For a single such region the diffuse intensity will occur for $\delta q < 10/L$. Here δq is measured from a Bragg spot q_0 and the numerical estimate derives from some previous work (KLEBAN et al. (4)). There will be scattering intensity for higher δq values as well, but it is not calculable by universal methods (the conformal calculations or in fact any universal results are only valid for $\delta q < 1/a$, where a is a substrate lattice spacing). However, the intensity will be small for these large δq values; the main peak is very accurately given by the conformal theory (BARTELT and EINSTEIN, (5)).

Now exactly at the phase transition temperature T_c, one has in general the property that the correlation length $\xi \propto L$. This is in fact a basic property of a second-order phase transition. By contrast, when one is not at a phase transition ξ is generally independent of and less than L. Its large size at T_c is an expression of the long-range correlations that exist there. Now, ξ, among other things, specifies the distance over which a disturbance is felt. Therefore, since $\xi \propto L$, the system always feels its boundary - a finite fraction of the system, no matter how large it is, remains influenced by the edge. Hence the scattering lineshape will reflect the size, shape and boundary conditions of the finite surface regions. By use of conformal invariance one may calculate the lineshape $S(\delta q)$ when the kinematic (single scattering) approximation is valid (KLEBAN et al (4)). But more is possible. To understand why we must give some background information. Exactly at T_c, there are only a finite number of critical operators (local thermodynamic quantities that exhibit long-range correlations. For the Ising model, the important ones are the (local) order parameter, which at low T measures the extent of long-range order, and the (local) energy density. When the overlayer or reconstruction ordering is not (1x1), these operators have different spatial symmetry, and are therefore associated with different Bragg peaks. For instance, only the order parameter-order parameter (or "spin-spin") correlation function will contribute to the diffuse intensity near q_0 in this case.

Now consider scattering probes for which multiple scattering is important. For definiteness we consider Low Energy Electron Diffraction (LEED). In this case the scattering intensity will involve multipoint correlation functions of high order, which are generally intractable. However, the diffuse intensity at T_c, by the arguments given above, must be governed by only a few two-point correlation functions of critical operators in the layer undergoing the transition; these are precisely the operators that give rise to long-range correlations. Hence the multipoint correlations collapse to two-point functions and the problem simplifies. Conformal field theory (CARDY (3), BELAVIN et al. (6)) provides the tools to make this idea explicit. If the electron mean free path $\lambda_e \ll L$, as is generally the case in surface experiments, the multipoint functions that contribute to the intensity at small δq always depend on two groups of points. Each group is of size $\sim\lambda_e$ but each group is separated from the other by a distance on the order of L. The operator product expansion and orthogonality of critical operators then allow one to explicitly reduce the multipoint functions. The result, under the conditions mentioned, is

$$I = f \, S(\delta q). \tag{1}$$

Here I is the LEED scattering intensity (up to an overall constant), S the

Fourier transform of the order parameter-order parameter correlation function, and f a function of the scattering parameters, which also includes the influence of the substrate atoms on the scattering. f may be determined by suitably modifying a standard dynamical LEED calculation (HENTSCHKE and KLEBAN (2)). It samples the surface and substrate only over approximately one mean free path λ_e, and is therefore only weakly varying with δq. S, on the other hand, as mentioned varies over the range $\delta q < 10/L$, and strongly reflects the geometry of the finite size regions. Furthermore, since f is influenced by the atoms undergoing the phase transition, under suitable conditions its variation with scattering parameters is substantially different than in the ordered state. A simple example is shown in Fig. 1. The main conclusion is that a detailed prediction of the scattering intensity at T_C is possible.

3. Comparison with Experiment

A partial test of the predictions of (1) is possible by comparison with LEED data for the Au(110) (1x2) to (1x1) phase transition reported by CAMPUZANO et al. (7).

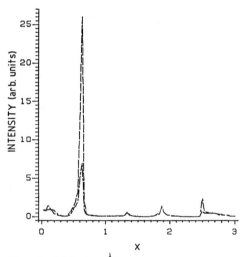

Fig. 1. LEED (o, $\frac{1}{2}$) beam intensity versus $x = \frac{1}{2}(a/\pi)^4 E$ for the (100) surface of an sc crystal with perfect (—·—) p(2x1) order or at the infinite plane critical point T_C (— —). Here E is the (normally incident) electron energy, a = 3 (a.u.) the lattice spacing and the real and imaginary part of the non-ion core potential are $V_0 = 0$ (a.u.) and V_{oi} = 0.2 (a.u.), respectively. For simplicity, only a constant s-wave phase shift $\eta_0 = \pi/2$ was used. The total scattering amplitude for perfect and critical ordering only differs to third order in the ion scattering amplitude (except for an unobservable overall constant). To capitalize on this, the single ion amplitude was multiplied by three, similar effects will occur for realistic cross-sections when large forward scattering enhances the higher-order multiple scattering terms. Comparable results may be obtained from a full-scale dynamical LEED calculation by straight forward modification of existing code.

At low temperatures, the Au(110) system reconstructs into a (1x2) state. The structure is described by a "missing row" model - alternate topmost rows of gold atoms on the unreconstructed surface are removed. In the sample employed for the scattering data analyzed here, the reconstructed regions were determined to have a size $L \sim 150\text{Å}$. Furthermore, the extra beams were approximately symmetrical at low temperatures, indicating that the reconstructed regions were not greatly elongated. Also, the width of the Gaussian part of the lineshape was independent of the temperature below the transition. Hence the finite size regions did not grow or shrink. As the temperature was raised, the (1x2) reconstruction was removed and the extra beams gradually disappeared. Detailed study of the lineshape and integrated intensity as a function of temperature by CAMPUZANO et al. (7) and CLARK et al. (8) show that the (1x2) to (1x1) phase transition is continuous and in the Ising universality class, as expected from symmetry analysis of BAK (9). Hence this system is a prime candidate for comparison with the conformal theory, which predicts the scattering lineshape at T_c. The value $L \sim 150\text{Å}$ indicates that finite size determines the shape of S for $\delta q < 0.07 \text{ Å}^{-1}$, which is exactly in the observed region. It should also be mentioned that resolution effects in this study are not important, since the instrumental coherence width greatly exceeded 150Å.

Comparison of experimental and conformal lineshapes for various temperatures are shown in Fig. 2. The data was previously fit (CAMPUZANO et al (7)) to a Gaussian plus a Lorentzian of variable heights and widths, to determine critical exponents. Here, the overall lineshape is compared with the conformal result by fitting the height and width of the latter to the experimental data at each temperature. Since neither the geometry of the finite regions nor the overall scattering intensity are exactly known, this procedure is appropriate. Some typical results are shown in Fig. 2. At temperatures of 635K or less, the lineshape is too Gaussian to agree with the conformal prediction. At temperature 687 or above, it is too Lorentzian. Lineshapes at intermediate temperatures fit well (658K is illustrated in Fig. 2). This good agreement is observed over a range of

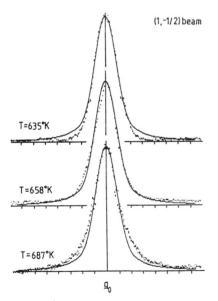

(1,-1/2) beam

T=635°K

T=658°K

T=687°K

q_0

Fig. 2. Comparison of experimental results for LEED lineshapes (squares) near the $(1, -\bar{2})$ beam with the conformal prediction (continuous curve) at three temperatures near the Au(110) (1x2) to (1x1) phase transition. q_0 is the parallel momentum transfer in the direction perpendicular to the rows in the (110) surface.

about 50K. More extensive data including directional dependence of the line shape on δq and dependence on other scattering parameters would allow a better comparison with (1).

The conformal prediction applies to scattering from a single region. In the experiment, more than one such region is included. However, the illuminated area on the surface was less than 0.1 μ across, so that only a few regions of size 150 Å were averaged over. The analysis assumes that the important regions did not differ greatly in size or shape and scattered incoherently with each other.

In general there may be a distribution of region sizes and shapes which is difficult to determine. Note however that the factor f in (1) is a very good approximation independent of these effects. This fact may prove important in surface characterization.

While the results reported here are very encouraging it is clear that a more detailed study needs to be done. The theory predicts the scattering intensity for all directions in reciprocal space, and its full variation with scattering parameters. The corresponding data is not yet available.

We thank M. Grunze for comments on the manuscript. R.H. acknowledges support from a University of Maine Graduate Fellowship.

References

1. P. Kleban, R. Hentschke and J.C. Campuzano: to be submitted.

2. R. Hentschke and P. Kleban: in preparation.

3. J.L. Cardy: in Phase Transitions and Critical Phenomena, Vol. 11, eds. C. Domb and J.L. Lebowitz (Academic Press, London, to appear).

4. P. Kleban, G. Akinci, R. Hentschke and K.R. Brownstein: J. Phys. A. 19, 437 (1986); Surface Science 166, 159 (1986); P. Kleban and R. Hentschke: Phys. Rev. B 34, 1980 (1986).

5. N.C. Bartelt and T.L. Einstein: J. Phys. A. 19, 1429 (1986).

6. A.A. Belavin, A.M. Polyakov and A.B. Zamolodchikov: Nucl. Phys. B 241, 333 (1984).

7. J.C. Campuzano, M.S. Foster, G. Jennings, R.F. Willis and W.N. Unertl: Phys. Rev. Letters 54, 2684 (1985).

8. D.E. Clark, W.N. Unertl and P. Kleban: Phys. Rev. B 34, 4379 (1986).

9. P. Bak: Solid State Commun. 32, 581 (1979).

Self-sustained Kinetic Oscillations in the Catalytic CO Oxidation on Platinum

Y.J. Chabal[1], *S.B. Christman*[1], *V.A. Burrows*[2], *N.A. Collins*[2], *and S. Sundaresan*[2]

[1]AT&T Bell Laboratories, 600 Mountain Ave., Murray Hill, NJ 07974, USA
[2]Princeton University, Princeton, NJ 08544, USA

1. INTRODUCTION

The observation of self-sustained oscillations in the rates of catalytic reactions under specific catalyst temperature is widespread [1]. It includes the oxidations of hydrocarbons, H_2, CO and other gases over Pt, Pd, Ir and Ni surfaces. In general, oscillations are observed on supported catalysts or foils in relatively high pressures (1-1000 Torr) and are characterized by *long* periods, ranging from several minutes to several hours. However, relatively rapid oscillations ($\approx$ seconds) have also been observed on *clean* Pt(100) crystals at low pressures ($\approx 10^{-4}$ Torr), and attributed to adsorbate-induced transformation of the surface structure [2]. Variations in surface reconstruction clearly cannot be involved for the majority of the systems under study since most do not reconstruct and all contain surface impurities under oscillation conditions. As a result, the driving mechanism for the majority of systems still remains poorly understood.

Gas phase effects were originally thought to account for the oscillations, partly because reaction rate oscillations have been observed in gas reactions and partly because gas composition and pressures directly entered in the kinetic equations and could be easily measured. However, the periods of the observed oscillations were far too long to be explained by gas transport phenomena or by the reaction kinetics alone. Attention was therefore turned to the catalyst surface itself where slow processes such as diffusion from the bulk could dramatically modify the reaction rates. Surface spectroscopic techniques had to be used to get insight on the elementary steps taking place at the catalyst surface.

In situ, real-time surface measurements are difficult at the pressures involved and early surface studies consisted mostly of indirect measurements made prior to and after oscillatory behavior. Based on such measurements, models were proposed which could account for the available set of observations. One such model, developed by SALES *et al.*[3], invokes the periodic oxidation and reduction of the catalyst surface. The experimental basis was the evidence for the presence of *strongly bound* oxide in addition to the reactive adsorbed oxygen [4]. The slow formation and reduction of this oxide was therefore a natural candidate to account for the long periods, with the attractive feature that it involved chemical species readily available from the ambient. The model, based on a Langmuir-Hinshelwood (L-H) reaction mechanism, provided satisfactory agreement with the observed time scales and reaction rates [3]. However, there have been few confirming real-time data to support it and the existence of the postulated oxides has recently been called into question [5].

The fundamental difficulty in understanding the process is common to many kinetic problems: while kinetic rate equations provide an accurate description of the reactant and product concentrations (i.e., easily observed quantities), they often give little insight into the detailed elementary steps taking place at the catalyst surface. In the specific case of self-sustained oscillations in the oxidation rate of CO on a catalyst surface, the basic mathematical description relates the time evolution of the surface coverages of the various species (oxygen, carbon monoxide and "impurity") to the available number of active sites by means of rate coefficients, assuming that both CO and oxygen need to be adsorbed before they can interact to form CO_2 (L-H mechanism). By so doing, a set of coupled *non-linear* differential equations is obtained. The nonlinearity, necessary for oscillatory behavior, comes from the dissociative chemisorption of oxygen requiring *two* adjacent empty sites for each O_2 molecule. Within this framework, two stable branches in reaction rate can be obtained for specific parameters and *any* "impurity" can induce a transition between these branches which can lead to an oscillatory behavior. The onus is therefore first on the experimentalists to identify the nature of the impurity which effectively modulates the number of active sites or the mechanism which modulates the activity of the reaction. Once established, the detailed mechanism can then be taken into account and the differential equations solved to verify that, under appropriate conditions, oscillatory behavior is indeed possible.

LEED and work function measurements were performed in the case of the rapid oscillations on Pt(100) [2]. The observation of temporal oscillations, under conditions giving a critical CO coverage of 0.5 monolayer (if this gas were present alone) led to the suggestion that this effect was associated with a periodic variation in the surface structure. The LEED pattern was indeed found to switch periodically between that of the reconstructed (hex) phase and that of the c(2x2) structure formed by 1/2 monolayer of adsorbed CO on the non-reconstructed (1x1) Pt(100) surface. To model these observations, the key consideration was that the sticking coefficient for oxygen on the hex phase is extremely low ($s_0 \approx 10^{-3}-10^{-4}$) compared to that on the 1x1 surface ($s_0 \approx 0.1$) at the reaction temperature. This strong variation in sticking coefficient provides a natural way to induce transitions between the two stable branches of a L-H reaction paths. Thus, instead of an impurity modulating the reactivity of the surface, specific changes in surface structure drive the observed oscillations. Mathematically the formalism is similar to that mentioned above except that, instead of rate coefficients for formation and removal of a surface impurity, there are now rate coefficients for growth and decrease of the 1x1 phase required for a high reaction rate.

While the latter mechanism describes well the observations obtained on Pt(100) under very clean conditions (no surface impurities and high gas purity), it cannot easily be reconciled with the results of high pressure and intrinsically dirtier experiments performed on a variety of substrates that do not reconstruct. The purpose of the present work was therefore to study experimentally those systems for which the traditional surface science techniques were inoperative because of high pressures. Since it is possible that surface or gas impurities, whatever their origins, play an important role in the oscillatory phenomenon, the approach was to understand the role of these impurities rather than to remove them.

To minimize possible ambiguities in the experimental search for the driving mechanism of these oscillations, we chose to study CO oxidation on a Pt *foil* for which real-time calorimetry *and* infrared studies (of both surface and gas phase species) could be performed simultaneously [6] and for which Auger spectroscopy and other surface science probes could also be useful [7]. In Section 2, where the experimental apparatus is described, we show that an accurate comparison between the state of the

Pt foil surface in the UHV chamber (for Auger analysis) and in the flow cell (for kinetic studies) could be made by means of surface infrared spectroscopy. In Section 3, we review the data obtained in the flow cell where oscillations could be induced. These include calorimetry and infrared (surface and gas phase) measurements *during* the oscillations ($T_{foil} \approx 500K$) and surface infrared measurements upon quenching of the foil to room temperature at different points of the oscillation cycle. The latter gave a measure of the number of active sites throughout the cycle. In Section 4, we review the data obtained in the UHV system (mostly Auger data) which help identify the nature and kinetics of the relevant surface impurities. In particular evidence is presented that carbon and not oxide can drive the oscillations. Finally, in Section 5, a simple model is proposed in which carbon blocks the active sites [8].

2. EXPERIMENTAL

The experimental apparatus is shown in Fig. 1. It consists of two main parts: (1) an infrared interferometer with a reaction cell constructed on its bench, and (2) an ultra-high-vacuum (UHV) chamber with infrared windows (CsI). Pieces of the *same* Pt foil were mounted in the reaction cell and the UHV chamber. In both vessels, the foils were subjected to similar temperature and gas exposure treatments and could be probed by calorimetry and infrared spectroscopy in identical manner.

In the reaction cell, the gas flow pattern was such that, for the range of pressures (1-1000 Torr) and gas composition (O_2, CO and He mixtures with P_{CO}/P_{O_2} varying from 1 to 3%), self-sustained reaction rate oscillations could be induced. *Real-time* calorimetry and infrared spectroscopy were performed.

In the UHV chamber, although the same range of pressure was used, the flow pattern prevented the establishment of periodic oscillations with detectable periods (< 2 hours). Both high and low reaction rate regimes could be well established, including

Fig. 1: Schematic diagram of the experimental apparatus. The interferometer bench and detector housing are purged with dry N_2 gas. Note the polarizers (POL.) used to discriminate between gas and surface absorptions.

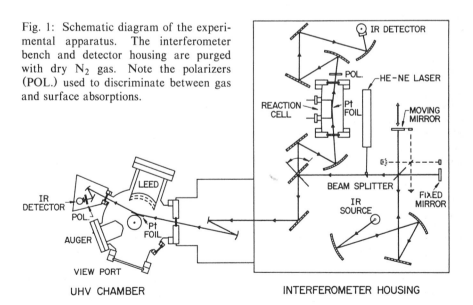

transient behavior between the two states, but the phenomena depending on gas flow had increased time-constants. Thus, after the foils in the UHV chamber were subjected to specific treatments reproducing those in the flow cell, the chamber was pumped below 10^{-6}Torr and electron spectroscopies were used.

2.1 Calorimetry

The heat generated by the exothermic CO oxidation reaction was monitored by measuring the current required to maintain (by resistive heating) the foil temperature constant. This was accomplished with a fast resistance controller in a balanced bridge configuration [6].

The thin foils (90 x 13 x 0.0127mm^3) had a typical resistance of $\approx 0.02\,\Omega$ at room temperature and required over 10A to be heated to the reaction temperature (≈ 500K) in several hundred Torr of gas mixture. The resistance controller was capable of delivering up to 20A while maintaining the foil resistance constant to within an equivalent average temperature of ± 1K with a 5ms response time [6].

Fig. 2a shows the resistance controller reading during a complete oscillation cycle. The abscissa is the negative of the voltage across the foil as the current is varied to maintain the resistance constant. During the spike, the least amount of current is needed because a large amount of heat is produced by the reaction.

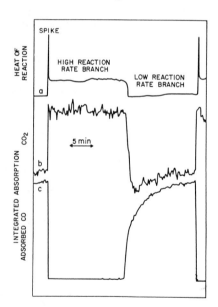

Fig. 2: (a) CO oxidation reaction rate measured by monitoring the Pt foil potential drop as a function of time for a *constant* foil *resistance*. (b) Gas phase CO_2 and (c) adsorbed CO integrated absorbances during an oscillation cycle.

2.2 Infrared Spectroscopy

As shown in Fig. 1, surface infrared spectroscopy was performed on both foils (in the cell and in the UHV chamber) using a grazing incidence ($\theta \approx 85°$) reflection geometry [6,9]. The infrared radiation was p-polarized (E-field in the plane of incidence) to maximize the electric field perpendicular to the surface. Vibrational spectra were obtained either by ratioing p-polarized spectra of the covered foil to s-polarized spectra

288

(insensitive to the surface) or to other p-polarized spectra of the clean foil. Gas phase absorption in the flow cell could readily be identified because it appeared both in the s-polarized and p-polarized spectra. Figure 2b and 2c show for example gas phase CO_2 and adsorbed CO absorptions as a function of time during a complete oscillation cycle. Gas phase CO was subtracted using S-pol. spectra.

The adsorbed CO and gas CO_2 spectra were obtained using an InSb infrared detector at a rate of 2 scans/s with a nominal resolution of $4cm^{-1}$. The noise level under these conditions was $\Delta R/R \approx 2 \times 10^{-3}$ per scan, which is adequate to detect a monolayer of CO ($\Delta R/R_{peak} \approx 6\%$). For better S/N, some of the spectra were signal-averaged from 10 scans (time resolved data) or from 2000 scans (static data).

The lower frequency region ($450cm^{-1}$-$1900cm^{-1}$) was investigated using several HgCdTe detectors. The only vibrational feature observed was a broad (FWHM> $50cm^{-1}$) and intense ($\Delta R/R \leq 8\%$) band in the 1000-$1400cm^{-1}$ region, which we have assigned to silicon oxide [7]. Adsorbed oxygen could not be detected because the Pt-O stretch mode is weak and occurs in a frequency range (400-$500cm^{-1}$) where the S/N is too low. However, by monitoring the broadband reflectance of the Pt foil (i.e. the electronic absorption), inference of the presence of adsorbed oxygen could be made as summarized in section 3.

2.3 UHV Studies

The Pt foil in the UHV chamber could be probed by calorimetry and infrared spectroscopy in identical manner (same dimension, resistance, and angle of incidence) as the foil in the flow cell. The main difference was that, in the UHV chamber, the ambient gas (typically at 100 Torr) was occasionally pumped down so that the surface could be investigated by Auger spectroscopy. The use of infrared reflection-absorption spectroscopy (IRRAS) was therefore crucial in comparing the surface composition of the two foils since the non real-time Auger data were only indirect measures of the surface composition *during* the oscillations. In particular, the amount of CO adsorbed on the surface at room temperature in a reaction mixture (O_2, CO, He) was found to give a reliable measure of the number of sites active towards oxidation reaction. Therefore, combining the IRRAS determination of the CO coverage with the Auger monitoring of the inactive sites (blocked by strongly bound impurities), a complete picture of the surface composition could be obtained. Another advantage of the UHV chamber was the ability to study such properties as broadband reflectance of a truly clean foil (obtained by Ar^+ bombardment and annealing in 10^{-11}Torr base pressure). Such Auger and IRRAS studies on clean foils were performed in UHV after the series of high pressure studies.

3. REAL-TIME DATA

3.1 During oscillations

The main results obtained from direct *vibrational* absorption spectra [6,7] are that:
(1) The production of CO_2 gas indeed oscillates in time for isothermal foil conditions and steady gas flow as shown in Fig. 2 (a) and (b). The high CO_2 production is correlated with high heat transfer to the foil by the CO oxidation reaction.
(2) During the high reaction-rate branch, no adsorbed CO was detected, indicating a low concentration and short residence time of CO at the surface (Fig. 2c).
(3) During the low reaction-rate branch, a large amount of linearly bonded CO (≤ 1 monolayer) was measured, the frequency of which was consistent with domain formation (Fig. 2c).

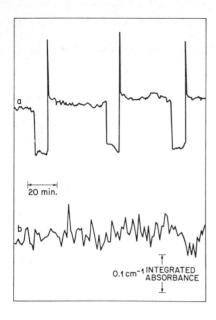

Fig. 3: (a) Three oscillation cycles measured by calorimetry. (b) Absorbance of the silicon oxide band, integrated from 1000 to 1400cm^{-1} (0.1cm^{-1} corresponds to 2% of the total silicon oxide signal).

20 min.

0.1 cm^{-1} INTEGRATED ABSORBANCE

(4) In both branches, substantial amounts of silicon oxide were evident with no detectable variation between the high and the low branch (Fig. 3).

Besides the vibrational studies, the reflectance of the *isothermal* foil, measured over the 1800-2050cm^{-1} region (devoid of vibrational bands), was found to be higher in the low reaction-rate branch than in the high reaction-rate branch by 0.1%. Studies performed in the UHV chamber on a clean foil with well controlled exposures of pure gases showed that only adsorbed oxygen would cause a decrease in reflectance by about 0.1% at room temperature. Both the clean foils and the CO covered foil had the same reflectance. This behavior was checked in other frequency regions (2150-2300cm^{-1} and 2400-4000cm^{-1}) where no discernible *vibrational* absorption features occurred. Hence, for a Pt foil in an O_2/ CO mixture, a decrease in reflectance was attributed to the presence of absorbed oxygen [7].

The real-time reflectance data taken in the flow cell during oscillation therefore indicate that there is a substantial oxygen coverage during the high reaction-rate branch where the rate of CO oxidation is limited by the rate of CO adsorption, and a very low coverage of oxygen in the low reaction-rate branch. The previous observation of substantial CO coverage in the low reaction-rate branch [6] confirms that, in this branch, the reaction rate is limited by oxygen adsorption.

In the low reaction-rate branch, the amount of CO adsorbed shown in Fig. 2c varies with two different time-scales. There is first a rapid rise in the amount of CO coverage which was associated with the removal of adsorbed oxygen. There is then a much slower rate of increase which cannot be associated with removal of oxygen (from reflectance data). This slow rise suggests, instead, that a surface impurity is slowly removed, freeing active sites onto which CO can adsorb. The next section shows more directly that there is indeed a large variation in the number of active sites during the oscillations.

3.2 Quench experiments

An oscillatory reaction state with a long period was established in the reaction cell. After the oscillation had completed at least one complete cycle, the reaction was quenched at various times (as measured from the start of a spike). The quench was accomplished by stopping the current supplied to the foil, while still maintaining the flow of the reaction gas mixture and reactor pressure. This would result in a cooling of the foil from $\approx$525K to room temperature (within seconds) and in an increase in the surface coverage of CO.

The variations in adsorbed CO were measured by time-resolved IR spectra collected from 30 seconds before the current interruption to 5 minutes after the interruption at a rate of 2 scans/sec. These spectra showed that in all cases the integrated infrared adsorption due to CO_{ad} reached a maximum and stabilized within one minute or less of the quench. It was found that, at the typical pressures used in this study and room temperature, virtually all the sites on the Pt sample active towards CO adsorption were populated by CO irrespective of whether the gas phase contained a CO/O_2 mixture (P_{CO}/P_{O_2}=0.011 or 0.030) or just CO. Thus the CO coverage (as integrated absorption) at room temperature provided an accurate measure of *the number of sites active towards oxidation reaction.*

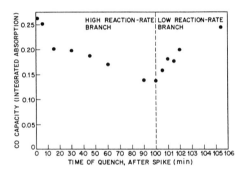

Fig. 4: Surface CO capacity during the course of an oscillation cycle. It is measured upon quenching to room temperature in the gas mixture with P_{CO}/P_{O_2}=0.03.

The CO coverage maxima at room temperature obtained at various quench times are presented in Fig. 4. The maximum CO coverage after quenching (henceforth referred to as the CO capacity of the surface) decreased approximately linearly with time during the high reaction-rate branch and recovered approximately linearly during the low reaction-rate branch. Thus, Fig. 4 clearly indicates a slow deactivation in the high reaction-rate branch of the surface sites which are active towards CO adsorption, and a slow recovery of these sites in the low reaction-rate branch. For the quench cycle shown, the CO adsorption capacity of the surface changes by 40% during the cycle. Such a change indicates that the concentration of the surface impurity should also vary substantially during the oscillation cycle.

4. AUGER ELECTRON AND INFRARED SPECTROSCOPIES IN UHV

To test whether periodic variations in the surface impurity concentration could be responsible for the periodic variation in the CO capacity of the surface, we subjected the Pt sample in the UHV chamber to a variety of treatments (temperatures and gas phase compositions) similar to those of the foil in the flow cell, characterized the surface region by AES, and measured the CO capacity of the surface after each of these treatments [7].

After baking the UHV chamber (and foil) to 470K, AES measurements on the "as received" Pt foil showed detectable amounts of C, O, Si, Ca, Cu, Fe, P and S. However, Cu, Fe, P and S were removed by a brief annealing in UHV (10^{-10}Torr) to 850K and never reappeared in detectable amounts ($\approx 10^{-3}$ monolayer). Calcium remained in very low concentration with no correlation to sample treatment.

Different sample treatments led to different concentrations of carbon and oxygen in the surface region, with [O/Pt] AES ratios varying between 0.19 and 0.40, and [C/Pt] AES ratios varying between 0.22 and 0.82. The CO adsorption capacity of the surface was measured by IRRAS after each sample treatment. The results are shown in Fig. 5. The figure reveals a strong inverse dependence of the CO adsorption capacity with [C/Pt], and no apparent correlation with [O/Pt].

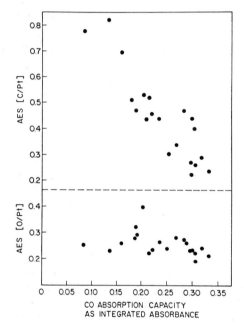

Fig. 5: AES carbon and oxygen ratios vs. CO capacity for many different sample treatments in the UHV chamber.

Quantitative analysis of a number of AES spectra (assuming all impurities to be at the surface) showed that the relative coverage O/Si was 1.3±0.5, indicating that essentially *all of the surface oxygen is in the form of silicon oxide*. Based on the knowledge that only silicon oxide and carbon are present at the Pt surface, the maximum variation in carbon coverage was found to correspond to ≈26% of a monolayer for the 75% change in CO capacity shown in Fig. 5 as opposed to ≈2% of a monolayer for the oxide [7].

Several transient experiments were also performed to investigate how the sample temperature and gas phase conditions affected the concentrations of carbon and oxygen in the surface region. The results are summarized in Table 1. The surface carbon

Table 1: AES Carbon and Oxygen After Heating to 550K

	in vacuum (10^{-10} Torr)		in 10^{-6} Torr O_2		in 10^{-6} Torr CO
	C/Pt	O/Pt	C/Pt	O/Pt	C/Pt
initial	0.13	0.38	0.44	0.26	0.43
15 min.	0.17	0.24	0.30	0.24	0.40
45 min.	0.37	0.23	0.22	0.23	0.40

level increased upon heating the foil in vacuum (10^{-10} Torr), decreased on heating in O_2 (10^{-6} Torr), and remained relatively constant upon heating in CO (10^{-6} Torr). The surface oxygen level decreased slightly upon heating in vacuum and remained relatively constant upon heating in O_2. Therefore, carbon must diffuse from the bulk to the surface, oxygen can remove surface carbon faster than carbon can diffuse out of the bulk, and adsorbed CO blocks the carbon diffusion from the bulk, *at the reaction temperature.*

In contrast, it was found that the silicon oxide vibrational band did not change during the above transient experiments. Large changes were observed only at very high temperature ($>1000K$) in pure CO or O_2 and CO mixtures. Therefore, as evidenced by the results shown in Fig. 3, there are no detectable variations in the amount of silicon oxide at the oscillation temperatures.

5. THE CARBON MODEL

Based on the above observations, a model in which carbon drives the oscillations was proposed [7,8]: active sites are deactivated by carbon and reactivated as oxygen removes it. In the high reaction-rate branch, the rate of deactivation is greater than that of regeneration, yielding a gradual loss of active sites (Fig. 4). When the concentration of active sites falls below a certain value, the transition to the low reaction rate branch results. In the low reaction-rate branch, virtually all the active sites are filled by CO. The transient results of Table 1 imply that this adsorbed CO inhibits further surface deactivation by carbon. Since the surface carbon continues to be depleted by oxygen, a net increase in the active-site concentration takes place. When this concentration is close to maximum, the reaction undergoes an ignition from the low to the high reaction-rate branch, thus completing the oscillation cycle.

This model was quantified by an analysis based on the Langmuir- Hinshelwood surface reaction model with deactivation/regeneration of the surface by surface carbon formation and removal [8]. Thus, in addition to the reactions:

$$CO + * \underset{k_{-1}}{\overset{k_1}{\rightleftarrows}} CO_{ads}; \quad O_2 + 2* \underset{k_{-2}}{\overset{k_2}{\rightleftarrows}} 2\,O_{ads}; \quad CO_{ads} + O_{ads} \overset{k_r}{\rightarrow} CO_2 + 2*$$

the slow processes of surface carbon formation and removal:

$$C + * \overset{k_d}{\rightarrow} C_{surface} \qquad C_{surface} + \begin{bmatrix} O_{ads} \\ O_2 \\ \frac{1}{2}O_2 \end{bmatrix} \overset{k_g}{\rightarrow} \begin{bmatrix} CO_{ads} + * \\ CO_2 + * \\ CO_{ads} \end{bmatrix}$$

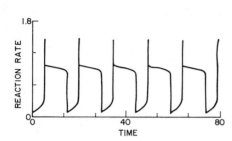

REACTION RATE

TIME

Fig. 6: Plot of the dimensionless reaction rate, $(X/X_T)^2 \theta_{CO} \theta_O$, versus time in units of k_d^{-1} obtained from ref. 8. Here X is the surface density of active sites, X_T the maximum surface site density and θ_i is the coverage of species i, normalized by X. The results are obtained for $k_g P_{O_2}/k_d = 0.042$, $P_{CO} = 0.76$ Torr and $(k_d/k_r X_T) = 0.0001$. Note that k_g includes only the process: $C_{surface} + O_2 \rightarrow CO_2 + *$ (see ref. 8).

are considered. The calculation is performed by means of a singular perturbation analysis following the description of Chang and Aluko [10].

It was found that, as long as carbon could be removed *directly* by oxygen, then long-period oscillations could occur as shown in Fig. 6. Comparison of these calculated oscillations with those observed experimentally (Fig. 2) shows remarkable agreement, including the reaction rate spike. This transition is found to occur when the fraction of active sites is at a maximum and the fractional coverages θ_O and θ_{CO} are at comparable values. Following this ignition, calculations show that the surface is primarily oxygen covered in the high reaction-rate branch during which the fraction of active sites decreases. When the fraction of active sites falls below a certain value, the reaction drops sharply to the low reaction-rate branch which is primarily covered with CO. In that branch, the fraction of active sites increases until ignition occurs and the cycle is repeated. The calculations are therefore in agreement with all of the observations [8].

It should be stressed that similar features can be obtained for an oxide model. Therefore, the calculations only indicate that the carbon model is possible and is consistent with all the observations such as carbon mobility, oxygen concentration, CO and carbon at the surface and gas phase composition. It suggests, in addition, direct removal of carbon by oxygen and possible experimental ways of discriminating between the carbon and oxide models [8].

6. CONCLUSIONS

This work demonstrates that infrared spectroscopy can help bridge the gap between gas kinetic measurements at high pressures and surface science studies in vacuum. It emphasizes that complex catalytic processes require *real-time, in situ* monitoring to identify surface impurities that can play an active role in the reaction kinetics. For CO oxidation on Pt at high pressures, the lack of in situ measurements and the indirect observation of surface oxides [4] had led to the oxide model [3], even though it is well established that carbon deactivates adsorption sites on metal surfaces.

In fact, the widespread occurrence of self sustained oscillations in the rate of catalytic oxidation, for catalysts with widely different reconstruction and oxidation properties, suggests that carbon is a good candidate as a general driving mechanism for most of the systems. Carbon is indeed a common contaminant in a typical reaction environment, i.e., not in an ultra-high-vacuum environment. Carbon diffusion rate is similar for most catalysts used (Pt, Pd, Ir, Ni) so that carbon movements at oscillation temperatures is expected. While *real-time, in situ* investigations for other oscillatory reactions are needed to establish this point, it is clear that carbon must be considered as a possible driving mechanism.

REFERENCES

1. See review by L. F. Razon and R. A. Schmitz, Cat. Rev.-Sci. Eng. *28*, 89 (1986) and references therein.

2. G. Ertl, P. R. Norton and J. Rustig, Phus. Rev. Lett. *49*, 177 (1983); R. Imbihl, M. P. Cox, H. Muller and W. Brenig, J. Chem. Phys. *83*, 1578 (1985) and *84*, 3519 (1986).

3. B. C. Sales, J. E. Turner and M. B. Maple, Surf. Sci. *112*, 272 (1982) and *114*, 381 (1982).

4. J. E. Turner, B. C. Sales and M. B. Maple, Surf. Sci. *103*, 54 (1981) ; *109*, 591 (1982) and *147*, 647 (1984).

5. H. P. Bonzel, A. M. Franken and N. G. Pring, Surf. Sci. *104*, 625 (1985).

6. V. A. Burrows, S. Sundaresan, Y. J. Chabal and S. B. Christman, Surf. Sci. *160*, 122 (1985).

7. V. A. Burrows, S. Sundaresan, Y. J. Chabal and S. B. Christman, Surf. Sci. (in press, 1987).

8. N. A. Collins, S. Sundaresan and Y. J. Chabal, Surf. Sci. (in press, 1987).

9. F. M. Hoffmann, Surf. Sci. Reports *3*, 107 (1983).

10. H.-C. Chang and M. Aluko, Chem. Eng. Sci. *36*, 1611 (1981); *39*, 37 (1984).

Kinetics of Surface Phase Transitions: Some Comments

P. Kleban

Laboratory for Surface Science and Technology and Department
of Physics and Astronomy, University of Maine, Orono, ME 04469, USA

It is a pleasure to have the opportunity to comment on two such excellent reviews concerning an emerging area in surface science, perhaps best denoted *collective kinetic effects*. The subject under consideration, in contrast to the others at this meeting, is not so much controversial as promising. Thus, in what follows we do not attempt to settle any disputes but rather discuss what are perhaps some important points in the articles, make a few suggestions that may prove helpful, and then end with some more general comments.

Klaus HEINZ's [1] review of the existing experimental situation, mainly for chemisorption and reconstruction systems, is exhaustive, provocative and well-balanced. One of the central issues here is clearly to determine whether the growth law for the average size of a domain

$$L \sim t^x \tag{1}$$

is valid, and if so for what systems, in what time domain and with what value of x. As emphasized in the article, a scattering lineshape measurement seems to be the least ambiguous way to do this. Here we would like to concur, and underline the fact that new developments in LEED (Low Energy Electron Diffraction) technology are very promising in this regard, in particular MEMLEED (Mirror Electron Microscopy LEED), as discussed by SHERN and UNERTL [2] or the very similar LEERM or LEEM (Low Energy Electron (Reflection) Microscopy) method described by TELIEPS and BAUER [3]. These instruments are well suited to determine the change of intensity with parallel momentum transfer, since the Bragg spot positions are independent of incoming electron energy. They are also capable of very high resolution, obviating the need for deconvolution and allowing one to follow domain growth for long times (assuming that surface defects do not intervene to limit L!). One can also obtain a *direct space* image of the domains if they are large enough (~200 Å with existing equipment). Finally, data collection using what are in many laboratories now standard video cameras will allow rapid measurement of the lineshape, so that its *evolution* may be determined. This promises to be a powerful way of attacking the problem. There will of course be the additional reward of determining the entire lineshape, not just its width. This allows comparison with, for instance, the kinetic scaling form of the lineshape described by Jim GUNTON in this volume [4]. In addition it should be mentioned that the strong anisotropy (along the surface) of many surface systems may provide some interesting new insights into "collective kinetics"--do domains really always grow with the same power in all directions? Is the kinetic scaling law for the scattering lineshape itself isotropic? Clearly some very fundamental questions remain to be answered.

We would also like to note that the basic questions addressed here mainly center around the kinetics of *interfaces*. In most if not all cases it appears that the surface *inside* the domain approaches its final (equilibrium) state rather quickly, and most of the subsequent changes are due to motion of the boundaries. It would be very interesting to find a way to *label* the interfaces, so they could be studied more directly. One possibility is *impurity doping*. It is known both theoretically (SELKE and HUSE [5], KARDAR and BERKER [6]) and experimentally (OLEFJORD [7]) that impurities often tend to segregate to domain or grain boundaries. A clever choice of systems and probes might provide some interesting information.

Another point worth mentioning is the Avrami or Kolmogrov growth law described by HEINZ [1]. This mixed power law and exponential behavior (also called "stretched exponential" -- see PALMER

[8]) describes the growth of domains at early times in some cases. For certain systems, the exponent n has what appears to be an anomalously low value. One possibility is that this is connected with *mass transport*. The low values occur in systems where transport apparently is taking place, e.g. due to an upquench from an unreconstructed state into a (2x1) *missing row* reconstruction state on an fcc (110) surface. For this to happen, atoms must be removed from the domain as the reconstructed region grows. If the diffusion of the missing row atoms away from the boundaries is rate limiting, clearly growth will be impeded. This suggests what might be termed a *snowplow* model: the advancing domain boundary piles up atoms in front of it, which must then diffuse away. The pile will necessarily grow with time, showing the boundary's motion, which could result in the low power in the growth law. Such a picture is also consistent with the observed failure to reach equilibrium after very long times, and by an amount that increases at lower temperatures. The removed atoms will run into step edges or other defects at long times, slowing the process further. This will take longer at lower temperatures, when the diffusion is slower. If all the domains nucleate simultaneously and grow according to (1), one can show that the exponent n (see HEINZ, section 3.2.b) satisfies n = 2x. Thus for *unimpeded* diffusion n = 2x = 1, while the *snowplow* model should result on an exponent n < 1. A complete quantitative treatment of this mechanism is not yet at hand, however.

Turning now explicitly to Jim GUNTON's [4] review of the theoretical situation, we find a most intriguing picture. The many interesting theoretical studies, which are very clearly explained there, allow some tantalizing glimpses of general principles -- especially Eq. (1) for domain growth and the related kinetic scaling law for the scattering lineshape. But the extent of their applicability (e.g. over what time range is (1) valid? when does x take on a given value?) and even whether they are the "correct" form is in some doubt! Certainly, as emphasized in the final paragraph of his article, careful experimental work will help clarify the situation. An example is the determination of the evolution of scattering lineshape mentioned above.

One interesting and apparently important distinction that emerges from this review is the difference between the time development of systems with "chemisorption" or "hopping" (Kawasaki) kinetics and those with "physisorption" or "spin-flip" (Glauber) kinetics, at least when the ordered state is sufficiently degenerate. One might be able to test this distinction with an experiment if a suitable *weakly chemisorbed* system can be found. What is required is a situation where the bonding energy and rates are such that the adsorbed species while moving between adsorption sites, spends most of its time *on the surface* at low temperatures, and, by contrast, most of its time *in the ambient gas* at higher temperatures. Then, for instance, studying domain growth via an upquench, as a function of the final temperature would allow one to see how the different mechanisms influence the time development.

Finally, we would like to step back a bit from the specific questions at issue in these two reviews and address some more general problems. These relate to the direction of research in the surface science area, especially for chemisorption or reconstruction systems. What needs to be emphasized is the importance of *collective effects*, such as the kinetic phenomena under consideration, or more simply thermodynamic phase transitions. Such effects are of great intrinsic interest, since they illustrate general types of behavior. In addition, they clearly play a major role in many important processes. Just to take one example, it will certainly be necessary to take phase transitions into account in extrapolating much of the information gleaned from UHV measurement to "real world" conditions (M. GRUNZE [9]). So it is not enough to dismiss collective effects as "too complicated" or "not practical." They can also lead to some surprising and powerful techniques, useful in specific cases--see the article by KLEBAN et al. in this volume [10], as a possible example. Such studies are therefore interesting for their illumination of general modes of behavior and also because "rules of thumb" may emerge that are very useful as tools for advancing understanding of other individual systems. Despite this, there seems to be an unfortunate "repulsive force" between many members of the community and the study of these phenomena. For instance, with some notable exceptions, the study of phase diagrams in these systems is in a state that borders the primitive. This is especially evident by contrast to the many careful and thorough experiments coupled with sophisticated theoretical work on physisorption systems. Surely it is time to move ahead!

References:

1. K. Heinz: this volume.

2. C.-S. Shern and W.N. Unertl: J. Vac. Sci. Tech., to appear.

3. W. Telieps and E. Bauer: Ultramicroscopy 17, 57 (1985); Surface Sci. 162, 163 (1985); Ber. Bunsenges. Phys. Chem. 90, 197 (1986).

4. J. Gunton: this volume.

5. W. Selke and D.A. Huse: Z. Phys. B 50, 113 (1983).

6. M. Kardar and A.N. Berker: Phys. Rev. Letters 48, 1552 (1982).

7. L. Olefjord: Intl. Metall. Rev. 4, 149 (1978).

8. R.G. Palmer, preprint.

9. M. Grunze, Surface Sci. 141, 455 (1984).

10. P. Kleban, R. Hentschke and J.C. Campuzano: this volume.

Index of Contributors